Series in BioEngineering

The Series in Bioengineering serves as an information source for a professional audience in science and technology as well as for advanced students. It covers all applications of the physical sciences and technology to medicine and the life sciences. Its scope ranges from bioengineering, biomedical and clinical engineering to biophysics, biomechanics, biomaterials, and bioinformatics.

More information about this series at http://www.springer.com/series/10358

Asim Bhatti · Kendall H. Lee
Hamid Garmestani · Chee Peng Lim
Editors

Emerging Trends in Neuro Engineering and Neural Computation

Springer

Editors
Asim Bhatti
Institute for Intelligent Systems Research and Innovation
Deakin University
Geelong, VIC
Australia

Kendall H. Lee
Mayo Clinic
Rochester, MN
USA

Hamid Garmestani
Materials Science and Engineering
Georgia Institute of Technology
Atlanta, GA
USA

Chee Peng Lim
Institute for Intelligent Systems Research and Innovation
Deakin University
Geelong, VIC
Australia

ISSN 2196-8861 ISSN 2196-887X (electronic)
Series in BioEngineering
ISBN 978-981-10-3955-3 ISBN 978-981-10-3957-7 (eBook)
DOI 10.1007/978-981-10-3957-7

Library of Congress Control Number: 2017932005

Printed on acid-free paper

This Springer imprint is published by Springer Nature
The registered company is Springer Nature Singapore Pte Ltd.
The registered company address is: 152 Beach Road, #21-01/04 Gateway East, Singapore 189721, Singapore

Preface

The understanding of how the brain works and enables intelligence, memory, learning and control of behaviour has been perhaps the most desired knowledge that mankind seeks. Improved understanding of the working principles of the brain could lead to accurate diagnosis and cure of neural diseases including seizure disorders (such as epilepsy), movement disorders (such as Parkinson's disease), migraine, delirium and dementia (such as Alzheimer's disease), as well as design of efficient neural prosthesis, such as bionic eye, ear, tongue and nose, to overcome the effects of structural, biochemical and/or electrical abnormalities.

Neurophysiological studies are of paramount importance in revealing the underlying behaviours and properties of neurons, providing a good understanding of the nervous system. These studies have paved the way for the development of neuro-prosthetics and brain–machine interface (BMI) devices. For instance, intra-neuronal recordings from the primary motor cortex have been investigated to develop neural decoders that can eventually drive artificial prostheses or machines. The contribution of these studies in understanding neurological disorders, such as the use of intracranial electrodes to gather information pertaining to epileptic patients, has played a key role in developing new therapies. Indeed, extracellular cellular recording technologies have been employed to understand the influence of gamma-protocadherine, which regulates the endurance of a neural network and the generation of new synapses.

One of the common factors behind all the aforementioned neuroscientific breakthroughs is the ability to access inter- and intra-neuronal functionality and communication, so as to decipher the neural networks' collective behaviours without disrupting their natural functioning. Extracellular recordings are the preferred techniques to aid in neurophysiological studies, and the recordings can be mainly grouped into two categories: in vivo (invasive) and in vitro (non-invasive).

In this book, we have made an attempt to explore the emerging trends of neuroengineering technologies and neural computation techniques to provide a general insight of the current research advancements and future research directions in this domain. New technologies and techniques appear on a regular basis, giving rise to the need of a unified information source that could keep readers up to date with the advancements in this field. This book will serve as a great resource of information in addressing the current trends and future prospects in the fields of neuroengineering and neural computation.

The book is divided into two major parts: neuroengineering and neural computation. The neuroengineering part covers emerging technological trends and novel techniques to interact with the brain. The neural computation part covers a variety of computing models and techniques to decipher useful information processing capabilities of the human brain.

Neuroengineering is an emerging multidisciplinary field of research that is about using scientific methods to understand and model the nervous system, and to use this knowledge to engineer systems that interact with, augment, or mimic the functionality of the nervous system. The dynamics, biophysical mechanisms, and information processing capabilities of individual neurons are fairly well understood; however, a detailed collection of the facts concerning the functionality of single neuron is insufficient to explain that of a large neuronal network. A quantitative and scalable explanation of how large recurrent neuronal assemblies develop and learn is the unsolved problem, and has attracted a great deal of attention in recent years.

Chapters "CMOS-Based High-Density Microelectrode Arrays: Technology and Applications" and "Microelectrode Arrays: Architecture, Challenges and Engineering Solutions" of this book highlight the technology trends, which include the architecture, challenges and material solutions, in developing highly efficient micro/nano electrodes as well as neural interfacing technologies such as CMOS-based microelectrodes for high resolution recording of the brain activity. Chapters "Revolutionizing Causal Circuitry Neurostimulation Utilizing the Optogenetic Technique Through Advanced Microsystems Development" and "Physiological Monitoring in Deep Brain Stimulation: Toward Closed-Loop Neuromodulation Therapies" highlight the emerging trends in opto-genetics and deep brain stimulations, which have attracted paramount attention from the neuroscience communities in recent years. Opto-genetics involves the use of light to control cells in living neurons, whereas deep brain stimulations have proven useful in treating a variety of neurological disorders, such as Parkinson's disease, tremors, rigidity, stiffness, slowed movement, and walking problems. Chapter "Mechanism of Docosahexaenoic Acid in the Enhancement of Neuronal Signalling" presents the use of extracellular recording and analysis to explore the effects of chemical stimulations. Chapter "Insects Neural Model: Potential Alternate to Mammals for Electrophysiological Studies" highlights the potential of electrophysiological studies of insects' neural model as an alternate to mammals, employing neuroengineering and extracellular analysis techniques. Chapter "Synchronization Criteria for Delay Coupled Izhikevich Neurons" describes the neural functionality of the brain employing mathematical techniques and stability analysis of the presented model.

Neural computation covers computing models and techniques that represent the characteristics and functionalities of the human brain, particularly the biological nervous system. As part of machine learning, these computing models attempt to imitate the information processing capabilities of the human brain through the combination of simple computational elements, or known as neurons, in a highly interconnected system. Some of the key characteristics of neural computation include input–output mapping (supervised learning), nonlinearity, adaptability, parallelism, and fault tolerance. While research in neural computation has been in existence for a few decades now, this area still attracts the attention of many scientists and researchers owing to the intrinsic importance of understanding and imitating the capabilities of the human brain in information processing.

Chapter "Capturing Cognition via EEG Based Functional Brain Networks" highlights the use of transfer entropy to analyse multi-channel electroencephalography (EEG) data and to examine the dynamics of functional brain networks with respect to cognitive activities. Chapter "Modelling of Tumour-Induced Angiogenesis Influenced by Haptotaxis" describes a mathematical model to simulate the influence of haptotaxis on angiogenesis, which shows that that migration of endothelial cells can be accelerated when the invasive tumour enhances haptotaxis. Chapter "Noise Reduction in ECG Signals Using Wavelet Transform and Dynamic Thresholding" presents a noise reduction technique to process noisy electrocardiogram (ECG) signals. Wavelet transform and dynamic thresholding are used to reduce different types of noise such that a high signal-to-noise ratio could be produced.

Chapters "Development of a Co-evolutionary Radial Basis Function Neural Classifier by a *k*-Random Opponents Topology"–"Monotone Data Samples Do Not Always Generate Monotone Fuzzy If-Then Rules" describe research in data analytics problems using artificial neural networks and related methods. A competitive co-evolutionary radial basis function neural network is presented in Chapter "Development of a Co-evolutionary Radial Basis Function Neural Classifier by a *k*-random Opponents Topology". During the co-evolutionary process, individual networks interact with each other in an intra-specific competition, while global and local search procedures are exploited to find the optimal solutions. In Chapter "Mining Outliers from Medical Datasets Using Neighbourhood Rough Set and Data Classification with Neural Network", a modified neighbourhood rough set is used as a pre-processing method to select representative samples for training the radial basis function network. An enhanced functional link neural network for data classification is described in Chapter "A Modified Functional Link Neural Network for Data Classification". The genetic algorithm is employed to optimize the functional link neural network by performing both weight tuning as well as selection of expanded input features. Chapter "Experimental Study of Elman Network in Temporal Classification" evaluates the usefulness of the Elman neural network as a temporal data classifier. In Chapter "Monotone Data Samples Do Not Always Generate Monotone Fuzzy If-Then Rules", a technique to generate fuzzy if-then rules is described. The results indicate that a set of multi-attribute monotone data may lead to non-monotone fuzzy rules, which is in agreement with the observation obtained from the adaptive neural fuzzy inference system that has neural learning capabilities.

The editors are grateful to all the authors who have contributed their articles, and to all the reviewers who have provided constructive comments and suggestions. Special thanks are due to the editorial team at Springer who have put together this book in a professional manner.

Geelong, VIC, Australia — Asim Bhatti
Rochester, MN, USA — Kendall H. Lee
Atlanta, GA, USA — Hamid Garmestani
Geelong, VIC, Australia — Chee Peng Lim

Contents

Part I Neuro Engineering

CMOS-Based High-Density Microelectrode Arrays: Technology and Applications 3
Marie Engelene J. Obien, Wei Gong, Urs Frey and Douglas James Bakkum

Microelectrode Arrays: Architecture, Challenges and Engineering Solutions 41
Dhurgham Khudhair, Saeid Nahavandi, Hamid Garmestani and Asim Bhatti

Revolutionizing Causal Circuitry Neurostimulation Utilizing the Optogenetic Technique Through Advanced Microsystems Development 61
R.P. Kale, S. Paek, S.J. Tye and A.Z. Kouzani

Physiological Monitoring in Deep Brain Stimulation: Toward Closed-Loop Neuromodulation Therapies 81
Seungleal (Brian) Paek, Rajas P. Kale, Katheryn M. Wininger and J. Luis Lujan

Mechanism of Docosahexaenoic Acid in the Enhancement of Neuronal Signalling 99
Md Ahsan Ul Bari, Julie Gaburro, Agnes Michalczyk, M. Leigh Ackland, Catherine Williams and Asim Bhatti

Insects Neural Model: Potential Alternate to Mammals for Electrophysiological Studies 119
Julie Gaburro, Saeid Nahavandi and Asim Bhatti

Synchronization Criteria for Delay Coupled Izhikevich Neurons 131
Imali T. Hettiarachchi, Lakshmanan Shanmugam, Asim Bhatti and Saeid Nahavandi

Part II Neural Computation

Capturing Cognition via EEG-Based Functional Brain Networks 147
Md. Hedayetul Islam Shovon, D. (Nanda) Nandagopal, Bernadine Cocks and Ramasamy Vijayalakshmi

Modelling of Tumour-Induced Angiogenesis Influenced by Haptotaxis 173
Wei Chen, Li Zhang, Chengyu Liu and Alamgir Hossain

Noise Reduction in ECG Signals Using Wavelet Transform and Dynamic Thresholding 193
Diptangshu Pandit, Li Zhang, Chengyu Liu, Nauman Aslam, Samiran Chattopadhyay and Chee Peng Lim

Development of a Co-evolutionary Radial Basis Function Neural Classifier by a *k*-Random Opponents Topology 207
Bee Yan Hiew, Shing Chiang Tan and Way Soong Lim

Mining Outliers from Medical Datasets Using Neighbourhood Rough Set and Data Classification with Neural Network 219
Pey Yun Goh, Shing Chiang Tan and Wooi Ping Cheah

A Modified Functional Link Neural Network for Data Classification 229
Toktam Babaei, Chee Peng Lim, Hamid Abdi and Saeid Nahavandi

Experimental Study of Elman Network in Temporal Classification 245
Shih Yin Ooi, Shing Chiang Tan and Wooi Ping Cheah

Monotone Data Samples Do Not Always Generate Monotone Fuzzy If-Then Rules 255
Chin Ying Teh and Kai Meng Tay

Part I
Neuro Engineering

CMOS-Based High-Density Microelectrode Arrays: Technology and Applications

Marie Engelene J. Obien, Wei Gong, Urs Frey and Douglas James Bakkum

Abstract Functional analysis of brain activity requires high-throughput and high-resolution tools for observation and manipulation. One approach is the use of microelectrode arrays (MEAs) for long-term in vitro extracellular recording of electrical activity from multiple neurons. Electrodes arranged on a planar substrate detect electric signals from surrounding neurons produced by ionic current flow through the cell membranes. Despite the advantages, MEA data analyses have been limited to extract parameters as a population average (e.g., firing rate). In order to extract information at the single-neuron or subcellular level, MEAs with high spatiotemporal resolution and good signal quality are required. In this chapter, we introduce the current trends on the technology and applications of complementary metal–oxide–semiconductor or CMOS-based high-density microelectrode arrays (HDMEAs). We review the recent HDMEA applications that facilitate single neuron and neuronal network studies and accelerate drug screening and biomarker discovery.

Keywords High-density microelectrode arrays (HDMEAs) · Extracellular recording · Action potential · Local field potential

1 Introduction

Increasingly, approaches for high-resolution activity mapping of neuronal networks are being developed through coordinated interdisciplinary efforts from various fields, including biology, genetics, biochemistry, optics, and engineering [1]. The main goal is to eventually map the whole brain and understand how the networks of

M.E.J. Obien (✉) · W. Gong · U. Frey · D.J. Bakkum
Bio Engineering Laboratory, Department of Biosystems Science and Engineering, ETH Zürich, Basel, Switzerland
e-mail: marie.obien@bsse.ethz.ch

M.E.J. Obien · U. Frey
MaxWell Biosystems, Basel, Switzerland

A. Bhatti et al. (eds.), *Emerging Trends in Neuro Engineering and Neural Computation*, Series in BioEngineering,
DOI 10.1007/978-981-10-3957-7_1

neurons are within the brain function [2]. This requires developing techniques for simultaneous recording of neuronal activity at multiple spatial and temporal scales and for manipulating the activity of neurons of interest. Complementary metal–oxide–semiconductor or CMOS-based high-density microelectrode arrays (HDMEAs) offer a promising platform to do so. Thousands of neurons can be simultaneously recorded and/or electrically stimulated over time scales of microseconds to months. Owing to the high-density feature, a single neuron can be recorded by hundreds of electrodes. This facilitates assigning recorded spikes to their source neurons, termed *spike-sorting*, and allows the subcellular mapping of a neuron's axonal arbor.

Electrical recording of neuronal activity has been the preferred means of analyzing single neuron and neuronal networks [3, 4]. Electrical signals produced by neurons can be detected at a distance from the source. Several recording tools apply to different spatial scales. At the mesoscale, where local neuronal populations can be analyzed, a popular method is extracellular recording using metal electrodes. An electrode placed inside a brain slice in vitro or inserted in the brain in vivo detects electrical signals produced by the surrounding cells. A wide range of neural phenomena can be observed, from the spiking activity of individual neurons (extracellular action potentials or EAPs; bandwidth: 300–3000 Hz) to the slower network oscillations of small populations (local field potentials or LFPs; bandwidth: 1–300 Hz), as shown in Fig. 1. Additionally, the same electrode can be used to deliver electrical stimulation to a local area in the brain. While applying this method for brain recording and stimulation is relatively straightforward, the challenge lies in

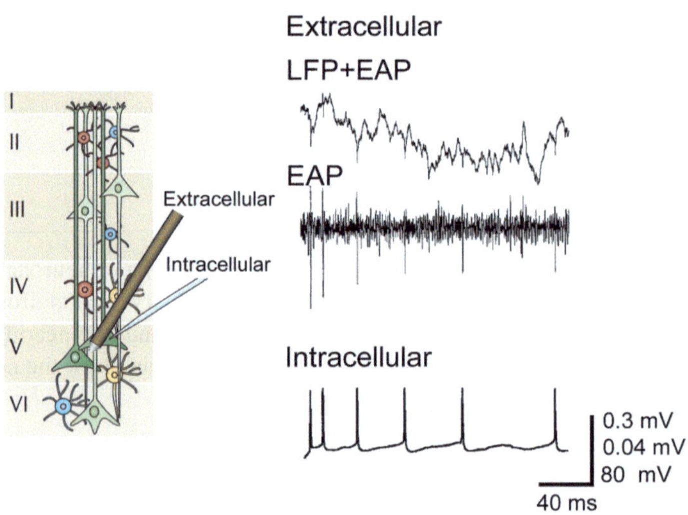

Fig. 1 Extracellular and intracellular recording. *Left* Illustration of cells across cortical layers modified with permission from [6]. *Right* Signals of simultaneous extracellular recording and intracellular whole-cell patch-clamp recording modified with permission from [7]

the analysis of recorded data. With hundreds of possible signal sources surrounding an electrode, the specificity and selectivity of such technique is poor. Thus, extracellular recording has been widely used for analyzing population activity. In contrast, intracellular recording by patch clamp has been the gold standard for analyzing single neuron. This method is powerful, yet necessitates intricate skill to perform. The viability of patched neurons lasts only up to a few hours. Moreover, current implementations of the experimental setup are bulky. All these limit the capability of the patch-clamp technique to scale for studying networks of neurons [5].

To achieve high-resolution activity mapping of neuronal networks, multiple electrical sensors tightly spaced in an array can be utilized. Microelectrode arrays (MEAs, also termed multielectrode arrays) allow simultaneous long-term recording of LFPs and EAPs from a population of neurons at sub-millisecond time scale. In order to increase spatial resolution, i.e., to place thousands of electrodes per square millimeter, the area taken up by wiring between electrodes and readout circuitry has to be reduced. This has been made possible using industrial CMOS technology to create high-density MEAs (HDMEAs). As an added benefit, readout circuitry, such as amplifiers and analog-to-digital converters, can be included on the same substrate as the electrodes in order to improve signal quality. The design of the on-chip signal conditioning circuitry should consider the electrode impedance and the possible sources of noise to ensure high-quality signals. HDMEAs with good signal-to-noise ratio (SNR) can be used to map single neuronal activity at subcellular resolution and to observe network activity at the same time [9, 10], as illustrated in Fig. 2.

This chapter focuses on the technology and applications of CMOS-based HDMEAs. As data obtained from MEA experiments done in vivo and in vitro are often very similar, the analytical concepts we include here apply to both. We discuss the current understanding of the MEA signal flow and the different factors affecting the shape of the recorded MEA signals from neurons. We also review recent works using MEAs, with emphasis on new techniques using HDMEAs, for neuroscience and medicine.

2 CMOS-Based HDMEA Technology

Advances in fabrication and semiconductor technologies allowed the miniaturization of modern biosensor devices. For MEAs, this meant integrating active electronic components in the same substrate, together with the electrodes. A technology for constructing integrated circuits is called complementary metal–oxide–semiconductor (CMOS), hence the term CMOS-based MEAs. CMOS technology enables increasing the electrode count in a single device, from dozens to tens of thousands of electrodes, in turn producing high-density microelectrode arrays (HDMEAs) while also improving the signal quality of recordings [11, 12]. Furthermore, tapping into the large and established CMOS production industry provides an economy of scale for HDMEA production.

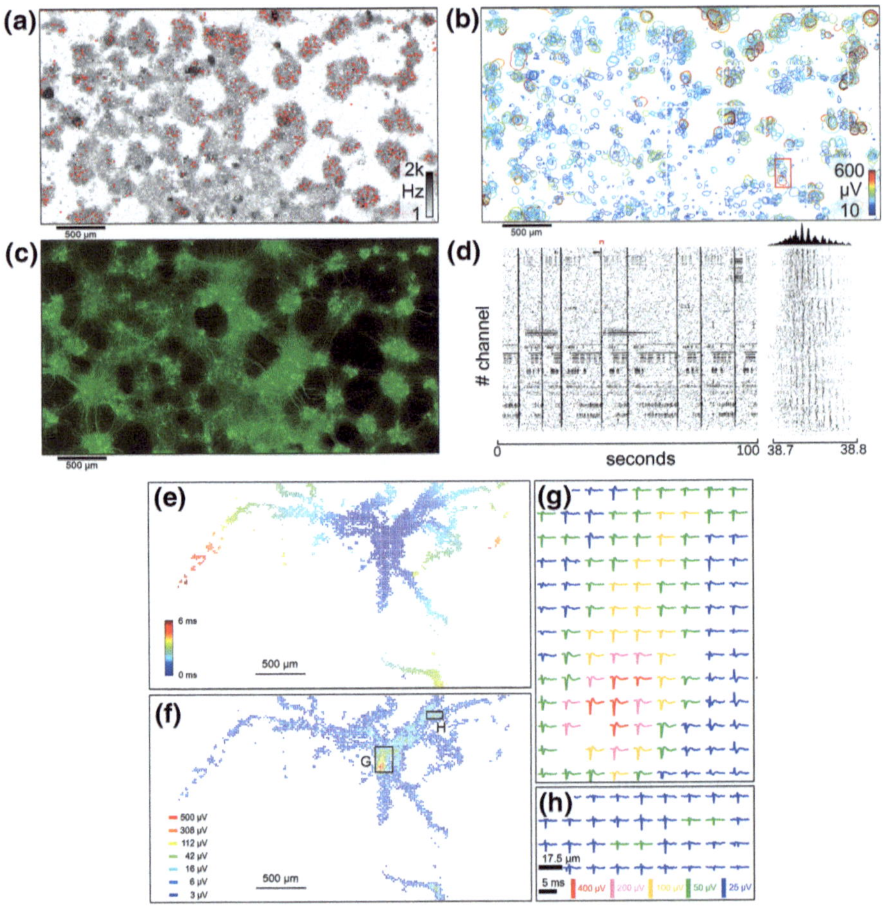

Fig. 2 Obtaining network-wide and single-neuron activity maps using CMOS-based HDMEAs. **a–d** Networks. **a** Average EAP firing rate as measured by each electrode (26,400 electrodes in total) shown as pixels colored with a logarithmic gray- scale between 1 Hz and 2 kHz. Red dots correspond to the electrodes selected for the raster plot in (**d**). **b** Representation of all 2000 individual neurons identified through spike-sorting signals. A circle represents each detectable cell; the edges indicate where the amplitude of the measured signals exceeds –4.5 standard deviations of the electrode noise. The colors correspond to the amplitude of the most negative peak detected by the electrodes within the circle. **c** Fluorescence image of transfected cells (around 5% of all cells in the culture). **d** Raster plot of 100 s of activity for 1024 electrodes recorded simultaneously. *Red marker* shows the time period in close-up view (bursting activity) on the right. Histogram at the *upper right* shows the number of spikes per time bin of the burst close-up. **e–h** Single-neuron electrical footprint. **e** All electrodes that captured activity attributed to a single neuron are colored according to the time of arrival of the AP at the electrode locations. **f** The same electrodes in **e** are *colored* according to the amplitude of the most negative peak detected. **g–h** Spike-triggered averages (30–50 averages) of the EAP electrical footprint from the two areas of the array as indicated by *black boxes* in (**f**). All figures modified with permission from [8]

2.1 Device Types

MEA architectures have evolved throughout the years. In general, the electrode-to-readout routing scheme can be divided into two types: fixed wiring (i.e., each electrode is directly wired to outside of the array, connecting to the signal conditioning circuit) and multiplexed array (i.e., routing from electrodes traverses switches before reaching the signal conditioning circuit). We further classify the MEA device types, as shown in Fig. 3.

Passive Conventional MEAs have fixed wiring and are passive (i.e., no active circuit elements, such as amplifiers). Each electrode connects directly to a signal pad outside the array through a wire. The pads are then connected to external equipment for signal conditioning. Passive MEAs are easy to fabricate and many different substrates and electrode materials can be used. The user has direct access to all electrodes simultaneously; however, wiring and electrode geometry limit the total number of electrodes that can fit in a given area. Examples of passive MEAs were developed and used by [13–28].

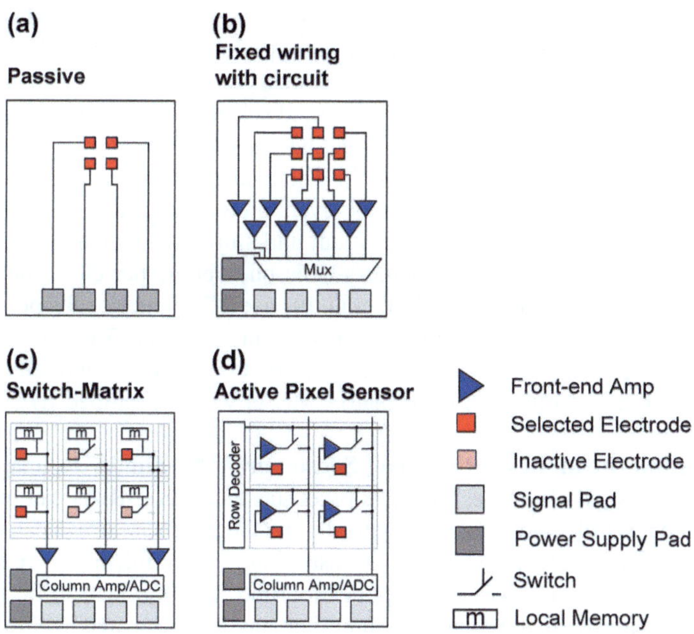

Fig. 3 MEA architectures. This table summarizes the different architectures used for MEAs. **a** Passive: Fixed wiring with electrodes directly connected to signal pads and no active circuitry. **b** Fixed wiring with electrodes directly connected to on-chip active circuitry for signal conditioning. **c** Switch matrix (SM): Multiplexed array with flexible addressing achieved by adding more routing resources within the array. **d** Active pixel sensor (APS): Multiplexed array with all electrodes sampled at fast speeds for a full-frame readout. Modified with permission from [12]

Fixed Wiring with On-chip Circuitry These types of MEAs have electrodes directly wired to on-chip active circuit elements that are used for signal conditioning, such as amplification and filtering. One variation employs multiplexers to allow readout of more electrodes despite a limited number of signal output pads. Multiplexing can be done only if the amplifiers and filters are before the multiplexer. Although this architecture allows for increased electrode count, the electrode density cannot be maximized (i.e., direct wiring of each electrode to signal conditioning circuitry occupies chip area that could have been used to add electrodes or readout channels). Selected fixed wirings with on-chip circuitry MEA references are [29–33].

Switch Matrix (SM) The switch matrix (SM) concept uses transistors to implement switches within the array to route signals from electrodes to readout circuitry placed outside the actual electrode array. In the SM concept, these routing means are operated in static mode, meaning that some electrodes are selected by opening or closing the switches and a recording is then started without changing the electrode selection. Typically, not all electrodes detect activity during an MEA experiment, thus choosing a subset of 'interesting' electrodes is possible. A common protocol is to first scan all the electrodes in successive recordings to determine which electrodes to later continuously record during an experiment. The advantage of this concept is that large, low-noise amplifiers can be implemented outside the actual electrode array, allowing to optimize amplifiers for best possible SNR. SM MEAs have been implemented and various degrees of flexibility that the routing means provide. Very simply row, column-based selectability has been implemented [34]. Increased degree of freedom in selecting subsets of electrodes was achieved for the following in vivo probes [35–37]. The availability of a large set of wires, switches, and local memory allows for even more complex routing paths that connect a subset of electrodes to the readout and stimulation channels in a flexible manner. Frey et al. [10] use 1.2 memory cells on average per electrode, allowing already fairly complex routing. Ballini et al. [9] use 2.2 memory cells per electrode, drastically increasing the possibilities in selecting subsets. Viswam et al. [38] increased the number of bits per electrode to more than 3, virtually allowing arbitrarily subset selections. Switch matrix MEAs were developed by [9, 10, 34–39].

Full-Frame Readout (Active Pixel Sensor or APS) Similar to image sensors used in cameras, all electrodes in active pixel sensor (APS) MEAs can be sampled at fast speeds in full-frame readout. Typically, rectangular sub-arrays can be chosen as regions of interest and sampled at faster rates than full-frame readout. For full-frame readout, the front-end amplification and filtering have to be before the multiplexing, meaning the front-end amplifier has to be located within the pixel itself. This is because the electrode exhibits high impedance and therefore, without an amplifier, cannot drive multiplexed readout lines at sufficient speed. The small pixel area (i.e., available area near each electrode) serves as a limitation to designing very low-noise circuitry for APS MEAs, since small-sized amplifiers inherently generate larger noise levels. Thus, while all electrodes can be recorded at the same time, only relatively large signals are detectable from noise. Examples of APS MEAs are [39–48].

Recent HDMEAs (SM and APS) aim to increase the total number of electrodes and the spatial resolution to allow for ever more demanding applications to be executed. The design of on-chip signal conditioning is crucial to achieve high-quality signals. However, due to area availability and power consumption

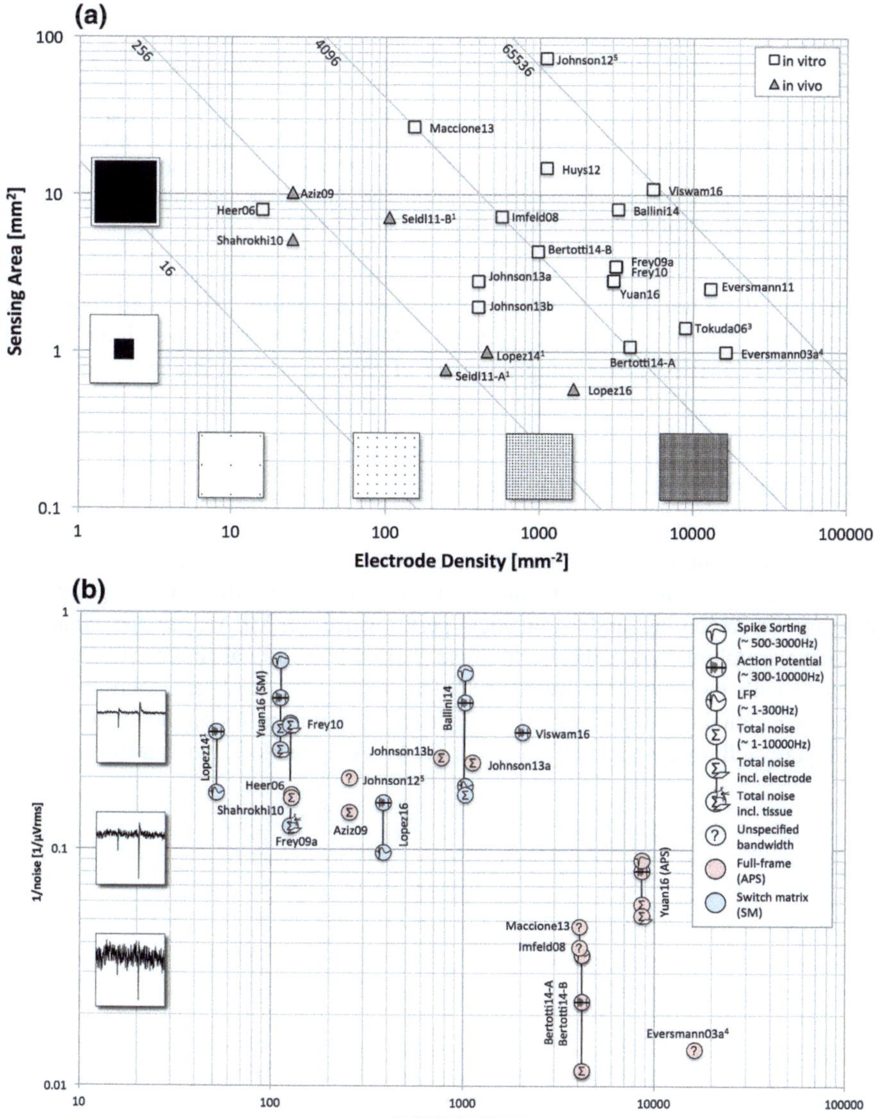

◀**Fig. 4** Device comparison. **a** HDMEA in vivo and in vitro implementations are shown according to the sensing area size and electrode density. For devices with a regular sensor pitch, such as most in vitro MEA devices, the total area is calculated as number of electrodes times the pixel area. For all devices, the number of electrode times the inverse of the electrode density matches the total area. The *light gray lines* illustrate the number of electrodes. **b** CMOS-based MEAs are compared with respect to parallel recording channel count and noise level. The noise values shown are approximated root-mean-square values stated in the respective citations. Note that the conditions under which these measurements were taken usually differ significantly (e.g., noise bandwidth, in- or exclusion of electrode noise, inclusion of ADC quantization noise, etc.). This graph only serves as a rough comparison, indicating noise values under both known and unknown conditions. The waveforms to illustrate the noise levels are simulated and have a spectrum typical for MEA recordings. The simulated spikes in the boxes (*left*) are typical spikes for acute brain slice measurements recorded with microelectrodes. The recorded amplitudes may vary significantly depending on preparation and sensor characteristics. Modified with permission from [12]. (See footnotes ([1]Area is calculated as total horizontal extent multiplied by the vertical extent. For probes, horizontal extent corresponds to shaft width. [2]Only one subarray of 16 electrodes is considered. [3]The frame rate for this design is 20 frames per second, significantly lower than other MEAs. [4]Noise values are taken from reference [49]. [5]The authors state that with a new acquisition board, the parallel channel count could be increased to 1024 at 9.3 kS/s.))

limitations, there remains a compromise between the quality of recorded signals and the number of parallel electrodes readout. SM HDMEAs prioritize signal quality, while APS HDMEAs target a high number of parallel readout channels, see Fig. 4.

2.2 Experimental Parameters

HDMEAs have been used for recording EAPs and LFPs from neurons and evoking neuronal activity through electrical stimulation. There are three main components affecting the recording and stimulation performance of MEAs: (a) the conductive extracellular volume where the electric field, caused by neuronal activity, forms; (b) the substrate with the embedded microelectrodes; and (c) the hardware connected to the electrodes, including amplifiers, filters, digitizer, data transmission, and stimulator [50, 51]. Figure 5 illustrates the components of the MEA signal flow. Noise affects the MEA

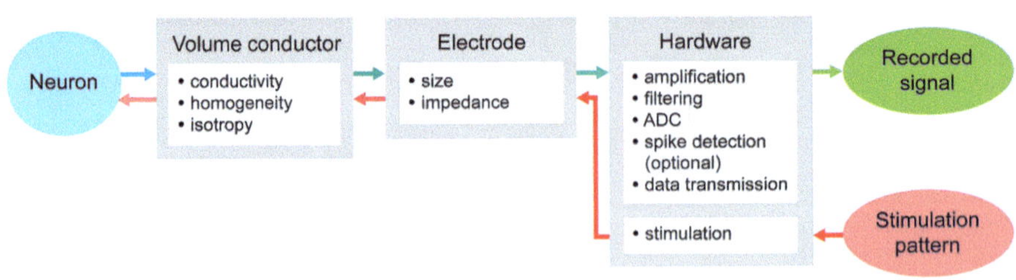

Fig. 5 MEA recording and stimulation system diagram. A neuron's signal, typically an action potential, is transduced through different components of the signal path into a digitally recorded trace. Similarly, a digital pattern generated from a computer or the MEA hardware applies current or voltage at the electrode during stimulation. Adapted with permission from [12]

signal across these components. Sources of noise include biological noise, electrode–electrolyte interface noise, and device noise [12].

2.2.1 Neuron–Electrode Interface

The early MEA neuron–electrode interface model assumed a tight seal between the neuron and the electrode [52]. However, extracellular microelectrodes can record EAPs and LFPs at a distance from active neurons, as observed in acute tissue and in vivo experiments. Likewise, for 2D neuronal networks grown on a MEA, EAPs can be detected from electrodes distant from the neuronal source. Thus, the neuron–electrode interface model can be separated into two parts: (1) the fluid side, which considers the effect of the volume conductor to the extracellular potential at the electrodes and (2) the metal side, which models the transformation of the extracellular potential through the electrode to the input of the front-end amplifier.

2.2.2 Volume Conductor

The distance and orientation of neurons with respect to measuring electrodes affect the amplitude and shape of the detected signals. The characteristics of the extracellular space, such as conductivity, anisotropy, and inhomogeneity, influence the spread of neuronal signals toward the electrodes. These effects can be estimated using the volume conductor theory illustrated in Fig. 6a. As a first-order approximation, the MEA surface can be considered as an infinite insulating plane, while the tissue and/or fluid in the MEA dish can be assumed to be infinite, homogeneous, and isotropic. A neuron's membrane current can be decomposed into several point current sources. The method of images can then be applied to Coulomb's law to solve the potential V_e at any given electrode e in a volume conductor with conductivity σ [12, 55]:

$$V_e = \frac{1}{2\pi\sigma}\sum\frac{I_n}{r_n},$$

where I_n represents the nth point current source and r_n represents the distance between the point source and the recording electrode e, with $n = 1, \ldots,$ N, where N is the number of individual point sources. For electrodes larger than an ideal point electrode, V_e can be solved at multiple locations of the electrode's surface area and then averaged. This equation can be extended to include the anisotropy and inhomogeneity of brain tissue, the saline layer above the tissue, and to use line sources instead of point sources to represent neuronal membrane currents [55].

2.2.3 Electrodes and Recording Hardware

Various materials have been used for fabricating planar microelectrodes for MEAs. The most crucial factors to be tested are biocompatibility and electrode impedance.

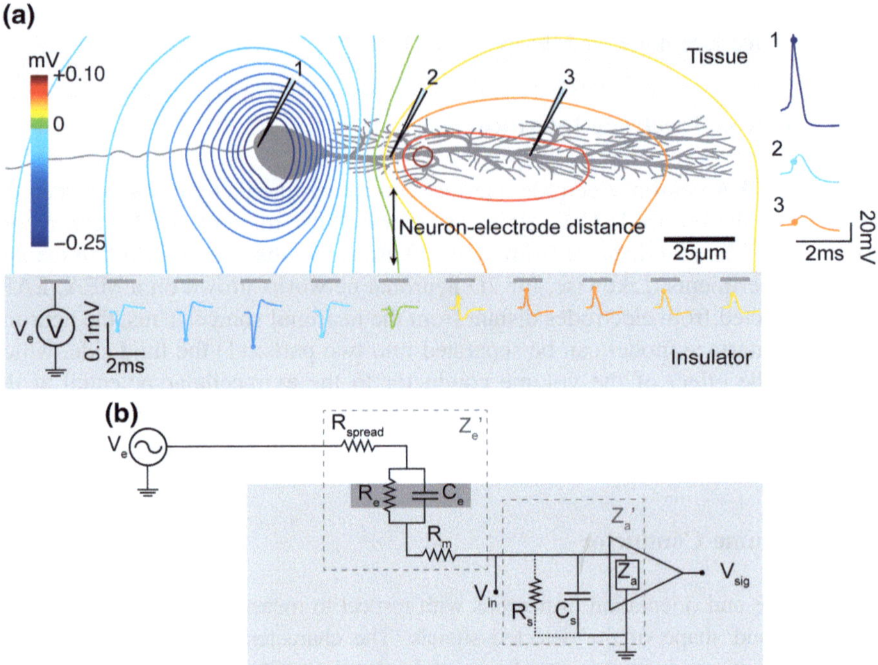

Fig. 6 MEA neuron–electrode interface divided into **a** fluid side and **b** metal side. **a** The potential at the electrode sites can be solved using the volume conductor theory. The MEA surface can be assumed as an insulator such that the method of images applies and can be used to solve the potential at any point on the MEA surface. The neuron–electrode distance and neuron orientation influences the signal amplitude and shape detected at the electrodes. High spatial resolution allows for recording EAPs at several locations of a single neuron, with large negative spikes at the perisomatic area and positive spikes at the dendritic area, i.e., return current. **b** The voltage measured at the electrode is transformed by the electrical parameters of the electrode–electrolyte interface, represented by Z_e' as the effective electrode impedance and Z_a' as the effective input impedance. This model is derived from [11, 53, 54]. R_{spread}–spreading resistance; R_e and C_e–resistance and capacitance of the electric double layer at the electrode–electrolyte interface; R_m—resistance of the metallic part of the electrode; R_s and C_s–shunt resistance and capacitance. Adapted with permission from [12]

Depending on the electrode material and fabrication method, the performance of electrodes may differ. Uniform impedance of all the electrodes in HDMEAs ensures consistency between recording sites.

A neuronal signal is transduced by an electrode into a current, and this process depends on the parameters of the effective electrode impedance Z_e' and effective input impedance Z_a'. We discuss this using the equivalent circuit of the electrode–electrolyte interface as shown in Fig. 6b. Noise (e.g., thermal noise and power line hum) can be injected into the recorded signal at the liquid–metal interface. Z_e' is the total impedance due to R_{spread}, R_e, C_e, and R_m. R_{spread} represents the effect of the electrode geometry and liquid conductivity. R_e and C_e are the resistance and capacitance of the electrode double layer formed at the electrode–electrolyte interface. R_m is the resistance of the metallic part of the electrode. Connected in series to Z_e' is Z_a', which is mostly influenced by the input impedance of the front-end amplifier Z_a and the shunt capacitance C_s. C_s includes the capacitances from connectors and wires from the liquid to the amplifier. The shunt resistance R_s is usually negligible. All these represent the metal side of the neuron–electrode interface. For more details on the circuit model, see [11, 53, 54].

Front-end amplifiers are designed to have large Z_a' in order to preserve signal quality. The ratio between Z_e' and Z_a' shows how to derive the voltage at the input of the amplifier as [53]:

$$V_{in}(\omega) = \frac{V_e(\omega)}{1 + \left(Z_e'(\omega)/Z_a'(\omega)\right)},$$

where $V_e(\omega)$ is the total extracellular potential at the electrode and $V_{\text{in}}(\omega)$ is the voltage at the input of the front-end amplifier. V_{in} will be smaller than V_e, i.e., the signal will be attenuated if Z_a' is not substantially larger than Z_e'.

Aside from the impedance effect to Z_e', electrode size plays an important role depending on the signal of interest. Large electrodes (>50 μm diameter) have a higher chance of being near a neuronal source [56]. However, the detected amplitude of a large EAP signal from a neuron is reduced as it is averaged out by nearby smaller amplitude signals. Sorting all the signals detected by a single large electrode to their respective individual sources can also be daunting when many neurons are nearby. For recording EAPs, especially for dissociated cell culture experiments, the use of small electrodes (<15 μm diameter) minimizes averaging. Small electrodes are inferior against large electrodes in terms of impedance, but this can be improved by surface modification. For example, the influence of electrode size (<10 μm diameter) on recorded signal amplitude significantly decreases by depositing Pt-black on platinum microelectrodes [57]. HDMEAs have small electrodes to allow the integration of a large number of sensors in an array. The dense grid of electrodes in HDMEAs, as shown in Fig. 7, increases the possibility of having an electrode 'at the right spot' while also allowing a single neuron to be recorded from multiple electrodes.

Novel 3D micro- and nano-structure electrodes aim to detect subthreshold neuronal signals (e.g., synaptic potentials, membrane oscillations) and membrane potentials [58]. These subthreshold signals are conventionally measured using invasive sharp microelectrodes or patch clamp. The 3D electrodes seek to achieve either a tight seal with the neuronal membrane or to temporarily puncture into the membrane and access the intracellular space of a neuron. With advancements in fabrication techniques, large-scale integration of 3D micro- and nano-electrodes will be feasible on HDMEAs.

HDMEA circuitry includes amplifiers, filters, and some sort of data transmitter of either the amplified analog signals or, more typically, of the already digitized data. As discussed in the previous section, the front-end amplifier needs high input impedance to ensure signal quality. The circuitry near the cells must be low

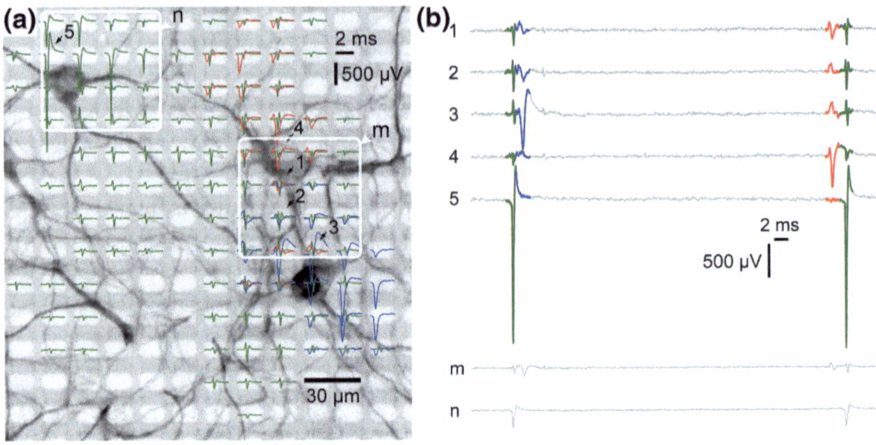

Fig. 7 Comparison of the recording capability of small electrodes at high-density and large electrodes. **a** The EAPs of three identified neurons (*green*, *red*, and *blue*) detected from each electrode site (*light gray rectangles*) are superimposed to a fluorescence image (MAP2 staining) of a cell culture on a HDMEA. Each spike represents the spike-triggered average over 50 trials. Spikes with amplitude below 50 μV are not shown. *White squares* represent the location of hypothetical large electrodes (60 μm × 60 μm) used for comparison of signals. **b** Raw traces from small electrodes (1–5) and large electrodes (m, n) indicated in (**a**). Signals for the large electrodes (m, n) are estimated by averaging the traces of the small electrodes within the area covered by the *white squares*. Estimated signals from (m, n) show reduced amplitudes due to the averaging effect only. The impedance effect due to electrode size differences may be neglected due to high input impedance at the first-end amplifier of the HDMEA, and is thus not considered. Modified with permission from [8]

powered in order to prevent heating that could damage the cells. Appropriate settings for gain and dynamic range of the readout depend on the preparation (e.g., maximal amplitudes of a few hundred microvolts in acute slice preparations and up to 10 mV in cardiomyocyte experiments). The recording bandwidth needs to be flexible to cover both LFP and EAP frequency ranges, depending on the experiment, in order to avoid filtering out signals of interest.

Analog-to-digital conversion (ADC) can be included in the device or done externally. Quantization noise due to the discretization error at the ADC can affect the recorded signals. Typically, a value of $1/\sqrt{12}$ times the magnitude of the least significant bit (LSB) approximates the quantization noise. Data transmission may also affect the quality of recorded signal, e.g., if lossy compression is used due to bandwidth constraints.

2.2.4 Stimulation

MEAs allow passive observation of neural activity but also can actively influence and control activity. Metal electrodes can deliver electrical stimuli directly. CMOS fabrication allows including electrical stimulation circuitry directly on-chip, in turn allowing a high degree of flexibility in generating spatiotemporal patterns of stimulation owing to dense and flexible wiring, higher spatial resolution for stimulation owing to densely packed electrodes, and room for on-chip circuitry to blank or suppress stimulation artifacts.

Electrical stimulation has been typically applied as a "trigger" for the so-called stimulus-triggered averaging [59]. By delivering electrical pulses through the microelectrode, action potentials (APs) can be triggered from nearby neurons, with an effective stimulation range depending on the neuron's distance from the stimulation site and the amplitude of the pulse. With HDMEAs, stimulus-triggered averaging reveals the electrical activity footprint of a single neuron, i.e., signals detected at the electrode sites corresponding to the EAPs from a single neuron, where negative spikes correspond to the AP initiated at the axonal initial segment and the positive spikes represent return current, including the propagation of APs in axons [60]. The stimulation amplitude has to be sufficient to consistently evoke an AP with small temporal jitter, e.g., a jitter of 160 μs [61]. Figure 8a shows how small axonal signals, typically undetectable from noise, become observable by increasing the number of trials averaged. The number of trials that must be averaged depends on the spike amplitude as shown in Fig. 8b.

One issue of electrical stimulation is the occurrence of artifacts in the recording channels. Stimulation pulses are typically three to four orders of magnitude larger than the recorded EAPs; the recording channels can pick up the artifacts through the wiring in the circuitry or through the media to neighboring electrodes. If the artifact amplitude is large, the amplification circuits may saturate and this prevents recording neuronal activity until the offset settles back to normal. Figure 8c provides an example of signal saturation due to stimulation in a SM HDMEA [10]. A recording electrode near the stimulation electrode (18 μm away) saturated for around 5 ms; another electrode located far from the stimulation site (1.5 mm away) did not saturate. Figure 8d presents the relationship between the distance from stimulation to recording electrode and the duration of saturation for an 11,011-electrode MEA [10], without employing any artifact suppression measures. As long as the amplifiers do not fully saturate, artifacts can be suppressed via software by subtracting the estimated artifact (based on templates, filters, or local curve fitting) from the data [62, 63]. To also allow recording from electrodes on which saturation would occur, counter measures in hardware have to be employed. One solution is to use a "reset" switch that can bring back the saturated amplifier into normal operation quickly, by resetting the high-pass filter of the front-end amplifier [10, 45].

Local delivery of stimulation pulses can be achieved by HDMEAs. Figure 8e, f shows stimuli-activated neuronal responses with high spatiotemporal precision. In a study to track axonal APs [60] several ten thousands of stimuli used for stimulus-triggered averaging did not damage the electrodes or the cells. Voltage-mode stimulation was used, although the stimulation hardware supported both current- and voltage-mode [64].

Combined recording and stimulation capabilities allow performing closed-loop experiments, whereby recorded signals are programmed to control the application of electrical stimuli. In such experiments, spike detection is performed online, typically through a dedicated hardware, e.g., a desktop with a real-time operating system or a field-programmable gate array (FPGA) [65, 66].

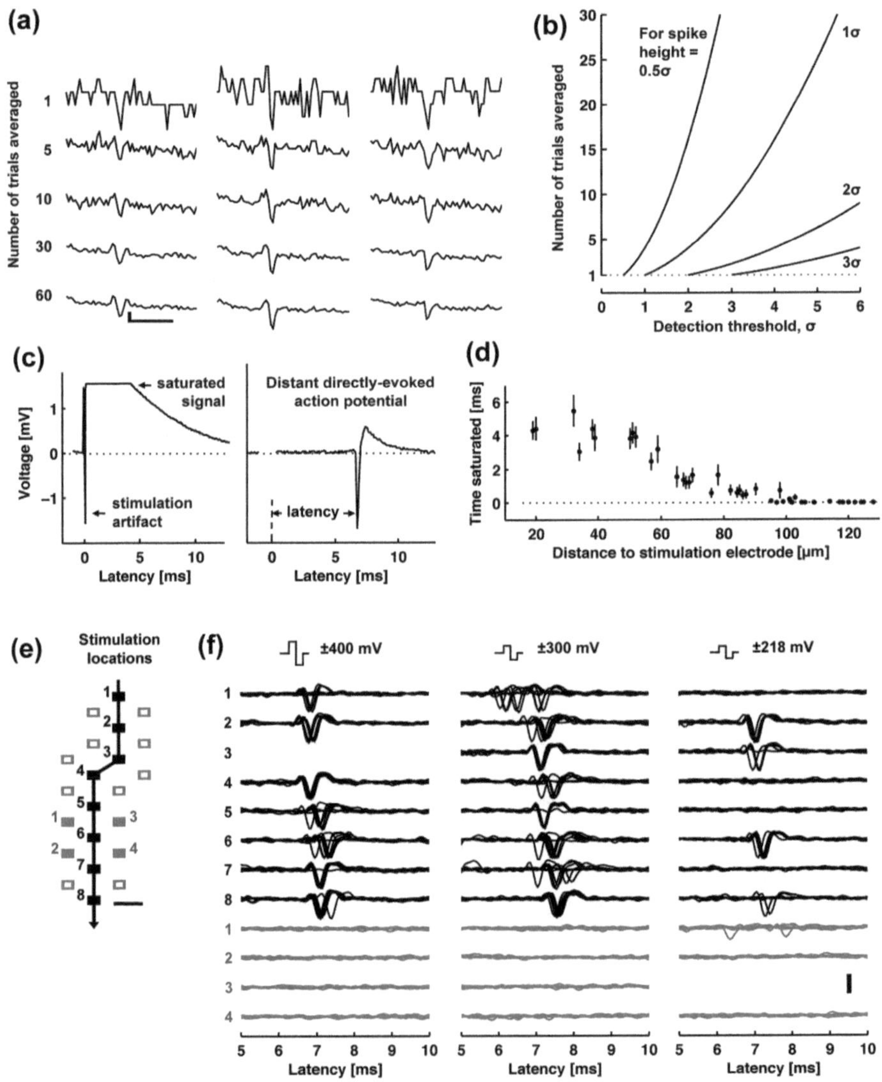

◀**Fig. 8** Stimulation capability of HDMEAs. **a–b** Stimulus-triggered averaging improves detection of axonal signals. **a** Evoked spikes detected at three chosen sites (columns) along the same axon. Each row shows individual traces obtained by increasing the number of averaged trials, from 1 to 60. Scale bars, 1 ms horizontal, 10 μV vertical. **b** The number of averaged trials necessary to detect a spike with a given height (0.5–3 times the standard deviation of the noise, σ) with respect to the detection threshold. **c–d** Electrical stimulation affects recorded signals of electrodes <100 μm away from the stimulation site. **c** *Left* A raw trace recorded at an electrode neighboring a stimulation electrode (18 μm away) saturated for about 4 ms (*flat line*). *Right* A raw trace recorded at an electrode located 1.46 mm away from a stimulation electrode did not saturate. **d** The duration of a saturated signal occurring after stimuli decreases with increasing distance from the stimulation electrode (mean ± s.e.m.; $N = 18$ stimulation electrodes from five HDMEAs). Stimuli consisted of biphasic voltage pulses between 100 and 200 ms duration per phase and between ± 400 and 800 mV amplitude. **e–f** Electrical stimulation can be delivered locally to axons and evoke action potentials. **e** Locations of stimulation electrodes that directly evoked (*black boxes*) or did not evoke (empty or *filled gray boxes*) APs detected at a soma located ~890 μm away. The *line arrow* indicates the orthodromic propagation direction. Scale bar, 20 μm. **f** Voltage traces of somatic APs elicited by biphasic voltage stimuli. Traces in response to eight stimuli are overlaid for each of three stimulation magnitudes (indicated at the *top*), plotted for all effective (*black*), and four ineffective stimulation sites (*gray at the bottom*). Stimulation electrode locations are represented as numbered boxes in (**e**). Scale bar, 200 μV. All panels and description adapted with permission from [60]

3 Applications in Neuroscience

In this section, we discuss specific neuroscience studies from selected experiments. Measurements done using passive MEAs can also be done using CMOS-based HDMEAs. However, the high spatiotemporal resolution of HDMEAs leads to novel types of data that were not possible to collect using conventional MEA devices.

In recent years, CMOS-based MEAs have been increasingly used for neuroscience and biomedical research. Figure 9 lists the currently available CMOS-based in vitro MEAs, their key specifications, and the experimental preparations for which they have been applied so far. The two most prominent preparations investigated using these devices are dissociated cell cultures from snails [44], rats [8, 49, 60, 65, 67–70] and chicken [65] and acute retina from mice [71–76], rats [77–79], rabbits [9, 80, 81], hamsters [82], guinea pigs [42, 83], and humans [84]. Additionally, data from acute slices of cerebellum [85, 86], cortex [87, 88], and olfactory bulb [46] have been presented. Cultured cardiomyocytes were also studied [30, 34, 89–91] and first results from mice organotypic hippocampal slices were presented [92]. This section reviews recent neuroscience applications of HDMEAs.

Ref	Micrograph	Key Specs	Published Recordings	Published Stimulation
APS (Eversmann et al., 2003) Other versions: (Eversmann et al., 2011)	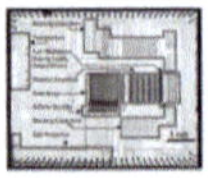	• Technology: 0.5μm • Chip area: 35mm^2 • Array area: 1.0mm^2 (2.9%) • Rec. sites: 16384 (OSFET) • Stim. sites: 0 • Channels: 16384 • Trans. density: 16384mm^{-2} • Power: 656mW	• Acute (Rr, Mr, Lr): (Menzler and Zeck, 2011; Stutzki et al., 2014; Zeck et al., 2011) • Cultures (S, Rc): (Eversmann et al., 2003; Lambacher et al., 2010)	• Acute (Lr): (Eickenscheidt et al., 2012)
APS (Heer et al., 2006) Other versions: (Heer et al., 2004)	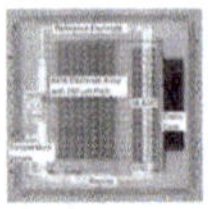	• Technology: 0.6μm • Chip area: 42mm^2 • Array area: 8.0mm^2 (19%) • Rec. sites: 128 (Pt) • Stim. sites: 128 (Pt) • Channels: 128 • Trans. density: 16mm^{-2} • Power: 120mW	• Cultures (Cch, Rc): (Hafizovic et al., 2007; Heer et al., 2007, 2004)	• Cultures (Rc): (Hafizovic et al., 2007)
APS (Imfeld et al., 2008) Other versions: (Berdondini et al., 2005;		• Technology: 0.35μm • Chip area: 29mm^2 • Array area: 7.2mm^2 (25%) • Rec. sites: 4096 • Stim. sites: 16 (newer versions) • Channels: 4096 • Trans. density: 567mm^{-2} • Power: 132mW	• Acute (Mcr): (Ferrea et al., 2012; Maccione et al., 2014; Medrihan et al., 2014) • Cultures (Rch): (Gandolfo et al., 2010; Imfeld et al., 2008)	• Cultures (Rc): (Maccione et al., 2013)
SM (Frey et al., 2009)	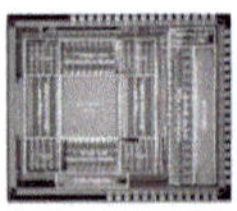	• Technology: 0.6μm • Chip area: 46mm^2 • Array area: 3.5mm^2 (7.6%) • Rec. sites: 11011 (Pt) • Stim. sites: 11011 (Pt) • Channels: 126 • Trans. density: 3150mm^{-2} • Power: 135mW	• Acute (Rpr, Lr, Hr, Mpr): (Frey et al., 2009; Fiscella et al., 2012, 2014, 2015; Jones et al., 2014, 2015; Obien et al., 2014; Reinhard et al., 2014; Franke et al., 2016) • Cultures (Rh, Rc): (Sanchez-Bustamante et al., 2008; Bakkum et al., 2013; Lewandowska et al., 2015, 2016) • Organotypic (Mc): (Gong et al., 2014)	• Acute (Rr): (Lloyd et al., 2014) • Cultures (Rc): (Bakkum et al., 2013)
SM (Huys et al., 2012)		• Technology: 0.18μm • Chip area: 64mm^2 • Array area: 14.7mm^2 (23%) • Rec. sites: 16384 (W) • Stim. sites: 16384 (W) • Channels: 1 • Trans. density: 1111mm^{-2} • Power: 35mW	• Cultures (Rh): (Huys et al., 2012)	• Cultures (Rh): (Huys et al., 2012)
APS (Johnson et al., 2013a) Other versions: (Johnson et al., 2013b)		• Technology: 0.18μm • Chip area: 4mm^2 • Array area: 2.8mm^2 (70%) • Rec. sites: 1120 (Al/Pt) • Stim. sites: 0 • Channels: 1120 • Trans. density: 400mm^{-2} • Power: 14.1mW	• Acute (Mo): (Johnson et al., 2013a)	
SM (Ballini et al., 2013)	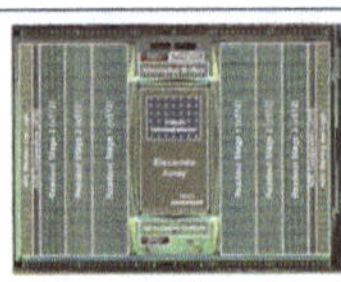	• Technology: 0.35μm • Chip area: 77mm^2 • Array area: 8.1mm^2 (11%) • Rec. sites: 26400 (Pt) • Stim. sites: 26400 (Pt) • Channels: 1024 • Trans. density: 3265mm^{-2} • Power: 75mW	• Cultures (Rc): (Ballini et al., 2014; Müller et al., 2015) • Acute (Mr): (Yonehara et al., 2016)	• Cultures (Rc): (Ballini et al., 2014; Müller et al., 2015)
APS (Bertotti et al., 2014)	Not available	• Technology: 0.18μm • Array area: 4.3; 1.1mm^2 • Rec. sites: 4225 (OSFET) • Stim. sites: 1024 (CAP) • Channels: 4225 • Trans. density: 977; 3906mm^{-2}	• Acute (Gr): (Bertotti et al., 2014)	• Acute (Gr): (Bertotti et al., 2014)
SM (Viswam et al., 2016)	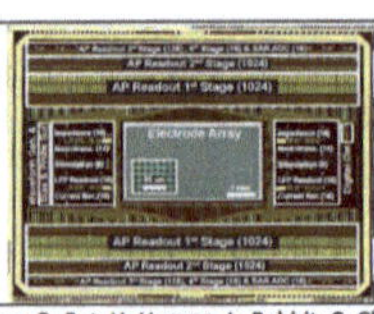	• Technology: 0.18μm • Chip area: 106.8mm^2 • Array area: 10.9mm^2 • Rec. sites: 59760 (Pt) • Stim. sites: 59760 (Pt) • Channels: 2048 (AP)+32(LFP) • Trans. density: 5487mm^{-2}	• Cultures (Rc): (Viswam et al., 2016)	• Cultures (Rc): (Viswam et al., 2016)

M: Mouse, R: Rat, H: Human, L: Rabbit, C: Chicken, G: Guinea pig, S: Snail
r: Retina, p: Cerebellum, o: Olfactory bulb, c: Cerebral cortex/hippocampus, h: Cardiomyocytes

Fig. 9 CMOS-based in vitro MEAs, their key specifications, and references to biological applications for recording and stimulation. The specifications may differ for other device versions. Modified with permission from [12]

3.1 Dissociated Cell Culture

3.1.1 Investigating Axonal Signals

HDMEAs with high SNR, such as SM HDMEAs [8, 10], allowed detection and tracking of APs propagating along a neuron's axon over days for the first time [60]. Axonal signals are difficult to measure using conventional methods—thin axons are challenging to patch and extracellular signal amplitudes are low compared to those from the soma and axon initial segment. In this work, the propagation of APs along the full arbor of a neuron has been electrically imaged, as shown in Fig. 10. Subsequently, axonal AP velocity was found to vary within single axon, hinting that axon velocity might contribute to temporal coding schemes of neuronal information. This capability can help expand new fields of research, such as axonal information processing and neuronal computation. Tracking the velocity of axonal signals also provides a new and promising parameter that can be used for analyzing the effect of different therapies, e.g., drugs and prolonged electrical stimulation, to the information transfer and signaling between neurons.

HDMEAs have also been used for precise microstimulation. By taking advantage of the high electrode density, the responses of neurons to different stimulation patterns and intensity can be investigated, including how best to selectively

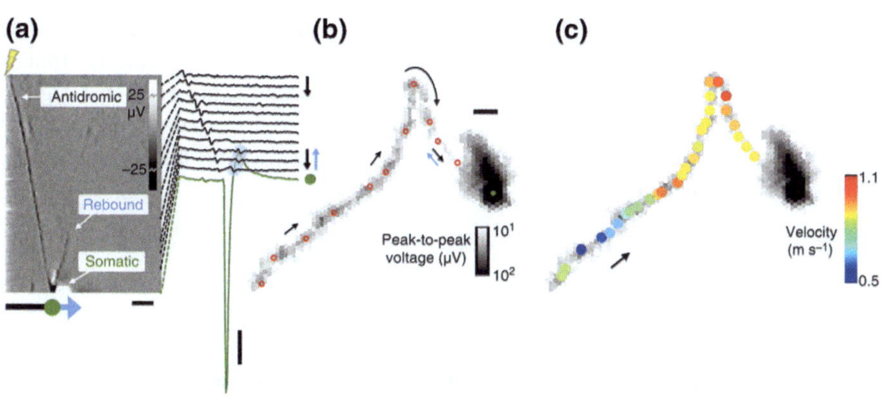

Fig. 10 Imaging axonal signal propagation with HDMEAs. **a** Antidromic action potential triggered by electrical stimulation at the axon. *Left* Heat map shows stimulation-triggered averages of 60 traces from 95 electrodes. *Dark colored line* from *top* to *bottom* indicates the antidromic propagation of an AP from the stimulation site. A subsequent rebound from the soma is also visible. *Right* A subset of averaged raw traces. Scale bars, 1 ms horizontal; 100 μV vertical. **b** The electrical footprint of the stimulated neuron recorded in (**a**). The *gray*-scale pixels indicate the maximum peak-to-peak amplitude of the APs detected at each electrode. The *red circles* denote the locations of the subset of traces in (**a**). The *black arrows* show the direction of the AP propagation along the axon, while the *blue arrow* indicates the subsequent rebound. The *green dot* is the location where the *green* trace in (**a**) was recorded from. Scale bar, 100 μm. **c** AP propagation velocity changes, as shown by the *colored dots* along the electrical image of the axon. Adapted with permission from [60]

stimulate single neuron [93]. Characterization of how stimulation affects neurons can benefit the design of stimulation therapies for clinical use.

3.1.2 Analyzing Networks

HDMEAs, especially devices with a readout channel count on the order of thousands, can be utilized to analyze neuronal networks in dissociated cell cultures. The spontaneous activity of neuronal networks varies from one culture to another. There can be behaviors, however, that are similar. For example, using parametric models to analyze activity patterns and their sensitivity to changes showed that neuronal networks are *sloppy*, i.e., insensitive to changes in many parameter combinations, but very sensitive to a few [94]. Sloppiness behavior was also observed in vivo by recording from monkey visual cortex. Another study investigated synchronized activities in spontaneously active cell cultures [95], see Fig. 11. The authors discovered that a repertoire of repeating spatiotemporal patterns exists during network bursts. Some patterns shared similar sequence of activation of subpopulations of neurons, which suggested that the network conserves certain spatiotemporal activation patterns. Different sets of activated subpopulations of neurons tended to appear repeatedly, indicating that such spatiotemporal patterns may be related to the state of the network.

For low-density cultures, functional connectivity of neurons in the network has been estimated [96]. Using low-density cultures, optical visualization of stained neurons provides an estimate of the structural connectivity of neurons. Imaged locations of neurons support the identification of neurons using HDMEA signals. Maccione et al. processed and analyzed the functional connectivity of neurons from HDMEA signals by ad hoc developed spatiotemporal filtering and by applying a cross-correlation-based method.

3.2 Acute Retina

3.2.1 Extracting Retina Cell Types by Light Stimulation

The retina encodes visual input and sends information to the brain via the optic nerve. Retinal ganglion cells (RGCs) serve as the final output of the retina; they encode visual information as complex patterns of APs. The planar arrangement of RGC bodies and axons is highly compatible with MEA recordings from retina explants. Acute retina tissue superfused with oxygenated Ringer's medium can survive and respond to light stimulations for several hours. A significant capability of applying HDMEAs to study the retina is the identification of different RGC types and the mapping of their receptive fields. Fiscella et al. developed a method to examine the AP responses of mice RGCs to different light stimulation patterns, e.g., moving bars at different angles. Different populations of ON–OFF direction-selective RGCs were identified by extracting the APs

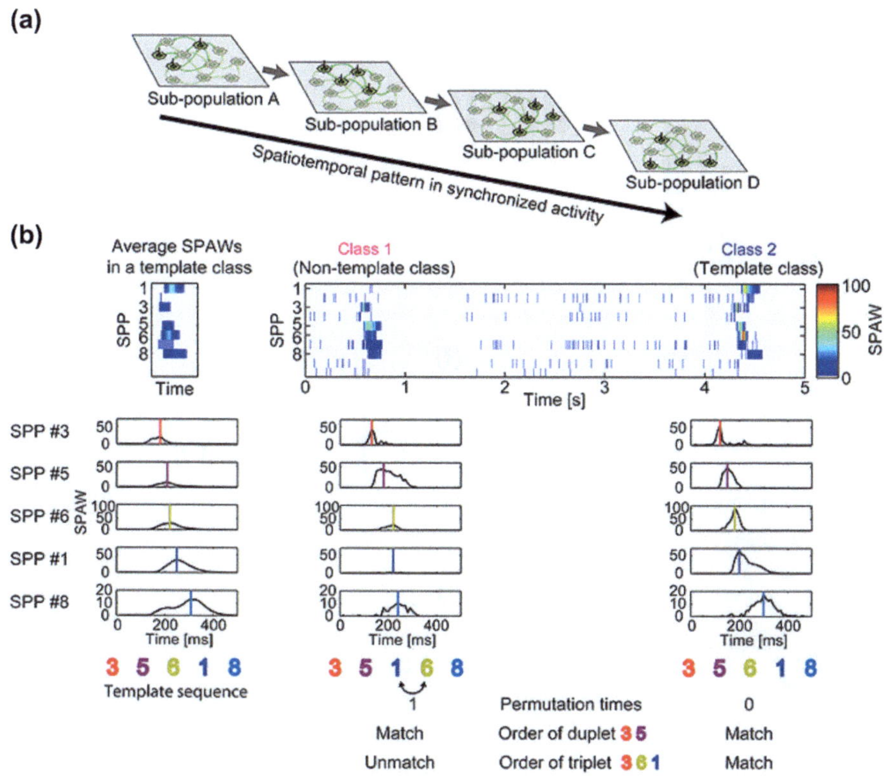

Fig. 11 Analysis of subpopulation activation during synchronized bursts. **a** Schematic illustration of the sequential activation of neuronal subpopulations during synchronized spontaneous activity in cell cultures. Such sequential activation of subpopulations is hypothesized to generate stable spatiotemporal patterns. **b** Illustration of the procedure to evaluate the sequences of subpopulation activation during bursts. *Upper right* A sample template of subpopulation activation weights (SPAWs) of five subpopulation patterns (SPPs) during a burst. *Lower right* SPAWs from the upper right image were converted into a sequence of their peaks. *Upper left* SPAWs computed for the chosen SPPs during a 5 s recording. *Lower middle* to *right* Subpopulation sequences of bursts were compared with the template sequence. Permutation times for matching refers to the way in which many pair permutations are required to match the template sequence. Duplet/triplet order matching indicates whether the order of two/three subpopulations matches the template sequence. Adapted with permission from [95]

synchronized with light stimulations and by spike sorting [72], see Fig. 12. Proper intensity and focal width adjustments of light stimulation aimed at the array area are important to prevent light-induced noise or fluctuations to the recording channels.

A similar technique was used to characterize hamster RGCs [82]. Light stimulation responses of different RGC populations were evaluated according to direction selectivity, speed tuning, width tuning, transience, and latency. Out of 262 RGCs recorded using HDMEA, seven distinct RGC groups were extracted using the evaluated parameters.

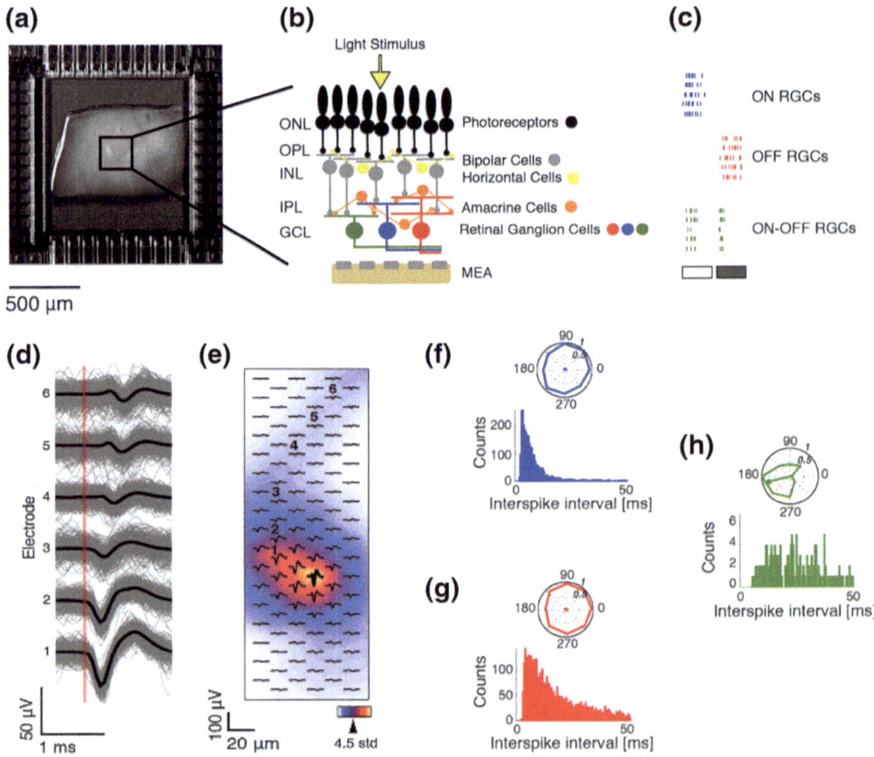

Fig. 12 Classification of ON–OFF retinal ganglion cells using HDMEAs and light stimulation. **a** Mouse retina patch placed on the HDMEA with the ganglion-cell-side down. **b** Schematic of the mammalian retina: ONL, outer nuclear layer; OPL, outer plexiform layer; INL, inner nuclear layer; IPL, inner plexiform layer; GCL, ganglion cell layer; MEA, microelectrode array. Light stimulus detected by the photoreceptors will be processed by the retinal circuitry and converted into action potentials, which are generated by the retinal ganglion cells (indicated by *green*, *blue,* and *red* cells) in the GCL. APs can be recorded by the MEA. **c** Samples of spike trains belonging to three different types of retinal ganglion cells (*Blue* ON type, *Red* OFF type, *Green* ON–OFF type) in response to the same light stimulus. The *white bar* represents a light stimulus brighter than the background light level. The *gray bar* represents a light stimulus darker than the background light level. **d** Superposition of 959 APs (*gray* traces) from six electrodes, indicated in (**e**). **e** Electrical footprint of a single RGC using averaged signals. The *thick-black* waveform denotes the highest peak-to-peak amplitude (central electrode). *Color code* (*right corner*) *yellow-red* indicates the region of maximum signal amplitude (active electrodes with APs –4.5 standard deviations above the noise level). **f–h** Physiological response of RGCs. *Top* Polar plot showing the responses of the RGC to motion of a bar in 8 directions at 45° radial intervals. *Bottom* Inter-spike interval distribution showing the time intervals between consecutive spikes. **f** *Blue* = ON RGC. **g** *Red* = OFF RGC. **h** *Green* = ON–OFF RGC. Adapted with permission from [72]

3.2.2 Network Oscillations in Degenerated Retinas

Aside from EAPs, LFPs have been detected from retinas using HDMEAs. In rod-degenerated (*rd1*) mice retinas, ON- and OFF-type RGCs exhibit spontaneous oscillatory spike activity due to the loss of photoreceptors [97]. Such oscillatory spiking was found to coincide with LFPs through HDMEA recordings [75], as shown in Fig. 13. LFPs initiated at random locations on the retina and propagated across the retina. LFPs also persisted in the presence of pharmacological blockers, such as TTX. The authors suggested that large-scale retinal network oscillations are caused by excitation of electrically coupled interneurons.

3.3 Acute Brain Slice

An acute brain slice can be placed on MEAs to monitor the electrical activity of neurons in a 3D environment. Cutting the brain into very thin slices allows access to neurons deep in the brain for imaging, i.e., mapping the anatomy. The same method can be used for recording the activity of neurons that are otherwise difficult to reach and identify in vivo. Perfusion of fresh oxygenated artificial cerebrospinal fluid keeps the neurons viable for up to 10 h. The neurons and network structure in slices are physiologically and biochemically similar to the in vivo situation.

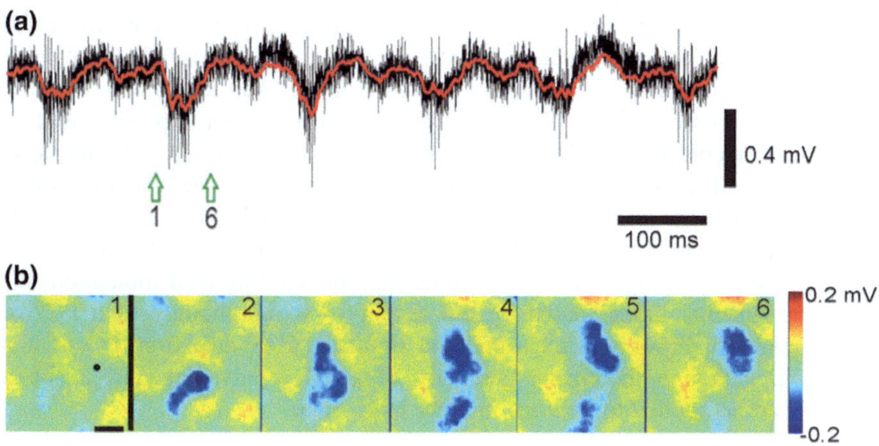

Fig. 13 RGC spiking and local field potential minima coincide in rod-degenerated mice retinas. **a** Extracellular recordings from an APS HDMEA electrode, indicated by a *black dot* in subplot (*b1*). Filter settings of 1 Hz to 3 kHz revealed single spike and a slow oscillatory extracellular potential. The superimposed red trace, low pass filtered at 1 to 60 Hz, reveals LFPs. The open *green arrows* mark the start point and end point of a six frame series of extracellular voltage maps, shown in (**b**). **b** Extracellular voltage maps or "electrical images" recorded at a spatial resolution of 8 × 16 μm captured the origin and propagation of LFPs. Each image represents the average extracellular voltage over 2 ms. Separation between images is 10 ms. Scale bar, 200 μm. Adapted with permission from [75]

3.3.1 Analyzing Local Network Activity in Brain Slices

LFPs and oscillations inherent in different states of the brain can be measured. Such recordings have been done for different brain areas, e.g., hippocampus, suprachiasmatic nucleus, etc. HDMEAs have the ability to extract electrical images of neuronal network activity in slices at high spatial resolution. For instance, functional imaging of the dentate gyrus has been demonstrated using HDMEAs [87]. Field excitatory postsynaptic potentials (fEPSPs) evoked by electrical stimulation were detected across different layers of the acute slice, as shown in Fig. 14a–c.

High-resolution imaging of epileptiform activity in slices has also been done. Not only can the propagation of activity be mapped using HDMEAs, but more data for analysis and statistics can be obtained compared to conventional MEAs. Medrihan et al. [88] showed that the absence of synapsin II (Syn II), a protein related to epilepsy, decreases tonic inhibition in mouse hippocampal slices, thus increasing synchronized bursts (see Fig. 14d, e). THIP (4, 5, 6, 7-tetrahydroisoxazolo [5,4-c] pyridin-3-ol; gaboxadol), a selective agonist of δ subunit-containing $GABA_A$ receptors, restores tonic inhibition.

3.3.2 Analyzing Single Neuron in Brain Slices

Depth recording of EAPs from neurons up to 100 μm distance from the MEA surface was also shown [98, 99]. Using HDMEAs, subcellular resolution recording from single Purkinje cell (PC) in acute cerebellar slices has been demonstrated [85]. To ensure the quality of recorded signals, proper tissue adhesion on the MEA surface has to be maintained throughout the experiment. Adhesion can be achieved by cellulose nitrate coating [98], but also by a slice anchor typically used for patch-clamp recordings. EAPs were observed along the PC layer and, after spike sorting, the EAP footprint of a single PC was analyzed. The negative spikes were recorded around the perisomatic area of the neuron, while positive spikes were obtained along the molecular layer corresponding to the dendrites of the PC. A comparison of the high spatiotemporal resolution recording with simulations of a full-compartmental model based on the stereotypical morphology of a PC was done. Figure 15 shows both measured and simulated EAP data from PCs at high resolution. Although the planar geometry of a PC is advantageous, a similar analysis can be applied to neurons in other brain areas.

3.4 Organotypic Brain Slice Culture

The organotypic brain slice culture method has been applied to study different brain areas, such as hippocampus [100], striatum [101], cortex, spinal cord [102], and cerebellum [103]. Moreover, several co-culture models study the interactions

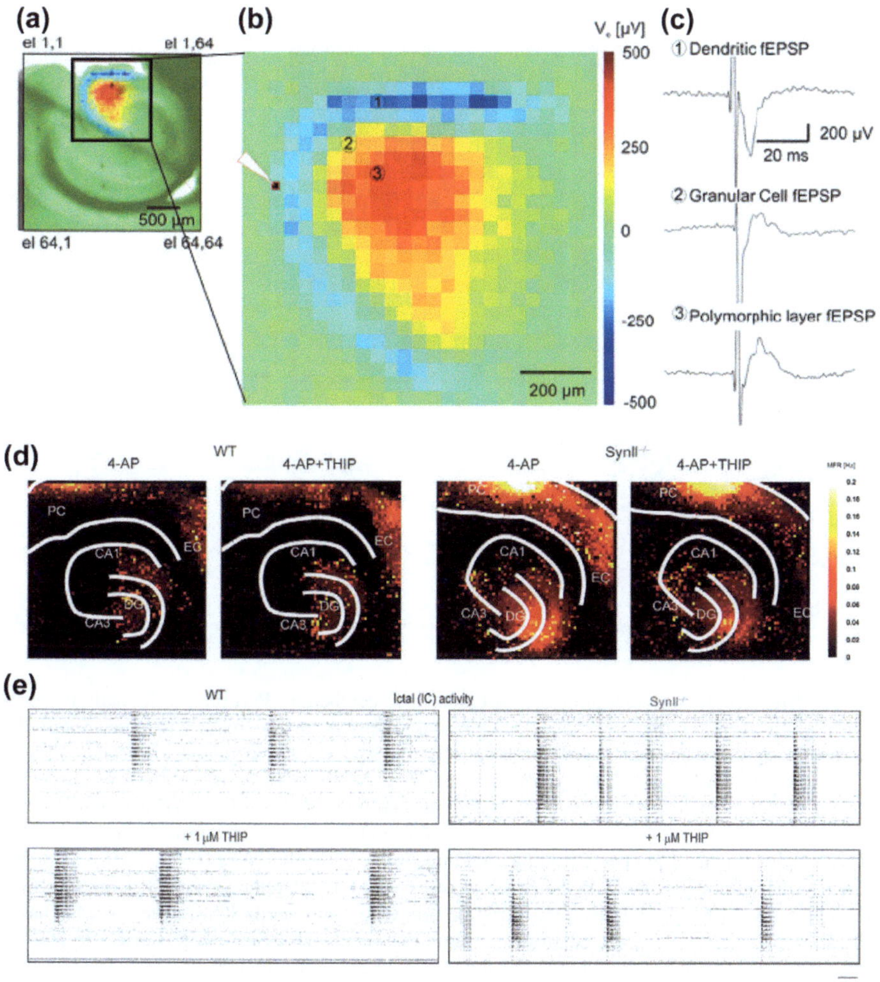

Fig. 14 Network waves in acute hippocampal slices. **a–c** Functional imaging of the dentate gyrus, adapted with permission from [87]. **a** A cortico-hippocampal slice placed atop an APS HDMEA, with superimposed color-coded fEPSP activity. **b** Close-up on the activated area in (**a**). The *white tip* indicates the site of stimulation using a patch pipette. **c** Recorded traces of fEPSPs from three chosen electrodes indicated in (**b**). Electrode 1 is located in the dendritic layer of the dentate gyrus, electrode 2 in the granular cell layer, and electrode 3 in the polymorphic layer. **d**, **e** Using HDMEAs to study the effect of deleting synapsin II (Syn II) on tonic inhibition in mouse hippocampal slices, adapted with permission from [88]. **d** Mean firing rate computed from each electrode from WT and Syn II knock-out hippocampal slices before and after THIP treatment. THIP: (4, 5, 6, 7-tetrahydroisoxazolo [5, 4-c] pyridin-3-ol; gaboxadol), a selective agonist of δ subunit-containing GABAA receptors. **e** Raster plots showing highly synchronized bursts, x-axis corresponds to time, y-axis corresponds to pixels (electrode). THIP reduced the high frequency bursts in Syn II knock-out hippocampus. Scale bar: 1 min

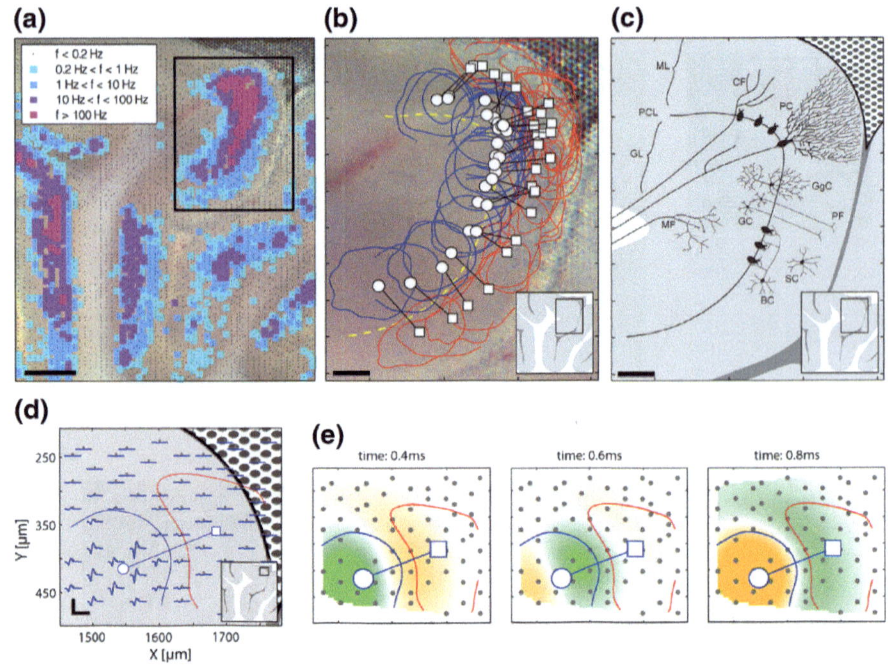

Fig. 15 High-resolution mapping of spontaneous cerebellar Purkinje cell activity using HDMEAs. **a** Activity map of the detectable spike activity in the recording area. *Small dots* correspond to the electrodes used for recording ($\sim$30% of the available electrodes). Events exceeding a threshold of ±36 μV were used to calculate the color-coded event rate. Scale bar: 0.3 mm. **b** Close-up of a region with high activity delimited in (**a**). All units identified by spike sorting are marked, i.e., the somatic region is *blue* and the dendritic region is *red*. Scale bar: 0.1 mm. **c** Schematic of the basic cellular structures in the cerebellar slice (*Gray*, 1918). Scale bar: 0.1 mm. ML, molecular layer; PCL, Purkinje cell layer; GL, granular layer; CF, climbing fiber; MF, mossy fiber; PF, parallel fiber; PC, Purkinje cell; GgC, Golgi cell; SC, stellate cell; BC, basket cell. **d** Footprint of a PC selected from the region shown in (**b**). Scale bar: vertical is 200 μV, horizontal is 1.9 ms. **e** Current source density (CSD) analysis for the cell shown in (**d**) at several points in time (*green* sink; *yellow* source). The sink moves from the soma at 0.4 ms to the proximal dendrites at 0.6 ms and covers the dendritic area, while the soma repolarizes. Frequency band: 180 Hz–3.5 kHz. **f–h** Matching simulated and measured EAP footprints. All panels and descriptions adapted with permission from [85]

between different brain networks, such as entorhino-hippocampal, cortico-spinal, and cortico-striatal preparations [104]. Compared with dissociated cell cultures, organotypic brain slice cultures better preserve the anatomy of brain regions, since the synaptic connections have been partially maintained. Advantages of organotypic slice cultures over acute slice cultures include a long (weeks) experimental time window with multiple measurement points. This allows observing effects from chronic manipulations on slice cultures.

3.4.1 Different Methods for Organotypic Slice Cultivation

The two most commonly used methods of organotypic slice preparation are the roller tube method [105] and the membrane interface method [106]. The roller tube method glues the slice culture on top of a glass coverslip using a drop of plasma and thrombin. The slice culture is continuously rotated inside a test tube, such that the slice cultures would be immersed in the culture media during half a cycle of the rotation, and exposed to air during the other half. On the other hand, the membrane interface method cultivates slice cultures on top of semipermeable membranes at the interface between culture medium below and air above. The slice cultures receive extra oxygen from the air, while still having contact with the culture medium. The membrane interface method is an easier preparation, while the roller tube method produces monolayer-thick slice cultures making it easier to image individual cells for analysis. The roller tube method can be directly integrated with MEAs by adhering slices on MEAs instead of glass slides. On the other hand, the membrane interface method cannot be directly implemented on MEAs because the slice would need to interface with the submerged MEA surface for recording. Instead, the slice culture can be cut off from the membrane insert and flipped upside down to face the electrode array in a one-time experiment. In this case, an anchor, similar to one used for acute slice experiments, is required on top of the slice culture to improve the contact between the slice culture and the electrode array.

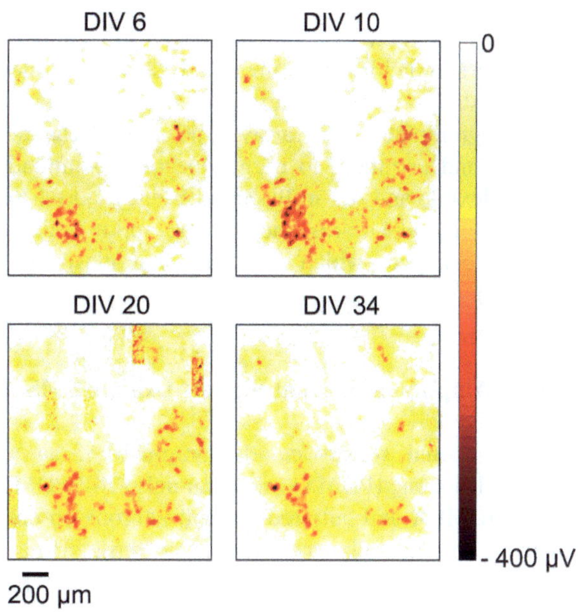

Fig. 16 Electrical activity maps of a hippocampal slice culture at four different time points (6–34 days in vitro, DIV). The *colors*, coded from *white* to *dark red*, indicate the largest amplitudes detected at each electrode. The neurons in the organotypic slice remained spontaneously active for over one month, with decreasing spiking amplitudes after DIV 10. Adapted with permission from [107]

Long-term HDMEA recordings of hippocampal slice cultures have been shown [107]. By developing a novel system based on the roller tube method, organotypic hippocampal slices were cultivated on HDMEAs for months. This system allowed continuous observation of neuronal spike activity at a single-unit and network levels at the same time, which was done over consecutive weeks at daily intervals, as presented in Fig. 16.

3.5 Other Advanced Techniques

HDMEAs allow easy integration of other methods and techniques for advanced measurements. Depending on the study, other equipment for neuronal recording and manipulation can be combined with the HDMEA setup, and the data from each tool can be synchronized. Microdevices, such as microfluidics, can also be readily attached on the HDMEA surface. Moreover, partial data analysis can be performed online, which can be useful to automatically deliver stimulation and control combined devices.

3.5.1 Investigating Neuronal Plasticity Through Closed-Loop Stimulation

SM HDMEAs enable simultaneous stimulation and recording of arbitrarily selected neurons in a network. By changing spike timing between sets of neurons via electrical simulations, the functional network connectivity was also changed [66]. In this study, a reprogrammable event engine unit was programmed into a field-programmable gate array. The system can detect arbitrary action potential patterns and use these to trigger electrical stimulations to arbitrary neurons, providing flexible and sub-millisecond latency closed-loop feedback. Cross-correlation analysis of spike trains showed the spike timing of the selectively stimulated neurons changed, which indicated that plasticity was induced in the network (Fig. 17).

3.5.2 Amplifying Axonal Signals Using PDMS Tunnels

Neuron axonal signals can be amplified if an axon is grown through polydimethylsiloxane (PDMS) micro-tunnels [70]. PDMS tunnels were attached on top of an HDMEA and cultured cortical neurons on each side of the tunnels. In time, many axons naturally enter and grow through the tunnels. While axonal signals outside of tunnels were also detectable, tunnels amplify the signals enough to avoid the need to average across trials and single axonal AP can be observed. By recording the spontaneous activities of the neurons, axonal signals were significantly amplified by a factor of 20–150.

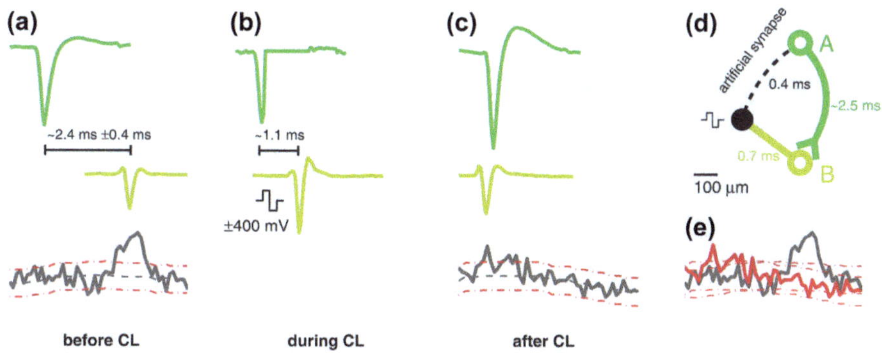

Fig. 17 Effect of closed-loop stimulation. **a** Spontaneous activity of two neurons before the application of the closed-loop stimulation. Spike traces are median waveforms of several spikes aligned at the negative peak. *Top* In *green*, spike trace from neuron A, the trigger neuron. *Middle* In *yellow*, spike trance of neuron B, a neuron with correlated spiking activity. *Bottom* Cross-correlation curve of spike times of neuron B with respect to neuron A. *Red-dotted lines* denote the 95% confidence intervals. Around 2000–3000 spikes were used to compute the cross-correlation. Elevated correlated activity of neuron B was observed around 2.4 ± 0.4 ms after neuron A fired an AP. **b** Same as (**a**), but with closed-loop feedback stimulation applied. The time delay of the spikes between neurons A and B was reduced to around 1.1 ms. Stimulation is applied upon detection of a spike from neuron A. During stimulation, the trace of neuron A was zeroed out. **c** Same as (**a**), but after application of the closed-loop stimulation. The cross-correlation plot changed after closed-loop stimulation. **d** Schematic of the synaptic connectivity between neurons A and B and the artificial synapse caused by closed-loop stimulation. **e** Comparison between the cross-correlation curves before (*black*) and after (*red*) the closed-loop stimulation. Adapted with permission from [66]

Stimulating the neurons at different stimulation frequencies caused changes in spike shape and velocity of axonal signals inside the channels [69]. Velocity and spike height decreased for increased stimulation frequency. Furthermore, potassium ion concentrations impacted axonal spike amplitude and propagation velocity. High potassium concentration improved signal fidelity while low concentrations caused axonal propagation failures (Fig. 18).

4 Applications for Medicine

HDMEAs provide a promising platform for drug screening and pre-clinical therapy diagnostics of electrogenic cells, e.g., brain, retina, and heart cells. In particular, human-induced pluripotent stem cells (hIPSCs) require an efficient readout for functional analysis, which can be achieved using HDMEAs. Culturing hIPSCs on HDMEAs has been proven feasible and the cells remained viable up to three months [108]. Spontaneous activity and responses to electrical stimulation were

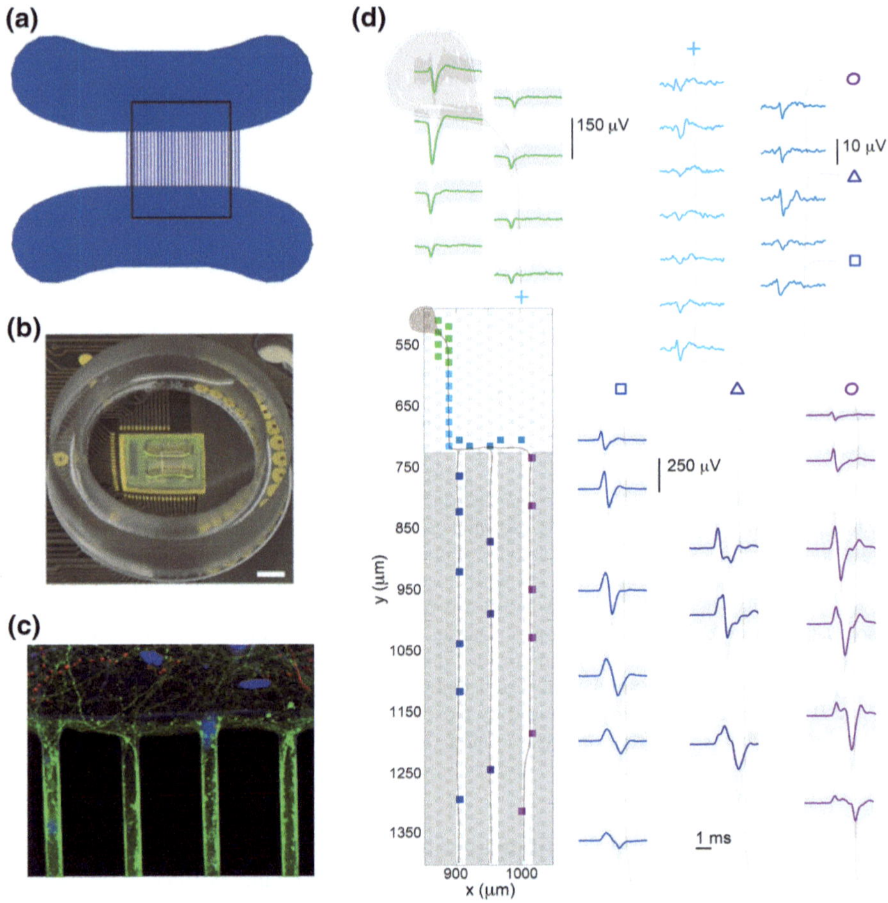

Fig. 18 PDMS axonal channel device on HDMEA amplifies axonal signals. Spontaneous spike propagation from a soma to a branched axon. **a** Illustration of the axonal channel device, with 12 μm channels, and neuronal culture chambers. The *black rectangle* shows dimensions of the HDMEA used (1.75 × 2.0 mm^2). **b** Photograph of a packaged HDMEA chip wire bonded to a printed circuit board and with the PDMS channel device on *top*. The plastic ring (18 mm diameter) holds the cell culture medium. Scale bar, 2 mm. **c** Close-up view of a fluorescence image of cultured neurons at the top chamber. Tau1 (axons) is shown in green, DAPI (nuclei) in blue, and GFAP (glia) in *red*. **d** Sample waveforms recorded in the PDMS channels. Individual spikes are shown in *gray* and spike-triggered averaged waveforms are indicated by *colors green* for signals from the perisomatic area and *blue* to *purple* from the axons under the channel device. A branching axon growing into three channels was observed. The small axonal signal outside of the channels is shown in *light blue* (averaged). Electrodes in three adjacent channels recorded spikes that were time aligned with that of the soma, and their positions and spike shapes are shown. The spike amplitudes detected under the channels were significantly larger compared with the signals from the axon outside the channels. A cartoon neuron was drawn over the traces to guide the eye. Adapted with permission from [70]

characterized. The authors found that spontaneous spiking activity of hIPSCs peaked around 81 DIV and that hIPSCs responded to electrical stimulation. Low-frequency electrical stimulations (0.2 Hz, biphasic current with peak-to-peak

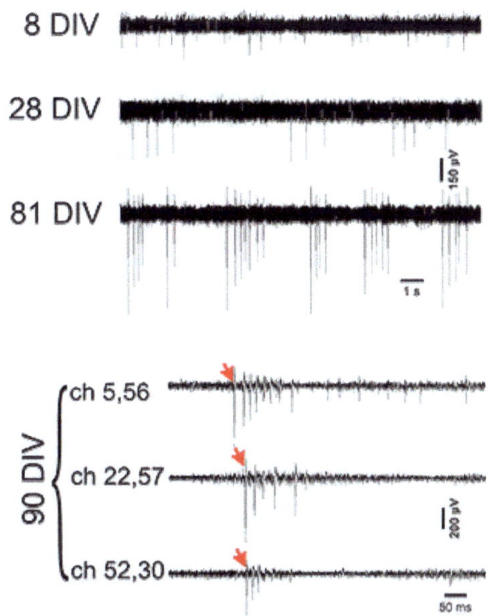

Fig. 19 Spontaneous activity of hIPSCs. Recorded extracellular signal traces show changes in firing rates during development. The activity develops from single spike (8 DIV), tonic firing (28 DIV) to bursting, and synchronized spikes (81, and 90 DIV). *Red arrows* denote the start of propagating burst. Adapted with permission from [108]

amplitude of 300 µA) led to an increase in the number of active electrodes, i.e., from 564 ± 28 to 688 ± 21, but decreased the mean firing rate, i.e., 0.66 ± 0.03–0.58 ± 0.03 spikes/s). Figure 19 summarizes these results.

The recent advent of CRISPR/Cas9-mediated genome editing has paved the way for fast development of disease models [109]. Mice models of human diseases can be used to characterize the functional differences of cells from different parts of the body compared to their healthy counterparts. HDMEAs can provide high-throughput and high-quality characterization of cells in culture and in acute preparations. One application of such characterization is biomarker identification, which has been done for a human retina disease called congenital nystagmus caused by FRMD7 gene mutation [76]. A mouse model of such disease was developed and light stimulation evoked responses of RGCs in the retina were recorded and analyzed in a high-throughput manner. Using HDMEAs, it was found that FRMD7 mutation leads to selective loss of horizontal selection selectivity of RGCs, as illustrated in Fig. 20.

5 Outlook

This chapter has shown the current state of CMOS-based HDMEA research in terms of technology and biomedical applications. Novel types of data can be obtained, which opens up new waves of possibilities for neuroscience discoveries and medical advancements. Potential future developments include device hardware improvements, advanced experimental methods, and new data analysis techniques.

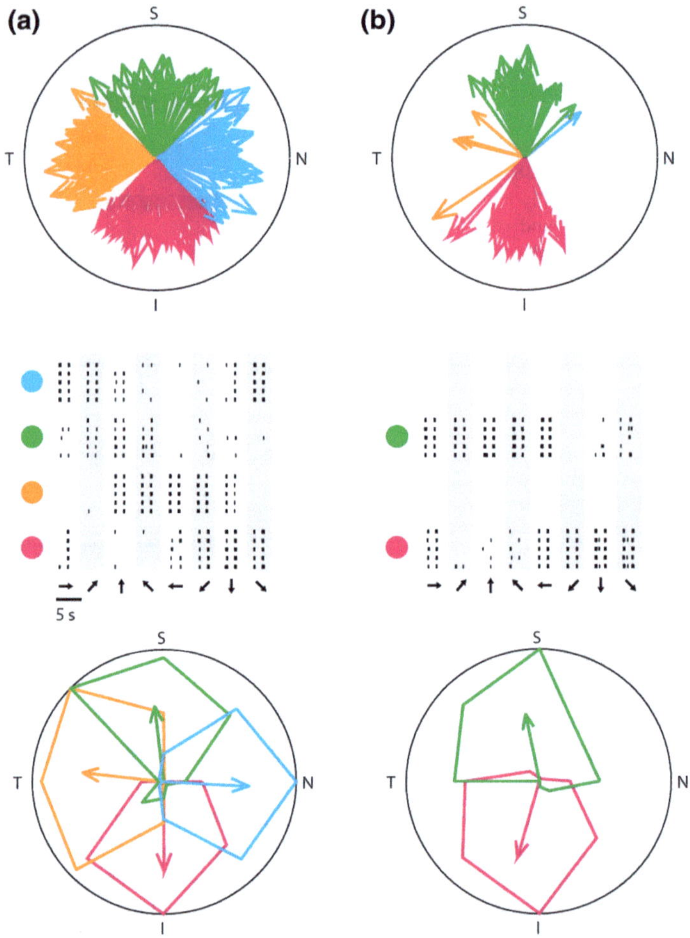

Fig. 20 Screening transgenic mouse models of human eye diseases with HDMEAs. *Top* Polar plots showing the preferred directions (directions of arrows) and direction-selectivity index (length of an arrow) of individual direction-selective retinal ganglion cells in **a** WT and **b** FRMD7^{tm} retinas. The *color code* shows the different preferred directions (*green* = superior, *blue* = nasal, *purple* = inferior, and *orange* = temporal). (*Middle* Raster plots showing the spike responses (each *black line* is a spike) of example DS cells in WT and FRMD7^{tm} retinas in response to motion in eight different directions, indicated by the arrows at the *bottom* of the plot. *Bottom* polar plots of the normalized mean spike numbers of cells shown in *middle* panes. The preferred direction and DSI of each cell are represented by the direction and length of the corresponding (*color-coded*) arrow. Adapted with permission from [76]

Next-generation HDMEAs may target increased array area, electrode density, and number of parallel recording/stimulation channels. A larger array area will extend the observable region of a sample, allowing simultaneous access to more neurons in cell cultures and to more distal brain areas in slices. This also enables opportunities for co-culturing different tissues or brain regions. Higher spatial resolution and more recording channels will assist spike-sorting accuracy and will potentially increase the number of detectable neurons per square millimeter. Multiple HDMEAs can also be combined in a multiwell-plate format, making HDMEAs compatible to current drug screening protocols.

Aside from improving the devices through resolution and scalability, adding new functionalities may also be done. Other readout circuitry may also be integrated in HDMEA devices, e.g., neurotransmitter and impedance measurement units [38]. Multi-modal measurement of neuronal activity will be helpful to understand the overall neuronal network function and the interplay between electrical activity and biochemical release.

Another promising route is the combination of HDMEA with different tools separate from the device, such as optical methods. Fluorescent calcium and voltage indicators, generic markers, and optogenetics have been used to map and manipulate brain activity. Similar to extracellular recordings, the presence of many molecules and compartments in the brain with different optical properties renders optical recording and analysis challenging. Of interest is to pinpoint the advantages and constraints of electrophysiological versus optical methods to determine how they can complement each other. For example, optogenetic manipulation of specific cellular subpopulations, while measuring the responses of the neurons using HDMEAs, will allow studying functional roles of different classes of neurons [110]. Additionally, the effect of different optogenetic therapies to compensate for neuronal dysfunction can be tested with HDMEAs.

Data obtained from next-generation HDMEAs and multi-modal experiments require advanced computational analysis and modeling techniques. Fast implementations of spike-sorting algorithms and parallel computing are needed to handle the large amounts of data produced during long-term HDMEA experiments. Multi-scale modeling, a system biology technique, may be employed to synchronize events recorded at different time and spatial scales. Overall, all data analysis methods need to be optimized to extract meaningful information within a feasible time from the massive amounts of data produced.

References

1. Alivisatos, A.P., Andrews, A.M., Boyden, E.S., et al.: Nanotools for neuroscience and brain activity mapping. ACS Nano **7**, 1850–1866 (2013). doi:10.1021/nn4012847
2. Marblestone, A.H., Zamft, B.M., Maguire, Y.G., et al.: Physical principles for scalable neural recording. Front. Comput. Neurosci. (2013). doi:10.3389/fncom.2013.00137
3. Contreras, D.: Electrophysiological classes of neocortical neurons. Neural Netw. **17**, 633–646 (2004). doi:10.1016/j.neunet.2004.04.003
4. Llinas, R.: The intrinsic electrophysiological properties of mammalian neurons: insights into central nervous system function. Science **242**, 1654–1664 (1988). doi:10.1126/science.3059497

5. Wood, C., Williams, C., Waldron, G.J.: Patch clamping by numbers. Drug Discov Today **9**, 434–441 (2004). doi:10.1016/S1359-6446(04)03064-8
6. Buzsáki, G., Anastassiou, C.A., Koch, C.: The origin of extracellular fields and currents—EEG, ECoG, LFP and spikes. Nat. Rev. Neurosci. **13**, 407–420 (2012). doi:10.1038/nrn3241
7. Henze, D.A., Borhegyi, Z., Csicsvari, J., et al.: Intracellular features predicted by extracellular recordings in the hippocampus in vivo. J. Neurophysiol. **84**, 390–400 (2000)
8. Müller, J., Ballini, M., Livi, P., et al.: High-resolution CMOS MEA platform to study neurons at subcellular, cellular, and network levels. Lab Chip **15**, 2767–2780 (2015). doi:10.1039/C5LC00133A
9. Ballini, M., Muller, J., Livi, P., et al.: A 1024-channel CMOS microelectrode array with 26,400 electrodes for recording and stimulation of electrogenic cells in vitro. IEEE J. Solid State Circuits **49**, 2705–2719 (2014). doi:10.1109/JSSC.2014.2359219
10. Frey, U., Sedivy, J., Heer, F., et al.: Switch-matrix-based high-density microelectrode array in CMOS technology. IEEE J. Solid State Circuits **45**, 467–482 (2010). doi:10.1109/JSSC.2009.2035196
11. Hierlemann, A., Frey, U., Hafizovic, S., Heer, F.: Growing cells atop microelectronic chips: interfacing electrogenic cells in vitro with CMOS-based microelectrode arrays. Proc. IEEE **99**, 252–284 (2011). doi:10.1109/JPROC.2010.2066532
12. Obien, M.E.J., Deligkaris, K., Bullmann, T., et al.: Revealing neuronal function through microelectrode array recordings. Front. Neurosci. (2015). doi:10.3389/fnins.2014.00423
13. Alpha MED Science Co., Ltd.: MED64—A low-noise and user-friendly multielectrode array system for in-vitro electrophysiology (2009). http://www.med64.com
14. Berényi, A., Somogyvari, Z., Nagy, A.J., et al.: Large-scale, high-density (up to 512 channels) recording of local circuits in behaving animals. J. Neurophysiol. **111**, 1132–1149 (2014). doi:10.1152/jn.00785.2013
15. Blanche, T.J.: Polytrodes: high-density silicon electrode arrays for large-scale multiunit recording. J. Neurophysiol. **93**, 2987–3000 (2005). doi:10.1152/jn.01023.2004
16. Du, J., Blanche, T.J., Harrison, R.R., et al.: Multiplexed, high density electrophysiology with nanofabricated neural probes. PLoS one **6**, e26204 (2011). doi:10.1371/journal.pone.0026204
17. Greschner, M., Field, G.D., Li, P.H., et al.: A polyaxonal amacrine cell population in the primate retina. J. Neurosci. **34**, 3597–3606 (2014). doi:10.1523/JNEUROSCI.3359-13.2014
18. Gross, G.W., Rieske, E., Kreutzberg, G.W., Meyer, A.: A new fixed-array multi-microelectrode system designed for long-term monitoring of extracellular single unit neuronal activity in vitro. Neurosci. Lett. **6**, 101–105 (1977). doi:10.1016/0304-3940(77)90003-9
19. Jones, K.E., Campbell, P.K., Normann, R.A.: A glass/silicon composite intracortical electrode array. Ann. Biomed. Eng. **20**, 423–437 (1992). doi:10.1007/BF02368134
20. Litke, A.M., Bezayiff, N., Chichilnisky, E.J., et al.: What does the eye tell the brain?: development of a system for the large-scale recording of retinal output activity. IEEE Trans. Nucl. Sci. **51**, 1434–1440 (2004). doi:10.1109/TNS.2004.832706
21. Multi Channel Systems, GmbH: Multi Channel Systems GmbH (2006). http://www.multichannelsystems.com
22. Nisch, W., Böck, J., Egert, U., et al.: A thin film microelectrode array for monitoring extracellular neuronal activity in vitro. Biosens. Bioelectron. **9**, 737–741 (1994). doi:10.1016/0956-5663(94)80072-3
23. Oka, H., Shimono, K., Ogawa, R., et al.: A new planar multielectrode array for extracellular recording: application to hippocampal acute slice. J. Neurosci. Methods **93**, 61–67 (1999). doi:10.1016/S0165-0270(99)00113-2
24. O'Keefe, J., Recce, M.L.: Phase relationship between hippocampal place units and the EEG theta rhythm. Hippocampus **3**, 317–330 (1993). doi:10.1002/hipo.450030307
25. Pine, J.: Recording action potentials from cultured neurons with extracellular microcircuit electrodes. J. Neurosci. Methods **2**, 19–31 (1980)

26. Regehr, W.G., Pine, J., Cohan, C.S., et al.: Sealing cultured invertebrate neurons to embedded dish electrodes facilitates long-term stimulation and recording. J. Neurosci. Methods **30**, 91–106 (1989). doi:10.1016/0165-0270(89)90055-1
27. Segev, R., Goodhouse, J., Puchalla, J., Berry, M.J.: Recording spikes from a large fraction of the ganglion cells in a retinal patch. Nat. Neurosci. **7**, 1155–1162 (2004). doi:10.1038/nn1323
28. Thomas Jr., C., Springer, P., Loeb, G., et al.: A miniature microelectrode array to monitor the bioelectric activity of cultured cells. Exp. Cell Res. **74**, 61–66 (1972). doi:10.1016/0014-4827(72)90481-8
29. Greve, F., Lichtenberg, J., Kirstein, K.-U., et al.: A perforated CMOS microchip for immobilization and activity monitoring of electrogenic cells. J Micromechanics Microengineering **17**, 462–471 (2007). doi:10.1088/0960-1317/17/3/007
30. DeBusschere, B.D., Kovacs, G.T.A.: Portable cell-based biosensor system using integrated CMOS cell-cartridges. Biosens. Bioelectron. **16**, 543–556 (2001). doi:10.1016/S0956-5663(01)00168-3
31. Olsson, R.H., Wise, K.D.: A three-dimensional neural recording microsystem with implantable data compression circuitry. IEEE J. Solid State Circuits **40**, 2796–2804 (2005). doi:10.1109/JSSC.2005.858479
32. Najafi, K., Wise, K.D.: An implantable multielectrode array with on-chip signal processing. IEEE J. Solid State Circuits **21**, 1035–1044 (1986). doi:10.1109/JSSC.1986.1052646
33. Bai, Q., Wise, K.D.: Single-unit neural recording with active microelectrode arrays. IEEE Trans. Biomed. Eng. **48**, 911–920 (2001). doi:10.1109/10.936367
34. Huys, R., Braeken, D., Jans, D., et al.: Single-cell recording and stimulation with a 16 k micro-nail electrode array integrated on a 0.18 μm CMOS chip. Lab Chip **12**, 1274 (2012). doi:10.1039/c2lc21037a
35. Lopez, C.M., Andrei, A., Mitra, S., et al.: An implantable 455-active-electrode 52-channel CMOS neural probe. IEEE J. Solid State Circuits **49**, 248–261 (2014). doi:10.1109/JSSC.2013.2284347
36. Seidl, K., Herwik, S., Torfs, T., et al.: CMOS-based high-density silicon microprobe arrays for electronic depth control in intracortical neural recording. J Microelectromechanical Syst. **20**, 1439–1448 (2011). doi:10.1109/JMEMS.2011.2167661
37. Lopez, C.M., Mitra, S., Putzeys, J., et al.: A 966-electrode neural probe with 384 configurable channels in 0.13 μm SOI CMOS. In: 2016 IEEE International Solid-State Circuits Conference ISSCC, pp. 392–393. IEEE (2016)
38. Viswam, V., Dragas, J., Shadmani, A., et al.: Multi-functional microelectrode array system featuring 59,760 electrodes, 2048 electrophysiology channels, impedance and neurotransmitter measurement units. In: 2016 IEEE International Solid State Circuits Conference ISSCC 2016, pp. 394–396. San Franc. CA USA Jan 31–Feb 4 2016. (2016)
39. Yuan, X., Kim, S., Juyon, J., et al.: A microelectrode array with 8,640 electrodes enabling simultaneous full-frame readout at 6.5 kfps and 112-channel switch-matrix readout at 20 kS/s (2016)
40. Aziz, J.N.Y., Abdelhalim, K., Shulyzki, R., et al.: 256-channel neural recording and delta compression microsystem with 3D electrodes. IEEE J. Solid State Circuits **44**, 995–1005 (2009). doi:10.1109/JSSC.2008.2010997
41. Berdondini, L., Imfeld, K., Maccione, A., et al.: Active pixel sensor array for high spatio-temporal resolution electrophysiological recordings from single cell to large scale neuronal networks. Lab Chip **9**, 2644 (2009). doi:10.1039/b907394a
42. Bertotti, G., Velychko, D., Dodel, N., et al.: A CMOS-based sensor array for in-vitro neural tissue interfacing with 4225 recording sites and 1024 stimulation sites, pp. 304–307. IEEE (2014)
43. Eversmann, B., Lambacher, A., Gerling, T., et al.: A neural tissue interfacing chip for in-vitro applications with 32 k recording/stimulation channels on an active area of 2.6 mm^2, pp. 211–214. IEEE (2011)

44. Eversmann, B., Jenkner, M., Hofmann, F., et al.: A 128 × 128 CMOS biosensor array for extracellular recording of neural activity. IEEE J. Solid State Circuits **38**, 2306–2317 (2003). doi:10.1109/JSSC.2003.819174
45. Heer, F., Hafizovic, S., Franks, W., et al.: CMOS microelectrode array for bidirectional interaction with neuronal networks. IEEE J. Solid State Circuits **41**, 1620–1629 (2006). doi:10.1109/JSSC.2006.873677
46. Johnson, B., Peace, S.T., Cleland, T.A., Molnar, A.: A 50 μm pitch, 1120-channel, 20 kHz frame rate microelectrode array for slice recording, pp. 109–112. IEEE (2013)
47. Johnson, B., Peace, S.T., Wang, A., et al.: A 768-channel CMOS microelectrode array with angle sensitive pixels for neuronal recording. IEEE Sens. J. **13**, 3211–3218 (2013). doi:10.1109/JSEN.2013.2266894
48. Maccione, A., Simi, A., Nieus, T., et al.: Sensing and actuating electrophysiological activity on brain tissue and neuronal cultures with a high-density CMOS-MEA, pp. 752–755, IEEE (2013)
49. Lambacher, A., Vitzthum, V., Zeitler, R., et al.: Identifying firing mammalian neurons in networks with high-resolution multi-transistor array (MTA). Appl. Phys. A **102**, 1–11 (2011). doi:10.1007/s00339-010-6046-9
50. Fejtl, M., Stett, A., Nisch, W., et al.: On micro-electrode array revival: its development, sophistication of recording, and stimulation. In: Taketani, M., Baudry, M. (eds.) Advances in Network Electrophysiology, pp. 24–37. Springer US (2006)
51. Stett, A., Egert, U., Guenther, E., et al.: Biological application of microelectrode arrays in drug discovery and basic research. Anal. Bioanal. Chem. **377**, 486–495 (2003). doi:10.1007/s00216-003-2149-x
52. Weis, R., Fromherz, P.: Frequency dependent signal transfer in neuron transistors. Phys. Rev. E **55**, 877–889 (1997). doi:10.1103/PhysRevE.55.877
53. Nelson, M.J., Pouget, P., Nilsen, E.A., et al.: Review of signal distortion through metal microelectrode recording circuits and filters. J. Neurosci. Methods **169**, 141–157 (2008). doi:10.1016/j.jneumeth.2007.12.010
54. Robinson, D.A.: The electrical properties of metal microelectrodes. Proc. IEEE **56**, 1065–1071 (1968). doi:10.1109/PROC.1968.6458
55. Ness, T.V., Chintaluri, C., Potworowski, J., et al.: Modelling and analysis of electrical potentials recorded in microelectrode arrays (MEAs). Neuroinformatics **13**, 403–426 (2015). doi:10.1007/s12021-015-9265-6
56. Camuñas-Mesa, L.A., Quiroga, R.Q.: A detailed and fast model of extracellular recordings. Neural Comput. **25**, 1191–1212 (2013). doi:10.1162/NECO_a_00433
57. Viswam, V., Jäckel, D., Jones, I., et al.: Effects of sub-10 μm electrode sizes on extracellular recording of neuronal cells. In: Proceedings 18th International Conference Miniaturized Systems for Chemistry Life Sciences MicroTAS. Chemical and Biological Microsystems Society, pp. 980–982. San Diego, California (2014)
58. Spira, M.E., Hai, A.: Multi-electrode array technologies for neuroscience and cardiology. Nat. Nanotechnol. **8**, 83–94 (2013). doi:10.1038/nnano.2012.265
59. Cheney, P.D., Fetz, E.E.: Comparable patterns of muscle facilitation evoked by individual corticomotoneuronal (CM) cells and by single intracortical microstimuli in primates: evidence for functional groups of CM cells. J. Neurophysiol. **53**, 786–804 (1985)
60. Bakkum, D.J., Frey, U., Radivojevic, M., et al.: Tracking axonal action potential propagation on a high-density microelectrode array across hundreds of sites. Nat. Commun. (2013). doi:10.1038/ncomms3181
61. Bakkum, D.J., Chao, Z.C., Potter, S.M.: Long-term activity-dependent plasticity of action potential propagation delay and amplitude in cortical networks. PLoS one **3**, e2088 (2008). doi:10.1371/journal.pone.0002088
62. Hashimoto, T., Elder, C.M., Vitek, J.L.: A template subtraction method for stimulus artifact removal in high-frequency deep brain stimulation. J. Neurosci. Methods **113**, 181–186 (2002). doi:10.1016/S0165-0270(01)00491-5

63. Wagenaar, D.A., Potter, S.M.: Real-time multi-channel stimulus artifact suppression by local curve fitting. J. Neurosci. Methods **120**, 113–120 (2002). doi:10.1016/S0165-0270(02)00149-8
64. Livi, P., Heer, F., Frey, U., et al.: Compact Voltage and Current Stimulation Buffer for High-Density Microelectrode Arrays. IEEE Trans. Biomed. Circuits Syst. **4**, 372–378 (2010). doi:10.1109/TBCAS.2010.2080676
65. Hafizovic, S., Heer, F., Ugniwenko, T., et al.: A CMOS-based microelectrode array for interaction with neuronal cultures. J. Neurosci. Methods **164**, 93–106 (2007). doi:10.1016/j.jneumeth.2007.04.006
66. Müller, J., Bakkum, D.J., Hierlemann, A.: Sub-millisecond closed-loop feedback stimulation between arbitrary sets of individual neurons. Front. Neural Circuits (2013). doi:10.3389/fncir.2012.00121
67. Gandolfo, M., Maccione, A., Tedesco, M., et al.: Tracking burst patterns in hippocampal cultures with high-density CMOS-MEAs. J. Neural Eng. **7**, 056001 (2010). doi:10.1088/1741-2560/7/5/056001
68. Heer, F., Hafizovic, S., Ugniwenko, T., et al.: Single-chip microelectronic system to interface with living cells. Biosens. Bioelectron. **22**, 2546–2553 (2007). doi:10.1016/j.bios.2006.10.003
69. Lewandowska, M.K., Radivojević, M., Jäckel, D., et al.: Cortical axons, isolated in channels, display activity-dependent signal modulation as a result of targeted stimulation. Front Neurosci. (2016). doi:10.3389/fnins.2016.00083
70. Lewandowska, M.K., Bakkum, D.J., Rompani, S.B., Hierlemann, A.: Recording large extracellular spikes in microchannels along many axonal sites from individual neurons. PLoS one **10**, e0118514 (2015). doi:10.1371/journal.pone.0118514
71. Fiscella, M., Franke, F., Farrow, K., et al.: Visual coding with a population of direction-selective neurons. J. Neurophysiol. **114**, 2485–2499 (2015). doi:10.1152/jn.00919.2014
72. Fiscella, M., Farrow, K., Jones, I.L., et al.: Recording from defined populations of retinal ganglion cells using a high-density CMOS-integrated microelectrode array with real-time switchable electrode selection. J. Neurosci. Methods **211**, 103–113 (2012). doi:10.1016/j.jneumeth.2012.08.017
73. Franke, F., Fiscella, M., Sevelev, M., et al.: Structures of neural correlation and how they favor coding. Neuron **89**, 409–422 (2016). doi:10.1016/j.neuron.2015.12.037
74. Maccione, A., Hennig, M.H., Gandolfo, M., et al.: Following the ontogeny of retinal waves: pan-retinal recordings of population dynamics in the neonatal mouse: Pan-retinal high-density retinal wave recordings. J. Physiol. **592**, 1545–1563 (2014). doi:10.1113/jphysiol.2013.262840
75. Menzler, J., Zeck, G.: Network oscillations in rod-degenerated mouse retinas. J. Neurosci. **31**, 2280–2291 (2011). doi:10.1523/JNEUROSCI.4238-10.2011
76. Yonehara, K., Fiscella, M., Drinnenberg, A., et al.: Congenital nystagmus gene FRMD7 is necessary for establishing a neuronal circuit asymmetry for direction selectivity. Neuron **89**, 177–193 (2016). doi:10.1016/j.neuron.2015.11.032
77. Eickenscheidt, M., Jenkner, M., Thewes, R., et al.: Electrical stimulation of retinal neurons in epiretinal and subretinal configuration using a multicapacitor array. J. Neurophysiol. **107**, 2742–2755 (2012). doi:10.1152/jn.00909.2011
78. Jones, I.L., Russell, T., Fiscella, M., et al.: Characterization of mammalian retinal ganglion cell response to voltage stimulus. In: Stett, A., Zeck, G. (eds.) Proceedings of MEA Meeting 2014 July 1–July 4 2014 Reutlingen, Germany. 9th International Meeting on Substrate-Integrated Microelectrode Arrays. NMI Natural and Medical Sciences Institute at the University of Tuebingen, Reutlingen, pp. 74–75 (2014)
79. Stutzki, H., Leibig, C., Andreadaki, A., et al.: Inflammatory stimulation preserves physiological properties of retinal ganglion cells after optic nerve injury. Front. Cell. Neurosci. (2014). doi:10.3389/fncel.2014.00038

80. Fiscella, M., Franke, F., Müller, J., et al.: Decoding of motion directions by direction-selective retina cells. In: Proceedings MEA Meeting 2014 July 1–July 4 2014 Reutlingen Germany. 9th International Meeting Substrate Integrated Microelectrode Arrays. NMI Natural and Medical Sciences Institute at the University of Tuebingen, Reutlingen, pp. 98–99 (2014)
81. Zeck, G., Lambacher, A., Fromherz, P.: Axonal transmission in the retina introduces a small dispersion of relative timing in the ganglion cell population response. PLoS one **6**, e20810 (2011). doi:10.1371/journal.pone.0020810
82. Jones, I.L., Russell, T.L., Farrow, K., et al.: A method for electrophysiological characterization of hamster retinal ganglion cells using a high-density CMOS microelectrode array. Front. Neurosci. (2015). doi:10.3389/fnins.2015.00360
83. Velychko, D., Eickenscheidt, M., Thewes, R., Zeck, G.: Simultaneous stimulation and recording of retinal action potentials using capacitively coupled high- density CMOS-based MEAs. In: Proceedings 9th International Meeting Substrate Integrated Microelectrode Arrays, pp. 78–79. Reutlingen, Germany (2014)
84. Reinhard, K., Mutter, M., Fiscella, M., et al.: Novel insights into visual information processing of human retina. In: Proceedings 9th International Meeting Substrate Integrated Microelectrode Arrays, p 102. Reutlingen, Germany (2014)
85. Frey, U., Egert, U., Heer, F., et al.: Microelectronic system for high-resolution mapping of extracellular electric fields applied to brain slices. Biosens. Bioelectron. **24**, 2191–2198 (2009). doi:10.1016/j.bios.2008.11.028
86. Obien, M.E.J., Hierlemann, A., Frey, U.: Technique for analysis of purkinje cell sub-cellular functional dynamics in acute cerebellar slices using a high-density microelectrode array. In: Proceedings 9th International Meeting Substrate Integrated Microelectrode Arrays, pp. 88–90. Reutlingen, Germany (2014)
87. Ferrea, E., Maccione, A., Medrihan, L., et al.: Large-scale, high-resolution electrophysiological imaging of field potentials in brain slices with microelectronic multielectrode arrays. Front. Neural Circuits (2012). doi:10.3389/fncir.2012.00080
88. Medrihan, L., Ferrea, E., Greco, B., et al.: Asynchronous GABA release is a key determinant of tonic inhibition and controls neuronal excitability: a study in the synapsin II-/-mouse. Cereb. Cortex **25**, 3356–3368 (2015). doi:10.1093/cercor/bhu141
89. Heer, F., Franks, W., Blau, A., et al.: CMOS microelectrode array for the monitoring of electrogenic cells. Biosens. Bioelectron. **20**, 358–366 (2004). doi:10.1016/j.bios.2004.02.006
90. Imfeld, K., Neukom, S., Maccione, A., et al.: Large-scale, high-resolution data acquisition system for extracellular recording of electrophysiological activity. IEEE Trans. Biomed. Eng. **55**, 2064–2073 (2008). doi:10.1109/TBME.2008.919139
91. Sanchez-Bustamante, C.D., Frey, U., Kelm, J.M., et al.: Modulation of cardiomyocyte electrical properties using regulated bone morphogenetic protein-2 expression. Tissue Eng. Part A **14**, 1969–1988 (2008). doi:10.1089/ten.tea.2007.0302
92. Gong, W., Jäckel, D., Müller, J., et al.: Long-term cultivation and recording from organo-typic brain slices on high-density micro-electrode arrays. In: Proceedings 9th International Meeting Substrate Integrated Microelectrode Arrays, pp. 335–336. Reutlingen, Germany (2014)
93. Radivojevic, M., Jäckel, D., Müller, J., et al.: Finding the most effective site for extracellular neuronal stimulation. In: Proceedings MEA Meeting 2014 July 1–July 4 2014, Reutlingen, Germany. 9th International Meeting Substrate Integrated Microelectrode Arrays. Reutlingen, Germany (2014)
94. Panas, D., Amin, H., Maccione, A., et al.: Sloppiness in spontaneously active neuronal networks. J. Neurosci. **35**, 8480–8492 (2015). doi:10.1523/JNEUROSCI.4421-14.2015
95. Yada, Y., Kanzaki, R., Takahashi, H.: State-Dependent propagation of neuronal sub-population in spontaneous synchronized bursts. Front. Syst. Neurosci. (2016). doi:10.3389/fnsys.2016.00028

96. Maccione, A., Garofalo, M., Nieus, T., et al.: Multiscale functional connectivity estimation on low-density neuronal cultures recorded by high-density CMOS micro electrode arrays. J. Neurosci. Methods **207**, 161–171 (2012). doi:10.1016/j.jneumeth.2012.04.002
97. Poria, D., Dhingra, N.K.: Spontaneous oscillatory activity in rd1 mouse retina is transferred from ON pathway to OFF pathway via glycinergic synapse. J. Neurophysiol. **113**, 420–425 (2015). doi:10.1152/jn.00702.2014
98. Egert, U., Heck, D., Aertsen, A.: Two-dimensional monitoring of spiking networks in acute brain slices. Exp. Brain Res. **142**, 268–274 (2002). doi:10.1007/s00221-001-0932-5
99. Frey, U., Egert, U., Jackel, D., et al.: Depth recording capabilities of planar high-density microelectrode arrays, pp. 207–210. IEEE (2009)
100. Newell, D.W., Barth, A., Papermaster, V., Malouf, A.T.: Glutamate and non-glutamate receptor mediated toxicity caused by oxygen and glucose deprivation in organotypic hippocampal cultures. J. Neurosci. Off. J. Soc. Neurosci. **15**, 7702–7711 (1995)
101. Ostergaard, K., Finsen, B., Zimmer, J.: Organotypic slice cultures of the rat striatum: an immunocytochemical, histochemical and in situ hybridization study of somatostatin, neuropeptide Y, nicotinamide adenine dinucleotide phosphate-diaphorase, and enkephalin. Exp. Brain Res. **103**, 70–84 (1995)
102. Krassioukov, A.V., Ackery, A., Schwartz, G., et al.: An in vitro model of neurotrauma in organotypic spinal cord cultures from adult mice. Brain Res. Brain Res. Protoc. **10**, 60–68 (2002)
103. Birgbauer, E., Rao, T.S., Webb, M.: Lysolecithin induces demyelination in vitro in a cerebellar slice culture system. J. Neurosci. Res. **78**, 157–166 (2004). doi:10.1002/jnr.20248
104. Cho, S., Wood, A., Bowlby, M.R.: Brain slices as models for neurodegenerative disease and screening platforms to identify novel therapeutics. Curr. Neuropharmacol. **5**, 19–33 (2007)
105. Gähwiler, B.H.: Organotypic monolayer cultures of nervous tissue. J. Neurosci. Methods **4**, 329–342 (1981)
106. Stoppini, L., Buchs, P.A., Muller, D.: A simple method for organotypic cultures of nervous tissue. J. Neurosci. Methods **37**, 173–182 (1991)
107. Gong, W., Sencar, J., Jackel, D., et al.: Long-term, high-spatiotemporal resolution recording from cultured organotypic slices with high-density microelectrode arrays, pp. 1037–1040. IEEE (2015)
108. Amin, H., Maccione, A., Marinaro, F., et al.: Electrical responses and spontaneous activity of human iPS-derived neuronal networks characterized for 3-month culture with 4096-electrode arrays. Front. Neurosci. (2016). doi:10.3389/fnins.2016.00121
109. Doudna, J.A., Charpentier, E.: The new frontier of genome engineering with CRISPR-Cas9. Science **346**, 1258096 (2014). doi:10.1126/science.1258096
110. El Hady, A., Afshar, G., Bröking, K., et al.: Optogenetic stimulation effectively enhances intrinsically generated network synchrony. Front. Neural Circuits (2013). doi:10.3389/fncir.2013.00167

Microelectrode Arrays: Architecture, Challenges and Engineering Solutions

Dhurgham Khudhair, Saeid Nahavandi, Hamid Garmestani and Asim Bhatti

Abstract Neural interfaces are connections linking the neuronal systems with electro-mechanical systems for information exchange. Microelectrodes of various designs have been fabricated utilizing both metallic and non-metallic materials so that they can be employed as neural interfaces. Recent studies have shown that the performance of microelectrodes can be enhanced significantly through structure controlling, surface chemistry and biotechnology. This review highlights the challenges including invasiveness, stability and selectivity associated with the employment of common electrodes as interfaces for neural recording and stimulation. It also includes controlling of electrode material and geometry as engineering solutions for the aforementioned challenges. Due to their high surface area, small size and high electrochemical properties, nanostructured electrodes show promise as electrodes that could be employed as neural interfaces for stable signal recording and stimulation. We hope this work will provide a concise picture of the evolution and the progress of current neural interfaces technology, the development of which is still in progress.

Keywords Microelectrode arrays · Neural interfaces · Microelectrodes architecture

---

D. Khudhair (✉) · S. Nahavandi · A. Bhatti
Institute for Intelligent Systems Research and Innovation (IISRI),
Deakin University, Waurn Ponds, Australia
e-mail: dkhudhai@deakin.edu.au

S. Nahavandi
e-mail: saeid.nahavandi@deakin.edu.au

A. Bhatti
e-mail: asim.bhatti@deakin.edu.au

H. Garmestani
School of Materials and Engineering, Georgia institute of Technology, Atlanta, GA 30332-0245, USA
e-mail: hamid.garmestani@mse.gatech.edu

A. Bhatti et al. (eds.), *Emerging Trends in Neuro Engineering and Neural Computation*, Series in BioEngineering,
DOI 10.1007/978-981-10-3957-7_2

1 Introduction

As neurons in the nervous system communicate by electrical signals, neuroscience has given great attention to understanding how they are electrically connected in their networks as a major key to exploring the physiological and pathological functions of neurons [1]. Neural interface is an effective tool that enables the information to be exchanged in two directions within the nervous system. Electrical stimulation is a process by which the introduced information to the nervous system can be employed as an external control. The contraction of muscle resulting in stimulation of motor nerves is an example of external control [2]. Recently, brain–machine interfaces have been used to control prosthetic limbs by enabling users to supervise and process the activities of neurons. Deep brain stimulation is an effective method to treat patients suffering from Parkinson's disease [3], mental illness [4], depression [5], obsessive compulsive disorder [6] and epilepsy as neurological disorders by applying pulses of voltage or current to particular profound regions in the brain. Furthermore, electronic devices have been successfully instilled in the brain, eyes and cochlea to elicit and iterate the missing or weakened sensory and muscle function. In such devices, neural interfaces are the most important components [7, 8]. To measure the action potential and record the neural activity, different methods have been applied. These methods include: (a) extracellular recording and stimulation, (b) intracellular recording and stimulation, (c) technologies of optical imaging and stimulation and (d) methods designed to record signals of in extensive neural populations such as electroencephalography, magnetoencephalography and functional magnetic resonance imaging and electrocardiography.

Intracellular and extracellular recording are two prime electrophysiology methods that have been employed for measuring action potential and realization of neural information processing in neural circuits. Intracellular recording is more suitable for sensitive recording, but it needs rending a part of plasma membrane to approach the cell abdomen directly. Therefore, intracellular recording is a highly invasive method and difficult to perform causing significant limitations for long-term or large-scale recording. Extracellular recording, however, is a non-invasive method and supports long-term recording, but the weakness in signal strength and poor quality of the recorded signals are significant limitations of this method [1, 8]. For recording of action potential, microelectrodes should be close to neurological target cells, the surface area of the electrode should not exceed 4000 μ^2 for single unit recording and the ratio of signal to noise of the recorded action potential should not be less than 5:1. Electrochemical impedance is one of the most important electrochemical properties of the recording electrode. Signal-to-noise ratio has an inverse relationship with the electrode impedance, as lower impedance is equivalent to having higher signal-to-noise ratio. Furthermore, amalgamation of high impedance of the electrode and the distribution of the capacitance between the recording amplifier and the electrode leads to reduction of high frequency response of the electrode [9].

Many materials such as platinum, gold, stainless steel, tungsten, iridium oxide and titanium nitride have been used to fabricate the recording electrodes [10]. Neural interfaces need to be implanted deeply in the brain to record the action potentials of neurons; therefore, the implantation process will be accompanied by significant clinical risks represented by tissue damage and infection of target sites. For less tissue damage and accurate long-term recording, implanted interfaces with smallest cross section and largest number of electrode sites are the ideal interfaces. Microelectrodes are used to stimulate neurons as well as signal recording. Current picoamperes range is required to excite single neurons by the technique of patch clamp while current in the microamperes and milliamps ranges are required for nerve and muscle stimulation respectively. Microelectrode arrays are the most important component in the brain–machine interfaces as they act in a way that facilitates direct contact between the neural tissue and the electrical sensor. To obtain harmonious signals recorded from small clusters of neurons with retention of micro-stimulation abilities, microelectrode arrays are fabricated in a way that enables them to provide a low impedance path for the charge movement represented by charge injection and charge transformation. For low impedance, microelectrode arrays were made of highly conductive materials and fabricated in specific geometries [11]. To decrease the impedance of the microelectrode and improve the neural recording, porous structures such as carbon nanotubes, Pt-black and high-conductive polymers were employed to increase the effective surface area of the exposed part of the electrode [12–15].

The use of electrical stimulation of neural tissue goes back to the time of the invention of electricity. Electrical shock with 400 V from catfish was used for pain relief and as a treatment of several diseases by Ancient Egyptians in 2500 BC [16]. Also, there is historical evidence that refers to the use of electricity in religious rituals as a means to influence the spirit [17]. For pain relief and stimulation of blood circulation, ancient Greeks employed electric eels to apply electrical pulses in foot baths. Benjamin Franklin [18, 19], in 1759 looked into the contraction of muscles as a result of an electrical shock. A fuzzy concept was introduced in 1791 by Luigi Galvani illustrating the electricity in an animal's body after electrically stimulating frogs legs [20]. In the year 1939, a major development in the design of neural interface occurred when Hodgkin and Huxley [21] studied the electrical signals recorded from single neural fibre by reduction of neural interface size. In 1960, an important attempt in use of the neural interface system was done by Evarts [22] when electrophysiological experiments were conducted on the primary motor cortex of springy monkeys. Evarts pointed out that the firing rate of solitary neurons highly corresponded with the force created by the joints of the moving arm. Another significant transitional step in neural interface design and application took place in 1985, when microwire array electrodes came into force when dealing with a large number of patients [23, 24].

Electrodes made of diverse materials with different shapes and geometries have been employed as interfaces for neural recording and stimulation. These electrodes can be classified into two main groups: metal-based electrodes and non-metal-based electrodes.

2 Metal-Based Electrodes

Metals of high electrochemical properties and good biocompatibility have been used to fabricate electrodes for neural interfaces. These electrodes have been designed in different geometries to match the biological and electrochemical requirements.

2.1 Microwire Arrays

Microwires were the first electrodes made of sharpened metal and were used to record electrical signals chronically from the individual neuron in the brain by implanting the electrodes inside the brain. The wires were totally insulated except their tips, which were left uninsulated to inject neurons with current pulses and record their extracellular potentials [25].

Figure 1 is an exemplary array of microwires. Nontoxic metals of high corrosion resistance such as gold, platinum, tungsten and stainless steel were used to fabricate microwire arrays [26]. Stainless steel microwires, with a diameter of 80 μm, were used to record the action potentials from stimulated neurons of animals that had been awake for more than week [27]. As a sharpened steel microwire electrode has insufficient rigidity, Hubel [28] used pencil-like tungsten microwire electrodes with a tip point of 0.5 μm diameter for recording signals receipted from mammalian nerves. The electrodes were insulated with a suitable varnish up to the tip. The measurements showed that tungsten microwire electrodes have a low signal-to-noise ratio and slow signals were lost when a high-pass filter was used to diminish the noise. To increase rigidity, elasticity and corrosion resistance of

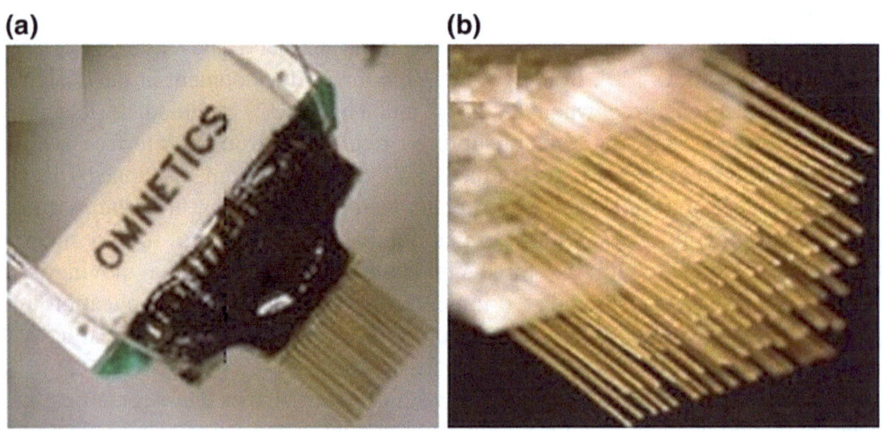

Fig. 1 Microwire arrays. **a** Arrays of microwires connected to the connector, **b** dense arrays of microwires [25]

microwire electrodes, iridium microwires were employed as microelectrodes for measurement of action potentials. It was reported that iridium has higher rigidity, corrosion resistance and elasticity than tungsten. Furthermore, increases in maximum charge density can be attained because it is possible to activate iridium surfaces electrochemically [26]. It was shown that an implemented array could consist of from 4 to over 100 wires. Nicolelis et al. used 704 microwires in 10 arrays to record 274 neurons individually in monkey cortex [29, 30]. An extracellular recording with spike amplitude of 60 μV was achieved by microwire tetrode from individual pyramidal cells distributed within a radius of 50 μm [31]. The ease of fabrication is an obvious advantage of microwire electrodes [32]. Another advantage of microwire electrodes is that these electrodes can access deeply in the brain to reach the target neurons [33]. The main obstacle of the microwires technique is that the bending of microelectrodes during implantation leads to loss of accuracy of the positions of the wire tips relative to each other [26].

2.2 *Planar-Type Microelectrode Array*

Planar-type microelectrodes made of nontoxic and high corrosion resistance metals such as gold, platinum, iridium and titanium nitride are a common electrode with a diameter of a few tens of micrometres. Planar microelectrodes are a cell culture dish used to study the activity and plasticity of the neural network [15]. To facilitate the observation of the cultured cells using conventional transmitted light microscopy, leads of one of the aforementioned metals are usually embedded in a glass wafer substrate. Figure 2 shows the planar-type electrode. It is usual to coat the glass substrate with laminin or polylysine to promote the cell adhesion and increase the sealing resistance between the substrate and the cultured cells. The substrate consists of 1–100 electrode sites spaced at inter distances of 100 μm. Organic or inorganic materials such as epoxy resin, polyimide, silicon oxide and silicon nitride

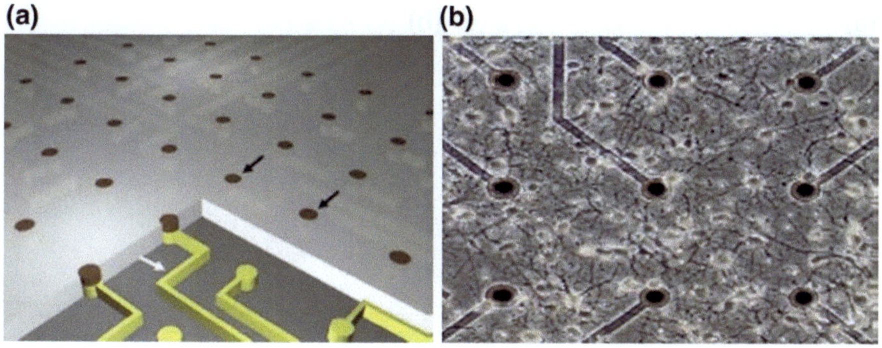

Fig. 2 **a** Planar-type electrode connected to contactor lines, **b** hippocampal neurons are cultured on the planar-type electrode [15]

are used to insulate the electrodes from each other. All electrodes are connected with a contact pad by thin contactors to transfer the captured signals to the amplifier [34]. It was reported that the impedance of a standard gold electrode with a diameter of 60 μm in electrolyte solution is 50 kΩ at 1 kHz. Individual sensing pads must be made smaller to match the size of the individual neurons. However, reduction of the surface area that accompanies the reduction of the sensing pad size results in a significant increase in the impedance and consequently decreases the signal-to-noise ratio. Nanostructures such as gold nanoflakes, gold nanopillers, carbon nanotubes or Ti_3N_4 compensate for the lack in the surface area resulting from the reduction of the electrode size [35].

2.3 *Mushroom-Shaped Microelectrode*

Although, micro patterned electrodes afford non-invasive and long-term extracellular recording, they suffer significantly in signal strength and quality. Recording understrength and quality signals restricts microelectrodes from sensitive recording such as, disinhibition, synaptic integration and under threshold oscillations [35].

To enable microelectrodes for such applications, Spira et al. used a micro-size protrusion made of gold with a mushroom shape as a sensing electrode (Fig. 3a). By increasing the electrical coupling coefficient between the neuron and the microelectrode to 50% recorded by a gold mushroom-shaped sensing electrode compared to 0.1% for the gold planar microelectrode array, it becomes possible to record synaptic potentials in addition to the action potentials. The enhancement in cell recording achieved by using gold mushroom-shaped electrodes was attributed to three main causes. The first is that the unique geometry of the electrode enables neurons to engulf the electrode as shown in Fig. 3b. The second is the high seal resistance between the cell membrane and the mushroom-shaped electrode and the third is because of the increase in the junctional membrane conductance [1].

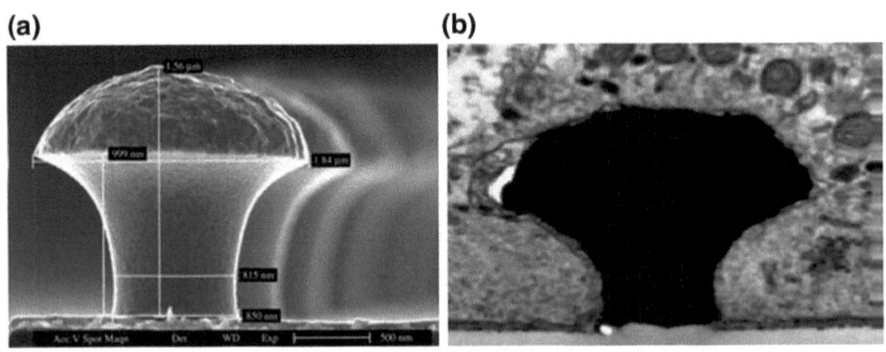

Fig. 3 **a** SEM image of gold mushroom-shaped microelectrode, **b** TEM image shows the Aplysia neurons engulf the mushroom-shaped electrode [1]

2.4 *Nanostructured Microelectrodes*

Great efforts have been exerted to overcome challenges such as inflammation and noise associated with the use of neural interfaces for neural stimulation and signal recording. Previous studies showed that the mechanical, physical and chemical properties have a crucial influence on the performance of the neural interfaces [36]. Ordinary interfaces in brain–machine interfaces utilize a small number of large electrodes for signal recording. It was reported that electrodes with diameter range of 10–100 microns were used in brain–machine interfaces for signal recording while electrodes with size of 4–8 mm were used for deep brain stimulation [37, 38]. As the human brain contains approximately a hundred billion neurons and each neuron has a diameter of about 10 microns, accurate monitoring and precise control on the neural circuit activities requires electrode arrays of high density and small size [39]. The efficiency of neural electrodes for neural signal recording and stimulation is greatly affected by the electrical coupling and the contact between the cell and the electrode surface, the electrochemical properties of the electrode and the biocompatibility of the material at the contact sites [11].

2.4.1 Nanowire Arrays

Due to their high aspect ratio, nanowires made of nontoxic and high corrosion resistance metals have attracted a lot of attention in different applications. Employing nanowires as neural interfaces is one of their important applications [40]. Platinum nanowires with a diameter of 150 nm and height of 1.5 μm were deposited on planer platinum electrodes by means of focused ion beams (Fig. 4a). A layer of Si_3N_4/SiO_2 with a thickness of 350 nm was deposited by plasma-enhanced chemical vapour deposition to insulate the substrate. Platinum nanowires were utilized to stimulate and record electrical signals from mitotic cardiac cells [40]. Bruggermann et al. [41] have fabricated gold nanowires with a

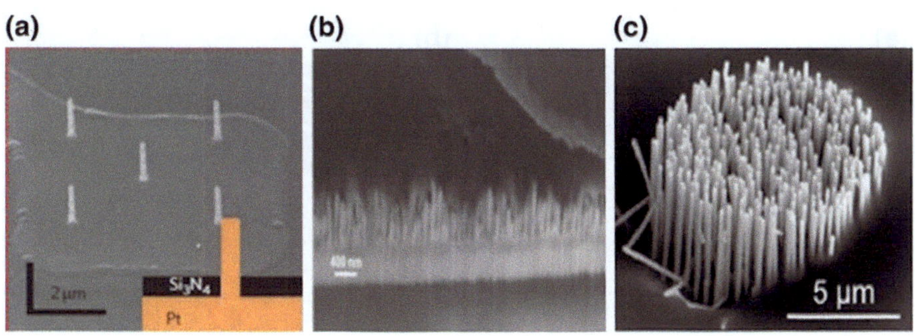

Fig. 4 SEM images of metal-based nanowire electrodes. **a** Platinum nanowire electrode [40]. **b** Gold nanowire electrode [41]. **c** gallium phosphide nanowire electrode [43]

diameter of 60 nm and height of 300–400 nm on pads with diameter of 2 μm (Fig. 4b). Extracellular potential recordings were taken from cardiac muscle cells (HL-1) using the gold nanowire electrode. The maximum-recorded amplitudes achieved by gold nanowire electrodes were 100% higher than those achieved by planar gold electrodes [42]. Gallium phosphide nanowires with 70 nm diameter and 5 μm length were fabricated by Suyatin et al. [43] on substrate of 12 μm diameter with a pitch size of 500 nm. A gold film was subsequently deposited on the top of the nanowires. Gallium phosphide nanowire electrodes are illustrated in Fig. 4c. Gallium phosphide nanowire based electrodes were employed to perform acute recordings in the rat cerebral cortex.

2.4.2 Nanotube Arrays

Previous studies have shown that vertical electrodes are able to achieve intracellular recording of action potential with high signal to noise ratio. These studies pointed out that the geometry of vertical electrodes has a crucial influence on the sensitivity and quality of recorded signals. Nanotube geometry was pointed out as an important factor influencing the performance of the vertical electrodes as the cell membrane wraps around and extends into the nanotube, thus the gap between the electrode and the membrane is significantly reduced [44–46].

Recently, Eick et al. [47] developed a new nanoelectrode consisting of iridium oxide nanotubes (Fig. 5a). After cardiomyocytes were cultured on the fabricated nanotube electrode, the images of scanning electron microscopy showed that the cell membrane wraps around and extends into the pore centres. Electrochemical measurements showed that iridium oxide nanotube electrodes possess lower electrochemical impedance and higher capacity of charge storage compared to gold nanopillar electrodes having the same surface area. The author demonstrated that the geometry of nanotubes promotes cell–electrode coupling and larger signals can be recorded compared with solid electrodes. Moreover, stable recording can be achieved by nanotube electrodes as they provide non-invasive and have longer

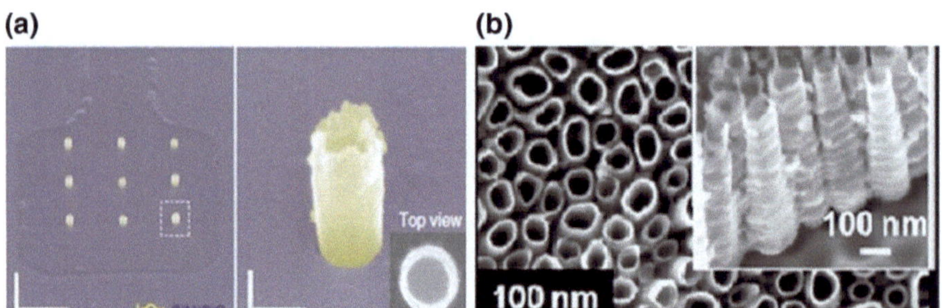

Fig. 5 **a** Iridium oxide nanotubes array on a platinum substrate (*left*). Iridium oxide nanotube (*right*) [47]. **b** TiO_2 nanotube array [48]

intracellular access. As a result, the author pointed out that nanotube geometry significantly enhances nanoelectrode performance.

In many studies, titania nanotubes were identified as a promising form of biological electrode as they possess high corrosion resistance, distinctive mechanical properties, elevated specific surface area in addition to excellent biocompatibility [44, 45, 48]. TiO_2 nanotube arrays are illustrated in Fig. 5b. It has been established that a titania tubular structure has a great potency for application in the medical area. Results have shown that TiO_2 nanotubes enhance the mineralization, proliferation and adhesion of osteoblasts and expedite the recovering of bone tissue. It has also been established that tube dimensions have a crucial influence in the differentiation of mesenchymal stem cell [44]. It was suggested that tube pores and the spacing in a titania tubular structure offer a substantial pathway for continuous supply of ions, nutrients and proteins required for healthy cell growth [45]. The anatase phase shows a better support for formation and growth of hydroxyapatite than rutile [46]. Park et al. showed that tube diameter has an important effect on adhesion as well as proliferation of mesenchymal stem cells. Authors have pointed out that tubes with a diameter of 15 nm improve the cell activity when the spacing between nanotubes is less than 30 nm while cells showed programmed death when they were cultured on tubes with a diameter larger than 50 nm [46]. Tube morphology has a critical effect on the electrical behaviour of a titania tubular structure. Tubes with wall thickness of 30–40 nm showed low impedance. Sun et al. attributed the electrical behaviour of these nanotubes to the thickness of their walls being sufficient to that being required for charge transformation [49].

Many approaches have been employed to fabricate TiO_2 nanotubes. These approaches include templating, sol-gel, photo-electrochemical etching and anodic oxidation. Low cost and the possibility of fabrication of TiO_2 nanotubes in a wide range of morphologies, anodic oxidation is generally used to fabricate titania tubular structures. Anodization parameters such as applied voltage, electrolyte type, electrolyte pH and anodization time have significant influence on the fabricated nanotube morphology [50]. Figure 6 shows TiO_2 nanotube arrays with different morphologies fabricated by controlling the anodization voltage. It was shown that subjecting the nanotubes to the annealing process at 450 °C enhances their

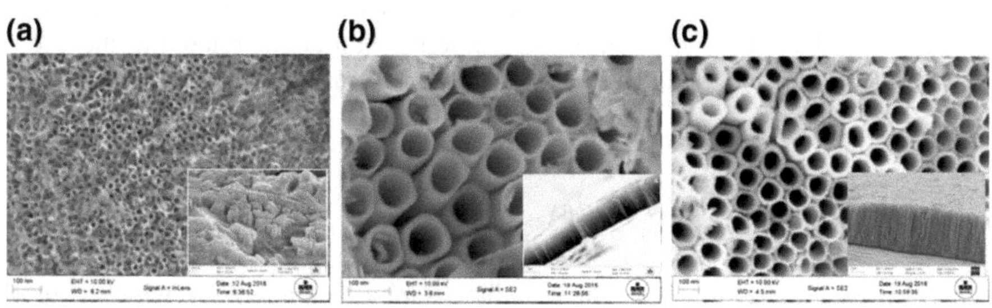

Fig. 6 TiO_2 nanotube arrays fabricated in ethylene glycol containing 0.5 wt% NH_4F and 4 vol.% of deionized water at anodization voltages of **a** 10 V, **b** 40 V and **c** 60 V

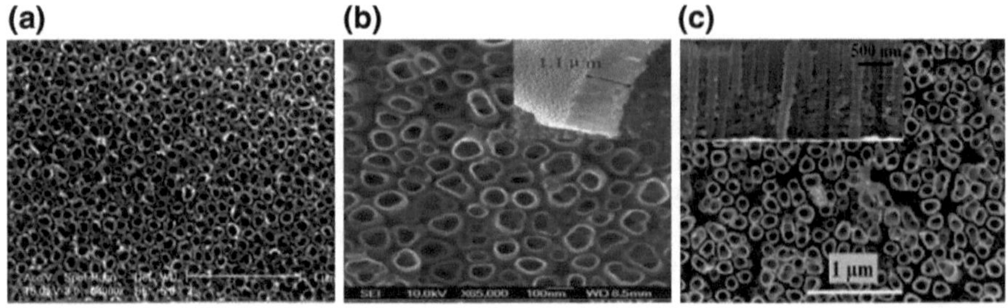

Fig. 7 TiO_2 arrays doped with **a** carbon [56], **b** nitrogen [55] and **c** tin [57]

electrochemical properties as a result of successful transformation from amorphous to anatase crystal structure [51]. As titania nanotubes inherently possesses low conductivity, which is highly desirable for neural interfacing, therefore electrochemical and physical methods have been employed to introduce metallic and non-metallic electrically active particles such as Pt, Ag, N and C into the titania lattice to enhance the electrical properties and biocompatibility of titania nanotubes [52–54]. It was reported that doping with nitrogen improved the electrochemical properties represented by electrochemical impedance and charge storage capacity [55]. To improve the capacitive properties of TiO_2 nanotubes, Zhang et al. have fabricated carbon-doped nanotubes. The authors show that the presence of carbon atoms in titania tubular structure improve the capacitance of the nanotube layer [56]. Kyeremateng et al. have investigated the effect of Sn doping on the electrochemical properties of TiO_2 nanotubes. The authors showed that Sn-doped nanotubes have higher capacitance than simple nanotube arrays [57]. Figure 7 illustrates TiO_2 nanotube arrays doped with different elements.

3 Non-metal-Based Electrodes

Due to the wide range of their structure, non-metals with their various physical, chemical and mechanical properties have been employed in different applications. One of their important applications is the neural interfaces. Neural electrodes made of non-metals have been utilized to overcome the limitations related to the use of metallic electrodes including inflammation, tissue damage and low flexibility.

3.1 Silicon-Based Electrodes

The emergence of technologies such as lithography techniques by which complicated structures can be fabricated has been followed by micromachined electrodes used to stimulate neurons and record potential signals [11, 58–60]. To overcome the

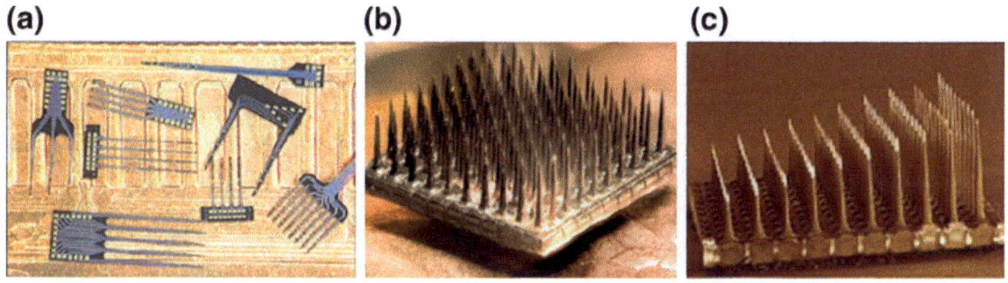

Fig. 8 Micromachined electrodes **a** different types of Michigan electrodes [62]. **b** Utah electrode (*flat array*) [63]. **c** Utah electrode (*incline array*) [63]

difficulties associated with the curvature of the wires used with the microwire technique, micromachining techniques utilizing silicon arrays have been used to generate a more rigid structure. As silicon arrays are smaller in size than that of microwire electrodes, larger number of sites in different layers of the cortex can be recorded by silicon arrays for the same quantity of tissue displacement [61]. There are two major models of micromachined electrode. The first model was fabricated in the University of Michigan and consists of a shank with silicon substrate having several electrode sites. Different types of Michigan electrodes are illustrated in Fig. 8a. The Utah array is the second model of micro machined electrodes. In Utah array, sharpened needles made of silicon with lengths up to 1.5 mm and diameter ranging from 1 to 100 μm were electrically insulated up to their tips using polymers with a good biocompatibility like parylene-C or polyimide. The bare tips of the silicon needles were coated with a conductive metal such as platinum or iridium. The architecture of Utah array provides the ability of signal recording from the individual neurons with high locative resolution as well as stimulation of the target neurons. Due to their biocompatibility, geometry and architecture, Utah arrays can be deeply inserted into the brain safely. Therefore, Utah arrays are widely used in neuroscience and medical researches [11, 62]. Figure 8b and c shows flat and inclined microelectrode arrays.

3.2 *Polymer-Based Electrodes*

Materials of rigid structures are usually used in conventional electrodes. Using stiff electrodes cannot provide vigorous interfaces with neurons and precludes long-term signal recording. The poor contact between the stiff interfaces and the soft neural tissue causes aggressive contact at the tissue-interface contact sites which leads to tissue damage and inflammation [63]. To avoid neural damage and inflammation of the implant sites, flexible arrays made of polymer-based electrodes have been developed to provide less invasive methods for neural stimulation and signal recording. Polymers such as polyimide, parylene-C, liquid crystal polymer, SU-8, benzocyclobutene and silk are common polymers used in fabrication of flexible

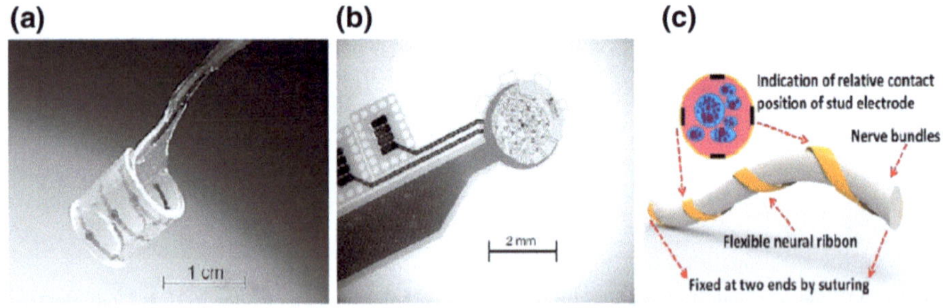

Fig. 9 Different types of flexible electrode. **a** Cuff electrode with polyimide substrate [74]. **b** Sieve electrode [74]. **c** Flexible ribbon electrode [76]

arrays [64–69]. Because of their good mechanical flexibility, high corrosion resistance and good biocompatibility, flexible arrays are widely used for long-term neural recording [70]. The variety of flexible electrode design was illustrated in Fig. 9. The low elastic modulus of polyimide and parylene-C electrodes (3.2 GPa) enable them to lessen the tissue damage and afford longer term recording than that of silicon [69–71]. In 1970s, Hoffer and Loeb successfully recorded neural signals by implanting cuff electrode and flexible platinum wire in the lumbar spine of a cat [36, 72]. The next attempt for flexible interfaces was made by Donaldson et al. to improve the flexibility and electrical conductivity by using the cuff interface with silicone rubber [35, 73]. As Parylene base electrodes have high flexibility, they show extremely consistent coverage of the tissue surface and this enables them to provide steady electrical contact and subsequently high signal to noise ratio [38].

A parylene-based microelectrode array has been employed for neural recording and drug delivery by utilizing electrode sites at different sides include the top, back and edges. Cuff electrodes made of hybrid polyimide entrenched in silicon guidance were fabricated to stimulate the peripheral nerves and record sensory signals. It was noticed that nerve trunks wrap around the cuff electrodes and furthermore, the muscles can be activated by stimulation of the motor fibres [74, 75]. However, although polymer electrodes are common neural interfaces, there are two main limitations for the use of these electrodes. The first is the difficulty of signal recording from neurons with different diameters and the second is the inability of communication with very small nerves (less than 300 μm in diameter). As a nerve size has a random distribution in the body, the predefined polymer electrodes lose the perfect match with target nerves [76, 77].

3.3 Nanostructured Electrodes

For many reasons nanomaterials are a promising technology in neural interfacing fabrication. Neurons are electroactive cells and the electrochemical properties of improved nanostructure can meet the requirements of the charge transformation. The distinctive chemical and mechanical properties of nanoscale structures strongly support the neural tissue for a long-term implantation. Furthermore, the high biocompatibility of improved nanomaterials solves the biological problems related to the implanted interfaces [78]. The discovery of nanomaterials with their unique properties of high electrical conductivity, exceptional chemical stability, excellent mechanical properties and high surface area moved neural interfaces to a new stage by enhancing the sensitivity and selectivity in addition of improving the biocompatibility and the response time [79].

3.3.1 Nanowire Arrays

Non-metallic nanowire used to record the electrical signals from the biological systems show higher sensitivity in observation of the changes in action potentials for cultured neurons than micro-size electrodes [80]. Nano wires made of semiconductor materials are a powerful technique and have a great impact on a wide range of scientific areas such as electronics, photonics, bioscience and healthcare. The good understanding of nanowire growth mechanism enables the production of nanowires with a homogeneous composition and diameter and, as a result, electrical and optical properties can be highly controlled. The correspondence in nanosize between the nanowires and the nano components in biological systems makes nanowires a promising candidate as a sensitive tool for investigation of biological systems [81]. As the bend of nanowires enables them to protrude between the lineaments of the cellular membrane, nanowires have a tight connection with the neural membrane. Decreasing the gap between the electrode surface and the cellular membrane enhances signal-to-noise ratio and increases the sensitivity of signal recording [82]. Two major techniques Langmuir–Blodgett and dry transfer techniques have been used to grow nanowire arrays on surfaces of silicone/silicone oxide substrate [83, 84].

Different strategies have been used to fabricate silicon-based vertical nanowires. Duan et al. [85] employed electron beam lithography to create gold islands on the gate of a nanowire field-effect transistor with nanoscale to use them as precursors for germanium nanowire growth by vapour–liquid–solid mechanism. By using the atomic layer deposition technique, germanium cores were coated with SiO_2 and subsequently etched by hydrogen peroxide and the result was vertical glass tubes (Fig. 10a). The author showed that these glass tubes have been successfully used for intracellular potential recording by penetrating plasma membranes of cardiomyocyte cells. To improve penetration, fabricated nanotubes were coated with phospholipid which makes it difficult to culture cells on these nanoelectrodes. Therefore, cells were cultured on separated substrate. As it is shown in Fig. 10 b, Robinson et al. [86] have fabricated silicon nanowire, using plasma etching techniques, which were insulated by thermally grown SiO_2 film. The silicon oxide film was removed from the tips of silicon nanowires followed by coating uncovered tips with vapourized platinum or gold film. The fabricated silicon nanowires were

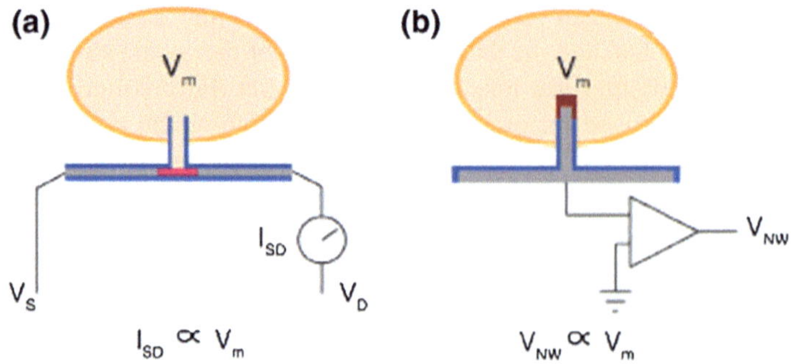

Fig. 10 Two strategies to fabricate nanowire electrode [8]. **a** Silicon nanotube on an insulated germanium nanowire. **b** Silicon nanowire with gold or platinum coated tip [8]

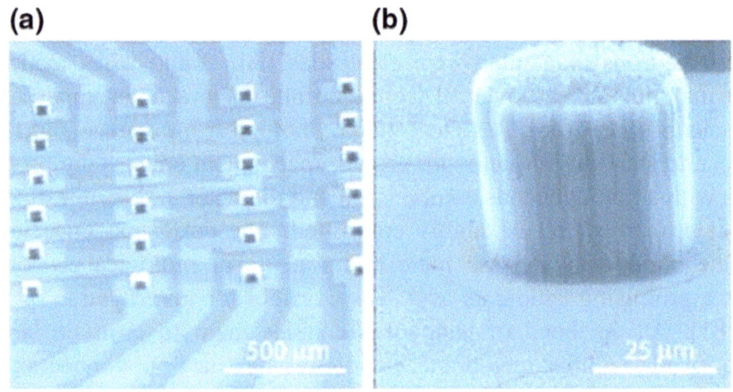

Fig. 11 SEM images of **a** Carbon nanotube electrode array [94]. **b** Carbon nanotubes electrode [94]

employed to stimulate individual neurons as well as record neural action potential. The main obstacle of the nanowire interfacing technique is the inherent high electrode impedance. It is worth mentioning that although a single pad in the interface contains a number of nanowires, the impedance of the electrodes is still too high to make it possible to record subthreshold potentials [40, 86]. Theoretically, increasing the number of the nanowires in a single pad may solve the problem of high impedance of electrodes. Nevertheless, high density of the nanowire electrodes precludes electrode to cell interior in intracellular recording.

3.3.2 Carbon Nanotube Arrays

Carbon nanotubes based electrodes are one of the most common nanotube electrodes because of their distinctive properties including very good electrical conductivity, high tensile strength, superior thermal conductivity and high aspect ratio [87–91]. Figure 11 shows an image of a carbon nanotube electrode array (a) and carbon nanotube electrode (b). Carbon nanotubes have been utilized in neuroscience research with one important application in recording of electrical activities of neurons [92, 93]. It was illustrated that carbon nanotubes enhance cell attachment and promote differentiation and growth of neurons [61]. Wang et al. [94] showed that vertically aligned carbon nanotubes offer a high level of charge injection (1–1.6 m C/cm^2). The authors demonstrated that carbon nanotubes increase the active surface area and decrease the impedance of the electrode–tissue interface. To improve the performance of single wall carbon nanotubes interface and achieve long-term recording, the nanotubes were electrochemically co-deposited with polypyrole. The modified electrode showed high injection level, good stability and low impedance [95].

4 Conclusion

In the past several decades, considerable effort has been made towards development of neural interfaces with many advantages and limitations of common interfaces being discovered. Various materials, utilizing different shapes and sizes, have been used to fabricate neural interfaces. Electrode design is important in that it has an influence on the cell–electrode coupling and tissue rehabilitation. Thin, sharp and flexible electrodes are desired for better performance in neural recording and stimulation. Lower electrochemical impedance and higher charge storage of interfaces are required to achieve good signal to noise ratio. Nanostructured electrodes are promising as their size matches the neuron size enabling signal recording for individual neurons and they can be shaped such that cells engulf the electrodes supporting enhancement of the cell–electrode coupling. As current neural interfaces have their limitations including high impedance, large size and high invasion, technology of interfaces fabrication is still in progress. This review focussed on the history and current status of neural interface technology.

References

1. Spira, M.E., Hai, A.: Multi-electrode array technologies for neuroscience and cardiology. Nat. Nanotechnol. **8**(2), 83–94 (2013)
2. Grill, W.M., Norman, S.E., Bellamkonda, R.V.: Implanted neural interfaces: biochallenges and engineered solutions. Annu. Rev. Biomed. Eng. **11**, 1–24 (2009)
3. Volkmann, J.: Deep brain stimulation for the treatment of Parkinson's disease. J. Clin. Neurophysiol. **21**(1), 6–17 (2004)
4. Robinson, A.J.: Clinical Electrophysiology: Electrotherapy and Electrophysiologic Testing. Lippincott Williams & Wilkins (2008)
5. Howland, R.H., et al.: The emerging use of technology for the treatment of depression and other neuropsychiatric disorders. Ann. Clin. Psychiatry **23**(1), 48–62 (2011)
6. Bourne, S.K., et al.: Mechanisms of deep brain stimulation for obsessive compulsive disorder: effects upon cells and circuits. Front. Integr. Neurosci. **6**(29), 1–12 (2012)

7. Nguyen-Vu, T.D., et al.: Vertically aligned carbon nanofiber arrays: an advance toward electrical-neural interfaces. Small **2**(1), 89–94 (2006)
8. Robinson, J.T., Jorgolli, M., Park, H.: Nanowire electrodes for high-density stimulation and measurement of neural circuits. Front. Neural Circ. **7**, 38 (2013)
9. Cogan, S.F.: Neural stimulation and recording electrodes. Annu. Rev. Biomed. Eng. **10**, 275–309 (2008)
10. Cogan, S.F., et al.: Over-pulsing degrades activated iridium oxide films used for intracortical neural stimulation. J. Neurosci. Methods **137**(2), 141–150 (2004)
11. Ghane-Motlagh, B., Sawan, M.: A review of microelectrode array technologies: design and implementation challenges. In: 2013 2nd International Conference on Advances in Biomedical Engineering (ICABME). IEEE (2013)
12. Cui, X., et al.: Surface modification of neural recording electrodes with conducting polymer/biomolecule blends. J. Biomed. Mater. Res. **56**(2), 261–272 (2001)
13. Franks, W., et al.: Impedance characterization and modeling of electrodes for biomedical applications. IEEE Trans. Biomed. Eng. **52**(7), 1295–1302 (2005)
14. Keefer, E.W., et al.: Carbon nanotube coating improves neuronal recordings. Nat. Nanotechnol. **3**(7), 434–439 (2008)
15. Nam, Y., Wheeler, B.C.: In vitro microelectrode array technology and neural recordings. Crit. Rev. Biomed. Eng. **39**(1) (2011)
16. Bresadola, M.: Medicine and science in the life of Luigi Galvani (1737–1798). Brain Res. Bull. **46**(5), 367–380 (1998)
17. Gordon, A.A.H.: The Quick and the Dead: Biomedical Theory in Ancient Egypt, vol. 4. Brill (2004)
18. Isaacson, W.: Benjamin Franklin: An American Life. Simon and Schuster (2003)
19. Finger, S.: Benjamin Franklin, electricity, and the palsies On the 300th anniversary of his birth. Neurology **66**(10), 1559–1563 (2006)
20. Gildenberg, P.L.: History of electrical neuromodulation for chronic pain. Pain Med. **7**(suppl 1), S7–S13 (2006)
21. Hodgkin, A.L., Huxley, A.F.: Action potentials recorded from inside a nerve fibre. Nature **144**(3651), 710–711 (1939)
22. Evarts, E.V.: Relation of pyramidal tract activity to force exerted during voluntary movement. J. Neurophysiol. **31**(1), 14–27 (1968)
23. Terzuolo, C.A., Araki, T.: An analysis of intra-versus extracellular potential changes associated with activity of single spinal motoneurons*. Ann. N.Y. Acad. Sci. **94**(2), 547–558 (1961)
24. Verzeano, M., Negishi, K.: Neuronal activity in cortical and thalamic networks a study with multiple microelectrodes. J. Gen. Physiol. **43**(6), 177–195 (1960)
25. Loeb, G., Peck, R., Martyniuk, J.: Toward the ultimate metal microelectrode. J. Neurosci. Methods **63**(1), 175–183 (1995)
26. Ghane-Motlagh, B., Sawan, M.: Design and implementation challenges of microelectrode arrays: a review. Mater. Sci. Appl. **04**(08), 483–495 (2013)
27. Strumwasser, F.: Long-term recording from single neurons in brain of unrestrained mammals. Science **127**(3296), 469–470 (1958)
28. Hubel, D.H.: Tungsten microelectrode for recording from single units. Science **125**(3247), 549–550 (1957)
29. Yuen, T., Agnew, W.: Histological evaluation of polyesterimide-insulated gold wires in brain. Biomaterials **16**(12), 951–956 (1995)
30. Nicolelis, M.A.: Actions from thoughts. Nature **409**(6818), 403–407 (2001)
31. Buzsáki, G.: Large-scale recording of neuronal ensembles. Nat. Neurosci. **7**(5), 446–451 (2004)
32. Polikov, V.S., Tresco, P.A., Reichert, W.M.: Response of brain tissue to chronically implanted neural electrodes. J. Neurosci. Methods **148**(1), 1–18 (2005)
33. Edel, D.J., et al.: Factors influencing the biocompatibility of insertable silicon microshafts in cerebral cortex. IEEE Trans. Biomed. Eng. **39**(6), 635–643 (1992)

34. Rutten, W.L.: Selective electrical interfaces with the nervous system. Annu. Rev. Biomed. Eng. **4**(1), 407–452 (2002)
35. Massobrio, P., Massobrio, G., Martinoia, S.: Interfacing cultured neurons to microtransducers arrays: a review of the neuro-electronic junction models. Front. Neurosci. **10** (2016)
36. Loeb, G., Bak, M., Duysens, J.: Long-term unit recording from somatosensory neurons in the spinal ganglia of the freely walking cat. Science **197**(4309), 1192–1194 (1977)
37. Donaldson, N.D.N., Donaldson, P.: When are actively balanced biphasic ('Lilly') stimulating pulses necessary in a neurological prosthesis? I Historical background; Pt resting potential; Q studies. Med. Biol. Eng. Comput. **24**(1), 41–49 (1986)
38. Campbell, P.K., et al.: A silicon-based, three-dimensional neural interface: manufacturing processes for an intracortical electrode array. IEEE Trans. Biomed. Eng. **38**(8), 758–768 (1991)
39. Williams, R.W., Herrup, K.: The control of neuron number. Annu. Rev. Neurosci. **11**(1), 423–453 (1988)
40. Xie, C., et al.: Intracellular recording of action potentials by nanopillar electroporation. Nat. Nanotechnol. **7**(3), 185–190 (2012)
41. Brüggemann, D., et al.: Nanostructured gold microelectrodes for extracellular recording from electrogenic cells. Nanotechnology **22**(26), 265104 (2011)
42. Suyatin, D.B., et al.: Nanowire-based electrode for acute in vivo neural recordings in the brain. PLoS ONE **8**(2), e56673 (2013)
43. Suyatin, D.B., et al.: Nanowire-based electrode for acute in vivo neural recordings in the brain. PLoS ONE **8**(2), e56673 (2013)
44. Xiao, X., Liu, R.: Synthesis and bioactivity of highly ordered TiO_2 nanotube arrays. Appl. Surf. Sci. **255**(2), 365–367 (2008)
45. Uchida, M., et al.: Structural dependence of apatite formation on titania gels in a simulated body fluid. J. Biomed. Mater. Res. Part A **64**(1), 164–170 (2003)
46. Park, J., et al.: Nanosize and vitality: TiO_2 nanotube diameter directs cell fate. Nano Lett. **7**(6), 1686–1691 (2007)
47. Eick, S., et al.: Iridium oxide microelectrode arrays for in vitro stimulation of individual rat neurons from dissociated cultures. Front. Neuroeng. **2**, 16 (2009)
48. Brammer, K.S., et al.: TiO2 nanotube structures for enhanced cell and biological functionality. JOM **62**(4), 50–55 (2010)
49. Sun, L., et al.: Effect of the geometry of the anodized titania nanotube array on the performance of dye-sensitized solar cells. J. Nanosci. Nanotechnol. **10**(7), 4551–4561 (2010)
50. Khudhair, D., et al.: Anodization parameters influencing the morphology and electrical properties of TiO 2 nanotubes for living cell interfacing and investigations. Mater. Sci. Eng. C **59**, 1125–1142 (2016)
51. Ghicov, A., et al.: Annealing effects on the photo response of TiO_2 nanotubes. Phys. Status Solidi **203**(4), R28–R30 (2006)
52. He, G., Hagiwara, M.: Ti alloy design strategy for biomedical applications. Mater. Sci. Eng. C **26**(1), 14–19 (2006)
53. Hamedani, H.A., et al.: Electrochemical fabrication of strontium-doped TiO2 nanotube array electrodes and investigation of their photoelectrochemical properties. J. Phys. Chem. C **115**(27), 13480–13486 (2011)
54. Huang, L., et al.: Wettability and in vitro bioactivity of doped TiO_2 nanotubes. 无机材料学报 **25**(7) (2010)
55. Xiao, P., et al.: Nanostructured electrode with titania nanotube arrays: fabrication, electrochemical properties, and applications for biosensing. J. Nanosci. Nanotechnol. **9**(4), 2426–2436 (2009)
56. Zhang, J., et al.: Fabrication and Capacitive Properties of C-doped TiO_2 Nanotube Arrays
57. Kyeremateng, N.A., et al.: Effect of Sn-doping on the electrochemical behaviour of TiO 2 nanotubes as potential negative electrode materials for 3D Li-ion micro batteries. J. Power Sources **224**, 269–277 (2013)

58. De Venuto, D., et al.: Impedance modeling of the intracortical microelectrode for a reliable design of a brain activity recording system. In: 2014 15th International Symposium on Quality Electronic Design (ISQED). IEEE (2014)
59. Reichert, W.M.: Indwelling Neural Implants: Strategies for Contending with the in vivo Environment. CRC Press (2007)
60. Ludwig, K.A., et al.: Chronic neural recordings using silicon microelectrode arrays electrochemically deposited with a poly (3, 4-ethylenedioxythiophene) (PEDOT) film. This work was supported by the Center for Wireless Integrated Microsystems NSF EEC-9986866 and the Whitaker Foundation. J. Neural Eng. **3**(1), 59 (2006)
61. Fattahi, P., et al.: A review of organic and inorganic biomaterials for neural interfaces. Adv. Mater. **26**(12), 1846–1885 (2014)
62. House, P.A., et al.: Acute microelectrode array implantation into human neocortex: preliminary technique and histological considerations. Neurosurg. Focus **20**(5), 1–4 (2006)
63. Turner, J., et al.: Cerebral astrocyte response to micromachined silicon implants. Exp. Neurol. **156**(1), 33–49 (1999)
64. Subbaroyan, J., Martin, D.C., Kipke, D.R.: A finite-element model of the mechanical effects of implantable microelectrodes in the cerebral cortex. J. Neural Eng. **2**(4), 103 (2005)
65. Metallo, C., White, R.D., Trimmer, B.A.: Flexible parylene-based microelectrode arrays for high resolution EMG recordings in freely moving small animals. J. Neurosci. Methods **195** (2), 176–184 (2011)
66. Lee, S.E., et al.: A flexible depth probe using liquid crystal polymer. IEEE Trans. Biomed. Eng. **59**(7), 2085–2094 (2012)
67. Cho, S.-H., et al.: Biocompatible SU-8-based microprobes for recording neural spike signals from regenerated peripheral nerve fibers. Sens. J. IEEE **8**(11), 1830–1836 (2008)
68. Lee, K., et al.: Biocompatible benzocyclobutene (BCB)-based neural implants with micro-fluidic channel. Biosens. Bioelectron. **20**(2), 404–407 (2004)
69. Kim, D.-H., et al.: Dissolvable films of silk fibroin for ultrathin conformal bio-integrated electronics. Nat. Mater. **9**(6), 511–517 (2010)
70. Green, R.A.: Unauthorised development and seismic hazard vulnerability: a study of squatters and engineers in Istanbul. Turk. Disasters **32**(3), 358–376 (2008)
71. Khodagholy, D., et al.: Highly conformable conducting polymer electrodes for in vivo recordings. Adv. Mater. **23**(36) (2011)
72. Hoffer, J., Loeb, G.: Implantable electrical and mechanical interfaces with nerve and muscle. Ann. Biomed. Eng. **8**(4–6), 351–360 (1980)
73. Donaldson, P.: Aspects of silicone rubber as encapsulant for neurological prostheses part 4: two-part rubbers. Med. Biol. Eng. Comput. **35**(3), 283–286 (1997)
74. Stieglitz, T., Schuetter, M., Koch, K.P.: Implantable biomedical microsystems for neural prostheses. Eng. Med. Biol. Mag. IEEE **24**(5), 58–65 (2005)
75. Stieglitz, T., Schuettler, M., Meyer, J.-U.: Micromachined, polyimide-based devices for flexible neural interfaces. Biomed. Microdev. **2**(4), 283–294 (2000)
76. Xiang, Z., et al.: Progress of flexible electronics in neural interfacing—a self-adaptive non-invasive neural ribbon electrode for small nerves recording. Adv. Mater. (2015)
77. Ordonez, J.S., et al.: Cuff electrodes for very small diameter nerves—prototyping and first recordings in vivo. In: Engineering in Medicine and Biology Society (EMBC), 2014 36th Annual International Conference of the IEEE. IEEE (2014)
78. Kotov, N.A., et al.: Nanomaterials for neural interfaces. Adv. Mater. **21**(40), 3970–4004 (2009)
79. Santhanam, G., et al.: A high-performance brain-computer interface. Nature **442**(7099), 195–198 (2006)
80. Reppel, M., et al.: Microelectrode arrays: a new tool to measure embryonic heart activity. J. Electrocardiol. **37**, 104–109 (2004)
81. Patolsky, F., et al.: Nanowire-based nanoelectronic devices in the life sciences. MRS Bull. **32**(02), 142–149 (2007)

82. Zhou, X., et al.: Supported lipid bilayer/carbon nanotube hybrids. Nat. Nanotechnol. **2**(3), 185–190 (2007)
83. Javey, A., et al.: Layer-by-layer assembly of nanowires for three-dimensional, multifunctional electronics. Nano Lett. **7**(3), 773–777 (2007)
84. Whang, D., et al.: Large-scale hierarchical organization of nanowire arrays for integrated nanosystems. Nano Lett. **3**(9), 1255–1259 (2003)
85. Duan, X., et al.: Intracellular recordings of action potentials by an extracellular nanoscale field-effect transistor. Nat. Nanotechnol. **7**(3), 174–179 (2012)
86. Robinson, J.T., et al.: Vertical nanowire electrode arrays as a scalable platform for intracellular interfacing to neuronal circuits. Nat. Nanotechnol. **7**(3), 180–184 (2012)
87. Ajayan, P.: Nanotubes from carbon. Chem. Rev. **99**(7), 1787 (1999)
88. De Volder, M.F., et al.: Carbon nanotubes: present and future commercial applications. Science **339**(6119), 535–539 (2013)
89. Wu, Y., et al.: Determination of the Young's modulus of structurally defined carbon nanotubes. Nano Lett. **8**(12), 4158–4161 (2008)
90. Kim, P., et al.: Thermal transport measurements of individual multiwalled nanotubes. Phys. Rev. Lett. **87**(21), 215502 (2001)
91. Lu, W., et al.: One-dimensional hole gas in germanium/silicon nanowire heterostructures. Proc. Natl. Acad. Sci. U.S.A. **102**(29), 10046–10051 (2005)
92. Cellot, G., et al.: Carbon nanotubes might improve neuronal performance by favouring electrical shortcuts. Nat. Nanotechnol. **4**(2), 126–133 (2009)
93. Lovat, V., et al.: Carbon nanotube substrates boost neuronal electrical signaling. Nano Lett. **5**(6), 1107–1110 (2005)
94. Wang, K., et al.: Neural stimulation with a carbon nanotube microelectrode array. Nano Lett. **6**(9), 2043–2048 (2006)
95. Lu, Y., et al.: Electrodeposited polypyrrole/carbon nanotubes composite films electrodes for neural interfaces. Biomaterials **31**(19), 5169–5181 (2010)

Revolutionizing Causal Circuitry Neurostimulation Utilizing the Optogenetic Technique Through Advanced Microsystems Development

R.P. Kale, S. Paek, S.J. Tye and A.Z. Kouzani

Abstract Optogenetics has recently emerged to become one of the most significant tools for in vivo causal analysis of neural networks. Developed through decades of pioneering work, the optogenetic toolbox has expanded utility to allow virtually total control over cellular actions. This article explores the emerging technologies that have been incorporated into making optogenetics a versatile technique in neuroscience research. Genetically engineered opsins continually evolve to directly activate or inhibit neuronal transmission with greater precision and functionality. A variety of light sources and fiber coupling methods employ unique photoactivation patterns and shapes. This article further explores the novel devices and systems that have been developed for the research setting and the technologies each system incorporates. These tethered systems, portable devices, and implantable microdevices have inherent benefits and detriments that are also discussed. Finally, emerging translational properties of optogenetics, particularly that for retinal pigmentosa, demonstrate how optogenetics may one day precipitate out of the research setting and into our healthcare practice.

R.P. Kale (✉) · A.Z. Kouzani
School of Engineering, Deakin University, Geelong, VIC 3216, Australia
e-mail: rajas.p.kale@gmail.com

R.P. Kale · S.J. Tye
Department of Psychiatry and Psychology, Mayo Clinic, 200 First Street SW, Rochester, MN 55905, USA

S. Paek
Department of Neurologic Surgery, Mayo Clinic, 200 First Street SW, Rochester, MN 55905, USA

S. Paek
Mayo Clinic Graduate School of Biomedical Sciences, Mayo Clinic, 200 First Street SW, Rochester, MN 55905, USA

A. Bhatti et al. (eds.), *Emerging Trends in Neuro Engineering and Neural Computation*, Series in BioEngineering,
DOI 10.1007/978-981-10-3957-7_3

1 Introduction

The implementation of targeted light stimulation has enabled neuroscientists to perform brain circuitry dissection with an unprecedented level of selectivity and specificity in spatial and temporal domains. This tool, known as optogenetics, utilizes light to control neurons that are genetically sensitized to specific types of light. To define optogenetics, Michael Häusser, a leading researcher in neural computation, states

> There's a broad definition and a narrow definition. The broad definition is rooted in etymology: any approach that combines optical interrogation with genetic targeting qualifies as 'optogenetic', and that includes the use of genetically encoded activity sensors. However, most people generally use the term optogenetics to mean the use of probes to manipulate activity, and (as is usual in English) usage normally wins [1].

Optogenetics exploits the function of genetically engineered light-sensitive membrane proteins that gate the ionic flux between internal and external space of the cell in a high temporal resolution. Ionic influx and efflux causes neuronal depolarization (activation) or hyperpolarization (inhibition), modulating the neuron's membrane potential. Furthermore, cell types can specifically be targeted to respond to light stimulation, as opposed to whole regions being vulnerable to the stimulation. The result is full control of neuronal firing patterns of preselected photosensitive neurons. Photosensitivity of neurons is possible through their induced expression of specialized proteins that are responsive to light. These proteins act as gated ion channels that open as wavelength-specific photons are absorbed. Activation of these channels floods the cell with either positive or negative ions as shown in Fig. 1. The type of ion and its charge is dependent on the type of light-sensitive protein used. Investigators utilize this technique to control neurons both in vivo and in vitro for a diverse range of experiments including neuronal connectivity mapping, functional connectivity, behavioral experiments, and induction of disease. These experiments can be carried out in cell cultures or in animal models, and one day may translate to clinical therapies for human patients.

Currently, optogenetic experiments are exclusive to preclinical settings. The first step in performing optogenetic studies is acquiring a genetically photosensitive cell group or animal. For example, murine optogenetic experimentation utilizes transgenic knock-in rodents. Knock-in rodent models incorporate a specific photosensitive protein into the animal's cellular DNA, and subsequently express this protein to become optogenetically viable. Transgenic animal lines are available almost exclusively within mouse models, but transgenic rats are becoming more available. The alternative to this is implementing viral vectors to specifically target neuronal cells of interest within the brain. Viral vectors offer more feasible transfection than transgenic cell lines in terms of resources, and can be performed in multiple types of animal models instead of genetically tractable rodents [80].

Photoactivation of cells was originally pioneered by Richard Fork in 1971 by stimulating native abdominal ganglia cells of marine mollusks at Bell Laboratories [23]. Decades later in 2005, Karl Deisseroth's lab first transfected mammalian cells

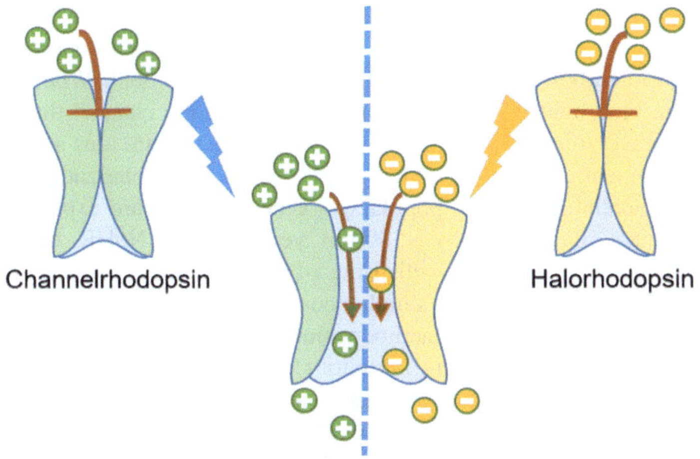

Fig. 1 Representation of channelrhodopsin and halorhodopsin in their closed conformational state. When activated by a light source, channelrhodopsin (at 460 nm wavelength) and halorhodopsin (at 590 nm wavelength) allow passage into the cellular membrane for cations (cellular depolarization) and anions (cellular hyperpolarization) respectively

with photosensitive proteins and performed millisecond-timescale activity control on these cells [8]. He would later coin the term 'optogenetics' and this technique would be hailed as nature's method of the year in 2010 [60]. A growing number of laboratories and research groups are continually make advances in optogenetics and the future of this technique seems endless in possibilities. We describe some of these advances in terms of photosensitive proteins, light sources, optical fibers and coupling, optogenetic microdevices and systems, and translational properties.

2 Opsins

Achieving optical neuronal control through optogenetics is often dependent on the photosensitive protein (known as an opsin) that is genetically encoded into the cell. Once expressed, opsins travel to the neuronal surface and are bound to the membrane. Naturally, opsins mediate photosynthetic functions by storing light energy into conjugated molecular bonds. Photon absorption results in conformational changes of the opsin into an open state to allow passage of ions across cell membranes. These opsins allow neuronal manipulation of the electrical potential in a high temporal precision (millisecond-scale) with light. Mediating light-driven depolarizations and hyperpolarizations of neurons with the opsins has proven to be facile [7].

The light-sensitive proteins channelrhodopsin-2 (ChR-2) from the alga *Chlamydomonas reinhardtii* and halorhodopsin (NpHR) from the archaeon *Natronomonas pharaonis* are two of the most commonly employed opsins in optogenetics [85]. Due to their robust temporarily precise control of neural activity in vitro [8] and in vivo [2, 3, 58, 63, 70], the use of these opsins, originally identified in microbes, has exploded in the neuroscience field. When illuminated by ~470 nm blue light [59], ChR-2 functions as a cation channel that evokes the influx of positive voltage change through ions such as sodium ions and protons but also potassium and calcium ions. Such influx results in depolarization of the basal membrane potential in specific population of neurons that are expressing ChR-2. Reversely, once illuminated by ~580 nm yellow light, NpHR functions as a chloride ion pump that inhibits cell firing of neurons that are expressing it [21]. ChR-2 and NpHR can be utilized simultaneously, allowing concurrent bidirectional control of neuronal activity. Archaerhodopsin-3 (Arch) is yet another opsin that has been identified to produce strong inhibitory photocurrents at yellow light wavelengths (566 nm) [16]. These photocurrents can be as high as 900 pA, dwarfing those of NpHR photocurrents of 38.9 pA [27]. These three opsins have originally allowed researchers to modulate cellular activity with some variety of optical parameters; however, their effective power densities (EPD50) are relatively quite low compared to today's bioengineered opsins. EPD50 is a measure of the light intensity required for 50% of the opsin population to become activated. The EPD50 of ChR-2 for example is 1.3 mW/mm^2, while the EPD50 of Arch is about 7.5 mW/mm^2 [53].

Opsins have been artificially advanced through genetic engineering throughout the years. These newer opsin versions expand the optogenetics toolbox, allowing researchers to stimulate or inhibit cells at optical wavelengths and intensities much different than from their native forms. For example, ChETA was derived from ChR-2 to produce sustained spike trains greater than 200 Hz, compared to only 40 Hz of ChR-2 [32]. ChETA was among the first class of bioengineered opsins that was produced through singe-amino-acid substitutions of ChR-2. A second class of opsins was produced through combining segments of ChR-1 and ChR-2. ChIEF is one of these opsins and was developed to produce more reliable photoactivation during persistent light stimulation [50]. At higher frequencies, ChIEF experiences only 33% inactivation compared to ChR-2's 77% inactivation. This allows ChIEF to produce action potentials that more closely resemble normal neuronal spiking patterns. Currently, there exists a variety of stimulating opsins that is available with an assortment of unique properties. ReaCHR for example is a ChR-2 variant that has been red-shifted to respond to orange or red light, which is completely on the other side of the optical spectrum compared to ChR-2's blue light optimal excitation [49]. ReaCHR opens up the possibility of manipulating retinal cells with wavelengths that do not cause damage to the eye, and allows the possibility of clinical optogenetic treatment for ocular conditions such as retinal pigmentosa (discussed at the end of this chapter). Interestingly, the amount of light that is absorbed by the tissue is dependent on the wavelength [2, 81]. Shorter wavelengths will tend to have higher absorption in the tissue and therefore have a lower volume of penetrance through the brain. Therefore, the wavelength at which the opsin responds can dictate the overall volume of illumination penetrance within the brain. ChR-2 has also been modified into forms that gate inhibitory chloride ions in order to hyperpolarize the cell, such as iClC2 and slow chloride-conducting channelrhodopsin (SloChloC) [6, 79]. Halorhodopsins have similarly observed evolutions with the implementation of eNpHR2.0 [27]. This original adaptation solved the issue of NpHR's poor expression in cells by promoting membrane localization and ER export. Photocurrents of this adapted NpHR therefore increased to 68.1 pA. This was followed by the production of eNpHR3.0, which had substantially increased light sensitivity and could respond to illuminations as weak as 3.5 mW/mm^2 [29]. Other inhibitory proton pumps, including ArchT [33] and Mac [16], have observed similar modifications as NpHR [53]. Table 1 offers a summary of opsins and their parameters that are typically used in optogenetics. Finally, an interesting class of opsins has also been bioengineered that acts on second messenger systems instead of gating ion channels. These opsins are referred to as OptoXRs and can be used to modulate cellular signaling molecules and cascades. At 473 nm light, Opto-α_1-AR has the ability to activate IP_3 or DAG recruiter GPCR, while Opto-β_2-AR activates cAMP recruiter GPCR [16]. These second messenger manipulations provide an even greater control of intricate molecular pathways of the cell to allow more refined modulations that mimic the actions of pharmaceutical drugs or specific diseased states.

Recently, a new method has come into the spotlight for its ability to edit, knockdown, or activate specific genes in localized or global cell populations. This genetic control can occur either at the germline level or in mature animal models. Known as CRISPR (clustered regularly interspaced short palindromic repeats), this method would allow researchers to take full control of genetic manipulations. Of the many fascinating possibilities CRISPR can unearth, one of the particular interests is its ability to genetically produce neurological disease models in animal research. However, the spatiotemporal precision of CRISPR lacks feasibility, especially considering the amount of toxic effects many of the chemicals used for CRISPR control can induce onto the cell and surrounding tissue during experimentation. This problem was alleviated with the production of photoactivatable-Cas9 (paCas9) [61].

Table 1 Examples of opsins used in optogenetics and their characteristics

Opsin	Subfamily	Gating type	Activation λ (nm)	EPD50 (mW/mm^2)	Photocurrent
Channelrhodopsin	ChR-2	Cation Channel	460	1.3	1-1.5 nA
Channelrhodopsin	$ChETA_A$	Cation Channel	470	3.1	<0.5 nA
Channelrhodopsin	ChIEF	Cation Channel	470	~1.5	390 pA
Halorhodopsin	NpHR	Chloride Pump	589	5.4	38.9 pA
Halorhodopsin	eNpHR3.0	Chloride Pump	566	5.5	450 pA
Archaerhodopsin	Arch	Proton Pump	566	~7.5	900 pA
Archaerhodopsin	ArchT1.0	Proton Pump	566	~6.5	~0.2 nA
Archaerhodopsin	eArch3.0	Proton Pump	566	~7.5	~1.2 nA
Leptosphaeria rhodopsins	eMac3.0	Proton Pump	540	~2.0	~0.3 nA

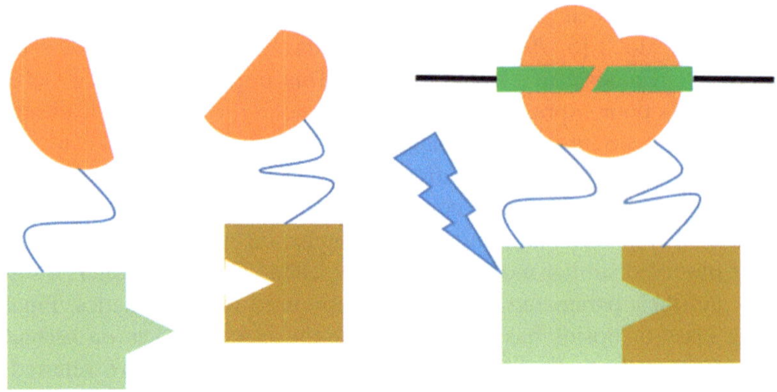

Fig. 2 Left: In the absence of *blue* light, nonfunctional paCas-9 is separated into two components that are each attached to magnets. Nonfunctional paCas-9 does not interact with DNA (*green bar*). *Right* with the induction of *blue* light, the magnets come together and form ionic bonds, allowing paCas-9 to dimerize into its functional form. The functional form is then able to produce DNA linkage breaks for targeted gene editing

Cas-9 is a vital control point in the CRISPR method as it produces DNA breakages to allow targeted genome editing. Thus, paCas9 allows researchers to turn this system on or off with precision by stimulating with blue light. This blue light activates a photoinducible dimerization system called magnets to bring the two components of paCas-9 together using ionic coupling. The functional paCas-9 can then produce targeted DNA breakage as shown in Fig. 2. Once the light is turned off, the magnets decouple and the CRISPR system shuts down, reversing the original effects. The potential of this system is far-reaching in biomedical sciences, with abundance in both exciting opportunities for research and ethical concerns for the human population [12].

3 Light Sources

Cellular performance of optogenetic stimulation relies on adequate delivery of photons from a light source. This light delivery has to occur at certain set parameters. These parameters include the specific wavelength of the photons needed for opsin activation, as well as the frequency and pulse width of the stimulation that govern the cell's response. As shown in Fig. 3, the needed wavelength of the produced light is dependent on the type of opsin and can range from near-infrared to near-ultraviolet light. Deviating from the optimal opsin-specific wavelength results in lower total activation. The frequency and pulse duration, however, can change based on the brain region and neuron type, as well as the behavior that is being sought.

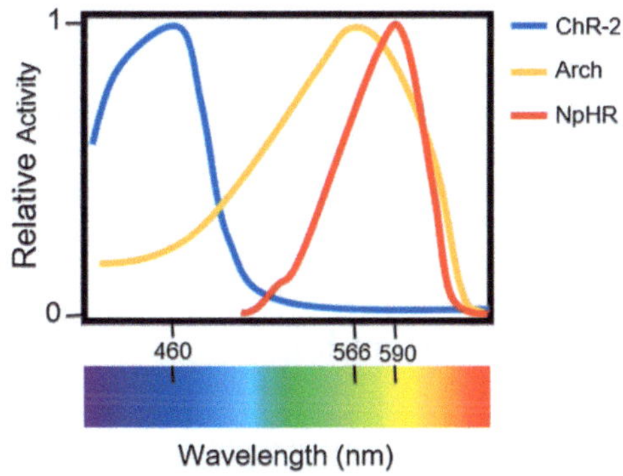

Fig. 3 Simplified activation profiles for ChR-2, Arch, and NpHR over the visible light spectrum. The greater the deviation from optimal wavelength, the less relative activity is expected from the opsins expressed by the cells

A variety of different light sources have been utilized to modulate neuronal activity in optogenetics experiments. These include lasers, liquid crystal displays (LCDs), arc lamps, halogen lights, light emitting diodes (LEDs), and organic LEDs (OLEDs). Previously, a majority of optogenetic systems have incorporated a laser-based system as the light source. However more recently, LEDs have become a primary source of photostimulation due to their benefits over laser-based systems in price, beam stability, instrumentation size, and high-frequency temporal precision [17]. In order to deliver a sufficient amount of light for ChR-2 activation, a relatively modest illumination irradiance is necessary. At 470 nm, these opsins necessitate between 0.1 and 1 mW/mm^2 [19]. Falling below this range results in less than 10% activation of ChR-2 [30]. Experiments have shown that LEDs [28, 40] can adequately supply this irradiance. Other experiments that use high intensity light sources, including lasers [2, 26, 34] and arc lamps [8, 84], have been performed. However, these powerful light sources promote the risk of damaging or killing the neural tissue by heating up the cell. Furthermore, these light sources produce only a single-point illumination, whereas OLEDs and LCDs can produce diverse patterns of two-dimensional arrays of illumination. However, their irradiance potential is around 1×10^{-4} mW/mm^2, which is 3 orders of magnitude too low for ChR-2 activation [31].

Neuromodulation experiments that utilize photostimulation, as opposed to electrical stimulation, beneficially produce very low-intensity stimulation artifacts. Having low artifacts facilitates better detection during simultaneous recording by increasing the signal-to-noise ratio. Various neural recording platforms have been incorporated into optogenetics experiments, including local-field potential [22], evoked-potential [20, 52], electroencephalogram [74], and single neuron recording [22]. Each study would have to consider all possible artifacts that could hinder their recording signal strength. Generally, there are three sources of artifacts occurring at the recording electrode level that are actively suppressed through signal filtering: capacitive coupling, common-mode conversion, and voltage gradients [44, 55]. The intensity of this noise generated by recording instruments (at the millivolt range) is orders of magnitude larger than the artifacts produced by optogenetics system (at the microvolts range) [17]. Therefore, the noise produced by optogenetic instruments can be relatively negligible and easy to minimize. These low-frequency artifacts are typically the product of the photoelectric effect, which is brought on by illumination-induced temperature kinetics directly changing the electrode conductance and releasing electrons picked up by the recorder [13]. Low-frequency artifacts produced by optogenetics can be remedied through high-pass filtering [34].

4 Fiber Coupling

The interface between light that is being produced and the neural region being modulated needs to retain efficiency in optogenetics experiments. Often this can be a limiting factor in the necessary power consumed by the system as well as the intensity of neural damage from optical fiber insertion into the brain. Optical fibers are typically inserted into the brain to transmit light into deep neural regions. Optical fibers are relatively inert and thin materials, which helps minimize the immune and inflammatory response of foreign body insertion as well as promote biocompatibility. Generally, a thinner optical fiber produces a smaller "kill-zone" in neural tissue [73]. Thinner optical fibers are, however, less efficient at transmitting light because of the smaller cross-sectional area for photons to enter the fiber. Optogenetics studies on larger animal models can tolerate thicker materials when inserted into the brain. Mice are able to tolerate optical fibers up to 300 μm, whereas rats can tolerate up to 400 μm [14].

Maximizing efficiency of optical fiber transmission is imperative, as it ensures sufficient photoactivation while simultaneously minimizing power consumption. Efficiency is influenced by the uniformity in direction that the photons travel once they exit the light source. Lasers inherently produce unidirectional light, but other light sources produce photons that are more scattered. The light source's viewing angle greatly affects the system's overall efficiency, as wider angled light sources result in a smaller percentage of photons that are able to enter the core of the optical fiber. Multimodal optical fiber cores, which have a greater core diameter compared to single-mode fibers, are preferred as the proportion of their cross-sectional core area is greater. A visual representation of viewing angle and fiber types is shown in Fig. 4. This directly relates to the numerical aperture (NA) of the fiber, which can be measured as

$$NA = \sqrt{n_{core}^2 - n_{cladding}^2},$$

where η_{core} represents the fiber core's refractive index and η_{ckadding} represents the fiber cladding's refractive index. Cladding is the material that lines the core and helps prevent the light from refracting out of the core as it reflects down the fiber. The NA is an important measure when comparing the relative light gathering capabilities of different fibers. Additionally, lenses can be used to focus light into the optical fiber's core, as long as the numerical aperture of the fiber is greater than

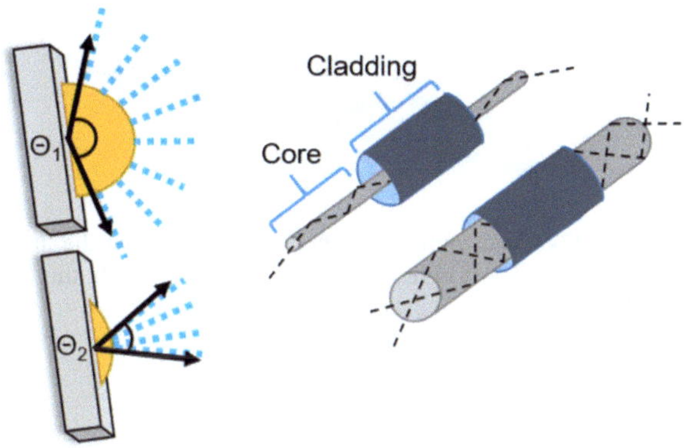

Fig. 4 *Left* Different viewing angles created by two LEDs. Although they produce the same amount of energy, Θ_2 viewing angle will have better efficiency when coupling to a fiber because the energy is more directed towards the fiber's core. *Right* Single-modal (*top*) and multimodal (*bottom*) optical fibers with the outer cladding stripped back to demonstrate light transmission. Multimodal fibers have larger core diameters, allowing for more light to enter the fiber and be delivered to the photoactivation region, thereby increasing efficiency

the numerical aperture of the lens. Typically, the lens' refractive index should be greater than 1.9 or else the maximum coupling efficiency is equivalent to simply butt-coupling the light source to the fiber [78].

More recently, there have been successful modifications to optical fibers for enhanced functionality during stimulation. For example, tapering the fiber produces a thinner fiber tip for focused activation while also minimizing neural damage at the proximal location to the site of stimulation. This helps the likelihood of maintaining the neural connections that may otherwise be damaged with a wider fiber penetrating the area of interest. Tapering can be accomplished chemically [35, 68] or thermo-mechanically [72]. Tapered fibers can also be coated with a thin metallic film, such as gold, to offer electrophysiological recording capabilities along the optical fiber [62]. In this case, the metallic film is inserted into a polyamide tube covered by a stainless steel tube. The multiple layers result in a larger diameter but are necessary for electrical insulation during electrophysiology recording and mechanical rigidity. Optical fibers have also been modified through focused ion beams to create optical windows on the fiber [64, 65]. These are essentially small windows that allow photons to escape at specific points along the fiber. Fabricating multiple windows and modifying the input angle can direct the light to exit through specific windows only. Therefore, researchers are able to target different neural sites using a single optical fiber that is easily changeable. These windows also allow the optical fiber to act as optical array by allowing the light to escape through multiple windows at once. This is demonstrated in Fig. 5. The implementation of optical windows on fibers further enhances the spatiotemporal resolution that optogenetics offers, and can be utilized with any optogenetic device that uses fibers.

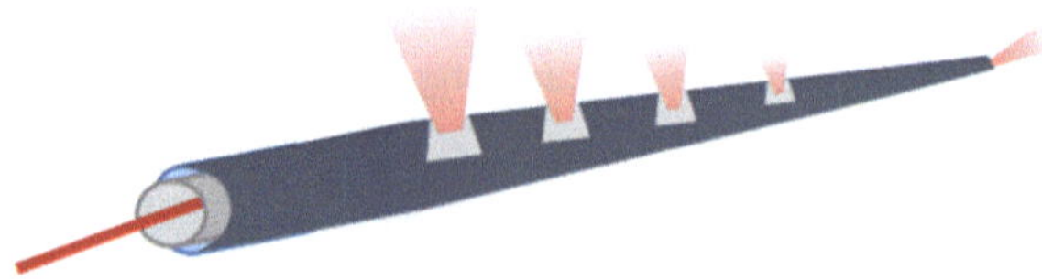

Fig. 5 A tapered optical fiber with optical windows. Depending on the angle of the incoming light source, the windows will project light out at different intensities. In this way, specific regions along the fiber can be targeted

5 Devices

The number of novel optogenetic systems that are being introduced into the research setting is vastly growing every year. Either for specific investigations or for commercial sales, these systems are becoming increasingly diverse in their functionality, size, cost, throughput, and more. The types of optogenetic systems that are being developed can be classified into three main categories: tethered systems, portable devices, and implantable microdevices [43]. Each of these has their own unique advantages and disadvantages, which will be discussed. Therefore, it is in the researcher's best interest to carefully determine which type of system suits their needs and would be optimal for their investigation.

5.1 Tethered Systems

Neuromodulation experiments typically require a power source to deliver stimulations to the intended neurons. In tethered systems, an optical fiber cord directly connects the animal to an external light source that is outside the animal's habitat. This light source is then connected to a wall power outlet, which delivers essentially an infinite supply of power to the system. Therefore, tethered systems can harness a very high current to power their light source, and deliver optical power with as much illuminance as needed. This process comes with limitations, particularly due to the animal being tethered to the light source. This setting potentiates the risk of the animal becoming tangled by its optical fiber cord. It would be difficult and dangerous to freely stimulate multiple animals during social behavior studies while preventing them from damaging each other's cords and implants. Additionally, behavioral tests in mood disorder models would be negatively affected by the cord as they deviate from the animal's naturalistic setting. The animal can become tangled by its own cord unless open-top home cage and behavioral apparatuses are used [48]. This issue may be alleviated with a rotary joint; however, these components inflict stressful rotational torques on the animal. Mice can typically withstand rotational pulling of up to 150 μN m, whereas rats can withstand upwards of 300 μN m [46]. Thus, selecting the ideal pulling strength is imperative—too soft and the cord could become tangled around the animal, too hard, and the force would cause stress and injury. Tethered optogenetic systems may be ideal for experiments that compare stimulation of motor pathways and the resulting muscle movements the animal performs. Recent studies have gained insights into how motor control is affected by optogenetic modulation of cells [36, 71].

A majority of tethered studies incorporate laser-based light sources because of their ability to harness high power demands. However, as previously discussed, LEDs can be advantageous to optogenetic studies [17]. With the incorporation of LEDs, one possibility is to eliminate the optical fiber entirely and simultaneously add a deep brain electrophysiological component onto a single optrode [11]. This is accomplished by incorporating the LED onto the tip of the optrode, so that the light source sits proximal to the site of photostimulation. Other tethered systems have used hybrid multielectrode recording and fiber optic arrays for both neuromodulations and measurements of cortical cells [47, 76]. Multifunctional systems can be adapted to a single platform and are not limited to electrophysiological recordings only. A novel device has combined a microfluidic channel to the recording and stimulating optrode [69]. This technology offers a combination of optogenetics, recording, and drug delivery, with the potential viral redelivery function to maintain opsin expression for long-term longitudinal studies.

5.2 Portable Devices

Most devices and technologies have a tendency to miniaturize over time. This concept is no exception in optogenetics. The detriments of tethering can pose a challenge to experimenters. Having the device mountable or wearable by the animal has exceptional value to the study, particularly in behavioral experiments. Tethered systems are capable of restricting behavioral experiments involved in complex paradigms such as wheel running, pain, and social interactions with other animals [54]. These restrictions can be minimized or eliminated with smaller portable devices when implemented in animal research. However, limitations inherently exist with portable devices too. The most obvious limitation is the lack of a power source capable of producing unlimited continuous power, which is not an issue with tethered systems. As a result, power consumption of the device needs to be minimized. This can come at the expense of limited functionality, lower optical power output, and potentially a higher cost of consumables.

Typically, portable devices can harness power from either a battery or through wireless energy harvesting. Batteries offer a reliable and cost-effective approach compared to wireless energy harvesters. For example, the amount of sophisticated components and circuitry is modest compared to wireless energy harvesters. However, batteries tend to be quite heavy, and this poses a problem for small animal research. Furthermore, inherent issues with thermodynamic chemical reactions that take place in batteries include slow recharging, limited charge and discharge cycles, and issues of being a biohazard if the internal chemicals are leaked

[4, 9]. Also considering that the animal may also destroy the battery and ingest the hazardous toxins, the risks of using batteries is often greater to animals in weakly constructed devices. At the same time, the battery would likely need to be easily accessible and replaced during the experiment. Therefore, there is a tradeoff between accessibility of the battery for the researcher and risks posed to the animal. Despite these risks, battery-powered microdevices can be relatively cost-effective compared to tethered systems and wireless energy harvesters. Teams that produce portable optogenetic devices claim to have a low-cost product that is easily manufactured using commercially available components [24, 41]. Alternative components have been utilized in bioelectronics to circumvent battery usage, such as piezoelectric, thermoelectric, biopotential harvesting, and glucose harvesting [51]. However, these components tend to produce energy densities much too low for portable devices.

A more sophisticated method for powering portable optogenetic devices is through wireless energy harvesting. This can be accomplished through radio-frequency (RF) transmission. Although this technology is relatively new, portable optogenetic devices have successfully utilized wireless energy harvesting [42]. They incorporate an antenna and rectifying circuitry to form a rectenna [39]. This rectenna supplies the circuit with DC power supplied from an RF transmitter some distance away. This harnessed current is subsequently used to power the light source. The light source, however, does not require constant power, but instead typically pulses at set frequencies and pulse widths that depend on the experimental parameters. Therefore, the energy must be stored within the circuit and pulsed in bursts to the light source. Two traditional components can accomplish this to some degree—batteries and capacitors. As described, there are inherent flaws with batteries such as size limitations and charge/discharge capacities. There is also little purpose utilizing RF transmission to power a bulky battery, as the benefits of size reduction are diminished. Furthermore, power densities of batteries are relatively low while energy densities are sufficiently high. On the other hand, capacitors usually have sufficient power densities but provide too low energy densities [56]. A fundamental middle ground is needed between batteries and capacitors. Fortunately, there exists a component that satisfies the sufficient energy and power densities to power an LED for optogenetics—supercapacitors. An electric double-layer capacitor, known as a supercapacitor, can be produced that measures only 1 mm^3 and can produce a high enough area of capacitance at 1.3 mF/mm^2 [56].

Supercapacitors have a tendency to be damaged by high current loads. This can be rectified through the use of a boost converter, which limits the current intensity on the circuit by increasing the voltage. An RF harnessing headstage was developed by Wentz et al. [77] that makes use of this boost converter in mice, with the headstage weight totaling about 2 grams [77]. This team has since formed a company (Kendall Research Systems) that produce wireless optogenetics headstages that concurrently transfer wireless neural data for closed-loop stimulation. Closed-loop systems can modify the stimulation parameters in real time to only stimulate when certain brain activity patterns are detected. Incorporating a recording electrode into the device has key benefits for closed-loop systems. With this, Ameli et al. (2013) produced a headstage that can wirelessly transfer the neural readouts to a base station over 2 meters away. A centralized base station offers control of multiple animals in parallel for high-throughput studies. Other devices have been developed that eliminate the optical fiber entirely by fitting the LED onto an optrode for direct stimulation deep into the brain [67]. A lower illumination intensity can be used when the LED is implanted into the site of stimulation because efficiency loss from fiber coupling is eliminated. This results in a lower power consumption of the overall headstage.

5.3 *Implantable Microdevices*

Newer technologies are allowing researchers to implement smaller components into their optogenetic systems, and in many cases these systems can become fully implantable. μLEDs, which can range in sizes from 625 to 10,000 μm^2, greatly reduce the overall dimensions of the implanted light source compared to traditional LEDs. Once implanted, they are able to direct the light to specific sites on the tissue with miniature lenses [54]. μLEDs can also be produced in an array format to deliver light distribution in complex shapes and designs, providing enhanced spatial resolution at the cellular scale [31]. Partially implanted microdevices have been produced that incorporate a rectenna system that is exposed outside the skull, while the majority of the components are implanted within the animal (Kim et al. [45, 54]. This device has the ability to optically stimulate while concurrently taking electrophysiological recordings using a Pt-contact pad, optical measurements using a μ-IPD, and temperature using a serpentine platinum resistor. Optical measurement provides assurance that light is being produced by the device since light can be difficult to see when implanted. The temperature sensor ensures animal safety as heat can be produced by any electrical and optical components within the implant.

Research groups more recently have been successful at producing microdevices that are small and safe enough to be fully implanted into the animal. Dr. Deisseroth's team at Stanford University have accomplished this by creating a chamber that uses resonance cavities to produce RF signals [57]. These signals target the optogenetics device that is implanted into the mouse and provide the energy to activate it. The mouse can move freely inside of an enclosure that is above the chamber. These devices are miniscule in size and weight, with the smallest version of the system being 10 mm^3 and weighing 20 mg. This same group had previously developed implanted microdevices that use direct inductive coupling between two coils to power their device [82]. However, electromagnetic fields produce stronger localization for energy in the low gigahertz frequencies, allowing for smaller coils to be utilized in RF systems compared to inductive coupling. RF-powered wireless technologies appear to be fruitful in the preclinical setting because they can more easily compensate for the different orientations that the animal can take, and can utilize sophisticated resonance chambers for the animal to reside in. On the other hand, midfield technologies may offer a unique method of coupling internal devices for humans to external power sources that are easily replaceable [38]. One important difference between human and animal implants is that the human device orientation is unlikely to move in relation to the power transmission coil. Midfield technologies may one day facilitate the translational properties of optogenetics to be incorporated into the clinical setting.

6 Translational Properties

Performing clinical optogenetic therapy to treat neurological diseases may be decades from becoming a reality because of existing concerns such as the invasiveness of this technique. Viral stability may also be an issue, and antibiotic-driven promoters may offer a solution for the future, as they provide precisely tuned gene expression and even complete shutoff at undesirable instances [15]. However, this task is hindered by regulatory agencies requiring full toxicology investigations before becoming clinically viable. Despite these setbacks, current therapeutic advances are being researched in the preclinical realm, making the translational properties of optogenetics a possible reality. One example worth noting is a type of blindness, and optogenetics may offer a remedy to many that are affected.

6.1 Retinal Pigmentosa

Retinal pigmentosa (RP) is a retinal degenerative disorder that currently affects roughly 100,000 patients in the US [15]. Investigations are occurring into activating ON bipolar or retinal ganglion cells of the eye using ChR-2. The light delivered to the eye from a light source transduces the photons into signals that can be processed by the visual system to enable sight [10]. Simultaneous inhibition of dysfunctional cells may resensitize retinal photoreceptors through inhibitory opsins. The eye, through intravitreal injections, can express these opsins. This is a relatively non-invasive procedure when compared to viral injections into the brain. The alternative to intravitreal viral injections would be nanoparticles or gene-gun approaches; however, viral vectors offer rapid delivery, high infectivity, and cell-specificity [37]. This offers a unique translational property of optogenetics to patient therapy before full-blown neural interfacing occurs. Currently, preclinical intravitreal transfection in primates remains difficult due to the poor transport of the virus through the inner limiting membrane [18]. Scaling viral delivery to primate eyes also proves to be difficult as a massive amount of particle numbers is necessary for whole eye coverage compared to regional coverage of the brain.

The noninvasiveness nature of optogenetic RP therapy demonstrates the important benefits of promoting translational research. However, many device and computational issues would need to be addressed. Ambient light would not sufficiently produce illumination intensities strong enough to activate opsins in the eye. Outerwear goggles can transmit the light at strong enough intensities with precise temporal and spatial locations onto the eye. The goggles would reproduce visual images from the external visual field, amplify this signal, and output the transformed visual images onto the retina through wavelengths that the ocular opsins can respond to [25]. Sophisticated components would be necessary to accomplish optogenetic RP therapy. One that stands out is the need for inward eye-tracking capabilities. Normally, eyes receive visual input from images that are produced from both a static retina and a moving retina. Exclusively static retinal images could produce visual stimulation different than what healthy individuals see, and therefore the goggles would need to compensate through altered arrays of projected intensities. Otherwise constant static images would result in adaptations to the eye, ultimately depressing visual response [66]. Additionally, computational compensation for ocular vergence would need to be performed for binocular vision of objects at different fixation points. Healthy-sighted individuals automatically adjust both eyes' positions simultaneously to maintain the site on an approaching object. This allows photons coming from the object to fall directly onto the focal point of the retina as it approaches or retracts away. Eye-tracking functionality could mediate this by actively adjusting the image's projected location so the retina can correct itself by radial stretch in the fovea, among other methods [5].

RP therapy has one of the highest translatable properties of any disease for optogenetics. However, many other disease states are being investigated by researchers. Animal models of disease coupled with optogenetics are providing these researchers with tools to further understand the underlying basis of disease genesis and progression, as well as novel methods to treat them. These diseases include but are not limited to anxiety, depression, schizophrenia, addiction, social dysfunction, Parkinson's disease, and epilepsy [75]. As scientists gain more knowledge of these diseases, better targeting drugs and personalized treatment options may be available as better treatment options for patients.

7 Conclusion

Advancing technologies in the optogenetic technique are emerging in this exciting and revolutionary field. The current trends towards technological sophistication and diverse functionality give this relatively new technique an exciting prospect to its future potentials both in preclinical and translational sciences. Clinical optogenetics may see a place in the future for patient healthcare once all of its caveats have been solved. For now optogenetics continues to inspire new research in complimentary areas, such as genetic engineering and neurostimulation. Novel methods are being investigated because of the excitement, such as focused ultrasound that offers noninvasive brain stimulation in deep regions of the brain [83]. Albeit the possibility exists that newer neurostimulation techniques may be adapted that are more translatable compared to optogenetics. But the high degree of causal information gained, coupled with the incredible spatiotemporal properties of this technique, will make optogenetics a viable option for uncovering the deep mysteries of the brain for years to come.

References

1. Adamantidis, A., Arber, S., Bains, J.S., Bamberg, E., Bonci, A., Buzsáki, G., Cardin, J.A., Costa, R.M., Dan, Y., Goda, Y.: Optogenetics: 10 years after ChR2 in neurons [mdash] views from the community. Nat. Neurosci. **18**(9), 1202–1212 (2015)
2. Aravanis, A.M., Wang, L.-P., Zhang, F., Meltzer, L.A., Mogri, M.Z., Schneider, M.B., Deisseroth, K.: An optical neural interface: in vivo control of rodent motor cortex with integrated fiberoptic and optogenetic technology. J. Neural Eng. **4**(3), S143 (2007)
3. Arenkiel, B.R., Peca, J., Davison, I.G., Feliciano, C., Deisseroth, K., Augustine, G.J., Ehlers, M.D., Feng, G.: In vivo light-induced activation of neural circuitry in transgenic mice expressing channelrhodopsin-2. Neuron **54**(2), 205–218 (2007). doi:10.1016/j.neuron.2007.03.005
4. Armand, M., Tarascon, J.-M.: Building better batteries. Nature **451**(7179), 652–657 (2008)
5. Asher, A., Segal, W., Baccus, S., Yaroslavsky, L.P., Palanker, D.V.: Image processing for a high-resolution optoelectronic retinal prosthesis. IEEE Trans. Biomed. Eng. **54**(6), 993–1004 (2007)
6. Berndt, A., Lee, S.Y., Ramakrishnan, C., Deisseroth, K.: Structure-guided transformation of channelrhodopsin into a light-activated chloride channel. Science **344**(6182), 420–424 (2014)
7. Boyden, E.S.: Optogenetics: using light to control the brain. Cerebrum: the Dana forum on brain science **2011**, 16 (2011)
8. Boyden, E.S., Zhang, F., Bamberg, E., Nagel, G., Deisseroth, K.: Millisecond-timescale, genetically targeted optical control of neural activity. Nat. Neurosci. **8**(9), 1263–1268 (2005)
9. Brodd, R.J., Bullock, K.R., Leising, R.A., Middaugh, R.L., Miller, J.R., Takeuchi, E.: Batteries, 1977 to 2002. J. Electrochem. Soc. **151**(3), K1–K11 (2004)
10. Busskamp, V., Roska, B.: Optogenetic approaches to restoring visual function in retinitis pigmentosa. Curr. Opin. Neurobiol. **21**(6), 942–946 (2011)
11. Cao, H., Gu, L., Mohanty, S.K., Chiao, J.: An integrated μLED optrode for optogenetic stimulation and electrical recording. IEEE Trans. Biomed. Eng. **60**(1), 225–229 (2013)
12. Caplan, A.L., Parent, B., Shen, M., Plunkett, C.: No time to waste—the ethical challenges created by CRISPR. EMBO Rep. **16**(11), 1421–1426 (2015)
13. Cardin, J.A., Carlén, M., Meletis, K., Knoblich, U., Zhang, F., Deisseroth, K., Tsai, L.-H., Moore, C.I.: Targeted optogenetic stimulation and recording of neurons in vivo using cell-type-specific expression of Channelrhodopsin-2. Nat. Protoc. **5**(2), 247–254 (2010)
14. Carter, M.E., de Lecea, L.: Optogenetic investigation of neural circuits in vivo. Trends Mol. Med. **17**(4), 197–206 (2011)
15. Chow, B.Y., Boyden, E.S.: Optogenetics and translational medicine. Science translational medicine **5**(177), 177ps175–177ps175 (2013)
16. Chow, B.Y., Han, X., Dobry, A.S., Qian, X., Chuong, A.S., Li, M., Henninger, M.A., Belfort, G.M., Lin, Y., Monahan, P.E.: High-performance genetically targetable optical neural silencing by light-driven proton pumps. Nature **463**(7277), 98–102 (2010)
17. Clements, I.P., Gnade, A.G., Rush, A.D., Patten, C.D., Twomey, M.C., Kravitz, A.V.: Miniaturized LED sources for in vivo optogenetic experimentation. In: SPIE BiOS, 2013. International Society for Optics and Photonics, pp. 85860X-85860X-85869
18. Dalkara, D., Kolstad, K.D., Caporale, N., Visel, M., Klimczak, R.R., Schaffer, D.V., Flannery, J.G.: Inner limiting membrane barriers to AAV-mediated retinal transduction from the vitreous. Mol. Ther. **17**(12), 2096–2102 (2009)

19. Degenaar, P., Grossman, N., Memon, M.A., Burrone, J., Dawson, M., Drakakis, E., Neil, M., Nikolic, K.: Optobionic vision—a new genetically enhanced light on retinal prosthesis. J. Neural Eng. **6**(3), 035007 (2009)
20. Dinner, D.S., Neme, S., Nair, D., Montgomery, E.B., Baker, K.B., Rezai, A., Lüders, H.O.: EEG and evoked potential recording from the subthalamic nucleus for deep brain stimulation of intractable epilepsy. Clin. Neurophysiol. **113**(9), 1391–1402 (2002)
21. Duschl, A., Lanyi, J.K., Zimanyi, L.: Properties and photochemistry of a halorhodopsin from the haloalkalophile, Natronobacterium pharaonis. J. Biol. Chem. **265**(3), 1261–1267 (1990)
22. Filali, M., Hutchison, W.D., Palter, V.N., Lozano, A.M., Dostrovsky, J.O.: Stimulation-induced inhibition of neuronal firing in human subthalamic nucleus. Exp. Brain Res. **156**(3), 274–281 (2004)
23. Fork, R.L.: Laser stimulation of nerve cells in Aplysia. Science **171**(3974), 907–908 (1971)
24. Gagnon-Turcotte, G., Kisomi, A.A., Ameli, R., Camaro, C.-O.D., LeChasseur, Y., Néron, J.-L., Bareil, P.B., Fortier, P., Bories, C., De Koninck, Y.: A wireless optogenetic headstage with multichannel electrophysiological recording capability. Sensors **15**(9), 22776–22797 (2015)
25. Goetz, G., Mandel, Y., Manivanh, R., Palanker, D., Čižmár, T.: Holographic display system for restoration of sight to the blind. J. Neural Eng. **10**(5), 056021 (2013)
26. Gradinaru, V., Mogri, M., Thompson, K.R., Henderson, J.M., Deisseroth, K.: Optical deconstruction of parkinsonian neural circuitry. Science **324**(5925), 354–359 (2009)
27. Gradinaru, V., Thompson, K.R., Deisseroth, K.: eNpHR: a Natronomonas halorhodopsin enhanced for optogenetic applications. Brain cell biology **36**(1–4), 129–139 (2008)
28. Gradinaru, V., Thompson, K.R., Zhang, F., Mogri, M., Kay, K., Schneider, M.B., Deisseroth, K.: Targeting and readout strategies for fast optical neural control in vitro and in vivo. J. Neurosci. **27**(52), 14231–14238 (2007)
29. Gradinaru, V., Zhang, F., Ramakrishnan, C., Mattis, J., Prakash, R., Diester, I., Goshen, I., Thompson, K.R., Deisseroth, K.: Molecular and cellular approaches for diversifying and extending optogenetics. Cell **141**(1), 154–165 (2010)
30. Grossman, N., Nikolic, K., Toumazou, C., Degenaar, P.: Modeling study of the light stimulation of a neuron cell with channelrhodopsin-2 mutants. IEEE Trans. Biomed. Eng. **58** (6), 1742–1751 (2011)
31. Grossman, N., Poher, V., Grubb, M.S., Kennedy, G.T., Nikolic, K., McGovern, B., Palmini, R.B., Gong, Z., Drakakis, E.M., Neil, M.A.: Multi-site optical excitation using ChR2 and micro-LED array. J. Neural Eng. **7**(1), 016004 (2010)
32. Gunaydin, L.A., Yizhar, O., Berndt, A., Sohal, V.S., Deisseroth, K., Hegemann, P.: Ultrafast optogenetic control. Nat. Neurosci. **13**(3), 387–392 (2010)
33. Han, X., Chow, B.Y., Zhou, H., Klapoetke, N.C., Chuong, A., Rajimehr, R., Yang, A., Baratta, M.V., Winkle, J., Desimone, R.: A high-light sensitivity optical neural silencer: development and application to optogenetic control of non-human primate cortex. Front. Syst. Neurosci. **5** (2011)
34. Han, X., Qian, X., Bernstein, J.G., H-H, Zhou, Franzesi, G.T., Stern, P., Bronson, R.T., Graybiel, A.M., Desimone, R., Boyden, E.S.: Millisecond-timescale optical control of neural dynamics in the nonhuman primate brain. Neuron **62**(2), 191–198 (2009)
35. Hanks, T.D., Kopec, C.D., Brunton, B.W., Duan, C.A., Erlich, J.C., Brody, C.D.: Distinct relationships of parietal and prefrontal cortices to evidence accumulation. Nature **520**(7546), 220–223 (2015)
36. Heiney, S.A., Kim, J., Augustine, G.J., Medina, J.F.: Precise control of movement kinematics by optogenetic inhibition of Purkinje cell activity. J. Neurosci. **34**(6), 2321–2330 (2014)
37. Henriksen, B.S., Marc, R.E., Bernstein, P.S.: Optogenetics for retinal disorders. J. Ophthal. Vision Res. **9**(3), 374 (2014)
38. Ho, J.S., Kim, S., Poon, A.S.: Midfield wireless powering for implantable systems. Proc. IEEE **101**(6), 1369–1378 (2013)

39. Hosain, M.K., Kouzani, A.Z., Tye, S.J., Abulseoud, O., Amiet, A., Galehdar, A., Kaynak, A., Berk, M.: Development of a compact rectenna for wireless powering of a head-mountable deep brain stimulation device. IEEE J. Transl. Eng. Health Med. **2**, 1–13 (2014)
40. Huber, D., Petreanu, L., Ghitani, N., Ranade, S., Hromádka, T., Mainen, Z., Svoboda, K.: Sparse optical microstimulation in barrel cortex drives learned behaviour in freely moving mice. Nature **451**(7174), 61–64 (2008)
41. Iwai, Y., Honda, S., Ozeki, H., Hashimoto, M., Hirase, H.: A simple head-mountable LED device for chronic stimulation of optogenetic molecules in freely moving mice. Neurosci. Res. **70**(1), 124–127 (2011)
42. Kale, R.P., Kouzani, A.Z., Berk, M., Walder, K., Berk, J., Tye, S.J.: Wireless optogenetics: an exploration of portable microdevices for small animal photostimulation. Proc. Technol. **20**, 225–230 (2015)
43. Kale, R.P., Kouzani, A.Z., Walder, K., Berk, M., Tye, S.J.: Evolution of optogenetic microdevices. Neurophotonics **2**(3), 031206–031206 (2015)
44. Kent, A., Grill, W.: Recording evoked potentials during deep brain stimulation: development and validation of instrumentation to suppress the stimulus artefact. J. Neural Eng. **9**(3), 036004 (2012)
45. Kim, T.-I., McCall, J.G., Jung, Y.H., Huang, X., Siuda, E.R., Li, Y., Song, J., Song, Y.M., Pao, H.A., Kim, R.-H.: Injectable, cellular-scale optoelectronics with applications for wireless optogenetics. Science **340**(6129), 211–216 (2013)
46. Klorig, D.C., Godwin, D.W.: A magnetic rotary optical fiber connector for optogenetic experiments in freely moving animals. J. Neurosci. Methods **227**, 132–139 (2014)
47. Kwon, K.Y., Sirowatka, B., Weber, A., Li, W.: Opto-array: a hybrid neural interface with transparent electrode array and integrated LEDs for optogenetics. IEEE Trans. Biomed. Circuits Syst. **7**(5), 593–600 (2013)
48. Land, B.B., Brayton, C.E., Furman, K.E., LaPalombara, Z., DiLeone, R.J.: Optogenetic inhibition of neurons by internal light production. Front. Behav. Neurosci. **8** (2014)
49. Lin, J.Y., Knutsen, P.M., Muller, A., Kleinfeld, D., Tsien, R.Y.: ReaChR: a red-shifted variant of channelrhodopsin enables deep transcranial optogenetic excitation. Nat. Neurosci. **16**(10), 1499–1508 (2013)
50. Lin, J.Y., Lin, M.Z., Steinbach, P., Tsien, R.Y.: Characterization of engineered channelrhodopsin variants with improved properties and kinetics. Biophys. J. **96**(5), 1803–1814 (2009)
51. Ma, A., Poon, A.S.: Midfield wireless power transfer for bioelectronics. IEEE Circuits Syst. Mag. **15**(2), 54–60 (2015)
52. Maegaki, Y., Najm, I., Terada, K., Morris, H.H., Bingaman, W.E., Kohaya, N., Takenobu, A., Kadonaga, Y., Lüders, H.O.: Somatosensory evoked high-frequency oscillations recorded directly from the human cerebral cortex. Clin. Neurophysiol. **111**(11), 1916–1926 (2000)
53. Mattis, J., Tye, K.M., Ferenczi, E.A., Ramakrishnan, C., O'Shea, D.J., Prakash, R., Gunaydin, L.A., Hyun, M., Fenno, L.E., Gradinaru, V.: Principles for applying optogenetic tools derived from direct comparative analysis of microbial opsins. Nat. Methods **9**(2), 159–172 (2012)
54. McCall, J.G., Kim, T-I., Shin, G., Huang, X., Jung, Y.H., Al-Hasani, R., Omenetto, F.G., Bruchas, M.R., Rogers, J.A.: Fabrication and application of flexible, multimodal light-emitting devices for wireless optogenetics. Nat. Protoc. **8**(12), 2413–2428 (2013)
55. McGill, K.C., Cummins, K.L., Dorfman, L.J., Berlizot, B.B., Luetkemeyer, K., Nishimura, D. G., Widrow, B.: On the nature and elimination of stimulus artifact in nerve signals evoked and recorded using surface electrodes. IEEE Trans. Biomed. Eng. **2**, 129–137 (1982)
56. Meng, C., Gall, O.Z., Irazoqui, P.P.: A flexible super-capacitive solid-state power supply for miniature implantable medical devices. Biomed. Microdevices **15**(6), 973–983 (2013)
57. Montgomery, K.L., Yeh, A.J., Ho, J.S., Tsao, V., Iyer, S.M., Grosenick, L., Ferenczi, E.A., Tanabe, Y., Deisseroth, K., Delp, S.L.: Wirelessly powered, fully internal optogenetics for brain, spinal and peripheral circuits in mice. Nat. Methods (2015)

58. Nagel, G., Brauner, M., Liewald, J.F., Adeishvili, N., Bamberg, E., Gottschalk, A.: Light activation of channelrhodopsin-2 in excitable cells of Caenorhabditis elegans triggers rapid behavioral responses. Curr. Biol. **15**(24), 2279–2284 (2005). doi:10.1016/j.cub.2005.11.032
59. Nagel, G., Szellas, T., Huhn, W., Kateriya, S., Adeishvili, N., Berthold, P., Ollig, D., Hegemann, P., Bamberg, E.: Channelrhodopsin-2, a directly light-gated cation-selective membrane channel. Proc. Natl. Acad. Sci. **100**(24), 13940–13945 (2003)
60. NatureMethods: Method of the Year 2010. Nat Meth, vol 8. Nature Publishing Group, a division of Macmillan Publishers Limited. All Rights Reserved (2011). doi:10.1038/nmeth.f.321
61. Nihongaki, Y., Kawano, F., Nakajima, T., Sato, M.: Photoactivatable CRISPR-Cas9 for optogenetic genome editing. Nat. Biotechnol. **33**(7), 755–760 (2015)
62. Ozden, I., Wang, J., Lu, Y., May, T., Lee, J., Goo, W., O'Shea, D.J., Kalanithi, P., Diester, I., Diagne, M.: A coaxial optrode as multifunction write-read probe for optogenetic studies in non-human primates. J. Neurosci. Methods **219**(1), 142–154 (2013)
63. Petreanu, L., Huber, D., Sobczyk, A., Svoboda, K.: Channelrhodopsin-2-assisted circuit mapping of long-range callosal projections. Nat. Neurosci. **10**(5), 663–668 (2007). doi:10.1038/nn1891
64. Pisanello, F., Sileo, L., Oldenburg, I.A., Pisanello, M., Martiradonna, L., Assad, J.A., Sabatini, B.L., De Vittorio, M.: Multipoint-emitting optical fibers for spatially addressable in vivo optogenetics. Neuron **82**(6), 1245–1254 (2014)
65. Pisanello, M., Della Patria, A., Sileo, L., Sabatini, B.L., De Vittorio, M., Pisanello, F.: Modal demultiplexing properties of tapered and nanostructured optical fibers for in vivo optogenetic control of neural activity. Biomed. Opt. Express **6**(10), 4014–4026 (2015)
66. Roska, B., Agboh, D.: Restoring Vision to the Blind: Optogenetics (2014)
67. Rossi, M.A., Go, V., Murphy, T., Fu, Q., Morizio, J., Yin, H.H.: A wirelessly controlled implantable LED system for deep brain optogenetic stimulation. Front. Integr. Neurosci. **9** (2015)
68. Royer, S., Zemelman, B.V., Barbic, M., Losonczy, A., Buzsáki, G., Magee, J.C.: Multi-array silicon probes with integrated optical fibers: light-assisted perturbation and recording of local neural circuits in the behaving animal. Eur. J. Neurosci. **31**(12), 2279–2291 (2010)
69. Rubehn, B., Wolff, S.B., Tovote, P., Lüthi, A., Stieglitz, T.: A polymer-based neural microimplant for optogenetic applications: design and first in vivo study. Lab Chip **13**(4), 579–588 (2013)
70. Schroll, C., Riemensperger, T., Bucher, D., Ehmer, J., Voller, T., Erbguth, K., Gerber, B., Hendel, T., Nagel, G., Buchner, E., Fiala, A.: Light-induced activation of distinct modulatory neurons triggers appetitive or aversive learning in Drosophila larvae. Curr. Biol. **16**(17), 1741–1747 (2006). doi:10.1016/j.cub.2006.07.023
71. Seeger-Armbruster, S., Bosch-Bouju, C., Little, S.T., Smither, R.A., Hughes, S.M., Hyland, B.I., Parr-Brownlie, L.C.: Patterned, but not tonic, optogenetic stimulation in motor thalamus improves reaching in acute drug-induced parkinsonian rats. J. Neurosci. **35**(3), 1211–1216 (2015)
72. Sileo, L., Pisanello, M., De Vittorio, M., Pisanello, F.: Fabrication of multipoint light emitting optical fibers for optogenetics. In: SPIE BiOS, 2015. International Society for Optics and Photonics, pp. 93052O-93052O-93057
73. Stensaas, S.S., Stensaas, L.: The reaction of the cerebral cortex to chronically implanted plastic needles. Acta Neuropathol. **35**(3), 187–203 (1975)
74. Swann, N., Poizner, H., Houser, M., Gould, S., Greenhouse, I., Cai, W., Strunk, J., George, J., Aron, A.R.: Deep brain stimulation of the subthalamic nucleus alters the cortical profile of response inhibition in the beta frequency band: a scalp EEG study in Parkinson's disease. J. Neurosci. **31**(15), 5721–5729 (2011)
75. Tye, K.M., Deisseroth, K.: Optogenetic investigation of neural circuits underlying brain disease in animal models. Nat. Rev. Neurosci. **13**(4), 251–266 (2012)
76. Wang, J., Wagner, F., Borton, D.A., Zhang, J., Ozden, I., Burwell, R.D., Nurmikko, A.V., van Wagenen, R., Diester, I., Deisseroth, K.: Integrated device for combined optical

neuromodulation and electrical recording for chronic in vivo applications. J. Neural Eng. **9**(1), 016001 (2012)
77. Wentz, C.T., Bernstein, J.G., Monahan, P., Guerra, A., Rodriguez, A., Boyden, E.S.: A wirelessly powered and controlled device for optical neural control of freely-behaving animals. J. Neural Eng. **8**(4), 046021 (2011)
78. Wetzel SJ (1993) Coupling light emitting diodes to multimode optical fibers
79. Wietek, J., Wiegert, J.S., Adeishvili, N., Schneider, F., Watanabe, H., Tsunoda, S.P., Vogt, A., Elstner, M., Oertner, T.G., Hegemann, P.: Conversion of channelrhodopsin into a light-gated chloride channel. Science **344**(6182), 409–412 (2014)
80. Witten, I.B., Steinberg, E.E., Lee, S.Y., Davidson, T.J., Zalocusky, K.A., Brodsky, M., Yizhar, O., Cho, S.L., Gong, S., Ramakrishnan, C.: Recombinase-driver rat lines: tools, techniques, and optogenetic application to dopamine-mediated reinforcement. Neuron **72**(5), 721–733 (2011)
81. Yaroslavsky, A., Schulze, P., Yaroslavsky, I., Schober, R., Ulrich, F., Schwarzmaier, H.: Optical properties of selected native and coagulated human brain tissues in vitro in the visible and near infrared spectral range. Phys. Med. Biol. **47**(12), 2059 (2002)
82. Yeh, A.J., Ho, J.S., Tanabe, Y., Neofytou, E., Beygui, R.E., Poon, A.S.: Wirelessly powering miniature implants for optogenetic stimulation. Appl. Phys. Lett. **103**(16), 163701 (2013)
83. Yoo, S.-S., Bystritsky, A., Lee, J.-H., Zhang, Y., Fischer, K., Min, B.-K., McDannold, N.J., Pascual-Leone, A., Jolesz, F.A.: Focused ultrasound modulates region-specific brain activity. Neuroimage **56**(3), 1267–1275 (2011)
84. Zhang, F., Wang, L.-P., Boyden, E.S., Deisseroth, K.: Channelrhodopsin-2 and optical control of excitable cells. Nat. Methods **3**(10), 785–792 (2006)
85. Zhang, F., Wang, L.P., Brauner, M., Liewald, J.F., Kay, K., Watzke, N., Wood, P.G., Bamberg, E., Nagel, G., Gottschalk, A., Deisseroth, K.: Multimodal fast optical interrogation of neural circuitry. Nature **446**(7136), 633–639 (2007). doi:10.1038/nature05744

Physiological Monitoring in Deep Brain Stimulation: Toward Closed-Loop Neuromodulation Therapies

Seungleal (Brian) Paek, Rajas P. Kale, Katheryn M. Wininger and J. Luis Lujan

Abstract Deep brain stimulation (DBS) is a widely used, efficacious neurosurgical treatment for neurological movement disorders. For example, electrical stimulation in the ventral intermediate thalamic nucleus drastically reduces tremor in patients with essential tremor. Likewise, stimulation in the subthalamic nucleus or the internal globus pallidus significantly attenuates tremor, rigidity, bradykinesia, and gait complications of Parkinson's disease. Its application is now rapidly expanding to a wide variety of conditions including epilepsy, neuropsychiatric disorders, Tourette syndrome, Alzheimer's disease, and intractable pain. However, the exact underlying therapeutic mechanisms of action of DBS remain unclear. Despite this lack of understanding, clinical utility of DBS cannot be underappreciated, and there is a great need for studies that can elucidate patient-specific optimization of DBS parameters and targets. This chapter explores recent approaches for studying the

S. (Brian) Paek
Mayo Clinic Graduate School of Biomedical Sciences, Mayo Clinic,
200 First Street SW, Rochester, MN 55905, USA

R.P. Kale
School of Engineering, Deakin University, Geelong, VIC 3216, Australia
e-mail: rajas.p.kale@gmail.com

S. (Brian) Paek · J.L. Lujan (✉)
Department of Neurologic Surgery, Mayo Clinic, Rochester, MN 55905, USA
e-mail: Lujan.Luis@mayo.edu

K.M. Wininger
Department of Molecular Pharmacology and Experimental Therapeutics,
Mayo Clinic, Rochester, MN 55905, USA

R.P. Kale
Department of Psychiatry and Psychology, Mayo Clinic, 200 First Street SW,
Rochester, MN 55905, USA

J.L. Lujan
Department of Physiology and Biomedical Engineering, Mayo Clinic,
Rochester, MN 55905, USA

A. Bhatti et al. (eds.), *Emerging Trends in Neuro Engineering and Neural Computation*, Series in BioEngineering,
DOI 10.1007/978-981-10-3957-7_4

underlying mechanisms of action of DBS. Additionally, it discusses the limitations of current open-loop approaches to DBS and accentuates the importance of developing a smart closed-loop DBS system that can optimize therapeutic parameters in real time to individual patients and symptoms.

1 Introduction

As recently as 30 years ago, surgical techniques for treating many neurologic disorders involved ablative procedures, potentially resulting in significant and sometimes generalized damage to the brain [26, 110]. The advent of thermal and cryogenic lesioning brought forth greater spatial selectivity during surgery; however, these procedures are irreversible and cannot be modulated if the treatment needs change [4, 10, 13]. In the early 1990's, deep brain stimulation (DBS) became a reversible alternative to lesioning procedures for the treatment of movement disorders such as Parkinson disease and essential tremor [8, 9, 87]. Since then, the use of DBS has become more widespread for the effective treatment of other neurologic disorders such as dystonia [55, 64, 87, 96], Tourette syndrome [24, 46, 87, 104, 108], epilepsy [27, 52, 103], depression [25, 37, 72, 74], neuropathic pain [31, 48, 64, 87], and obsessive-compulsive disorder (OCD) [1, 33]. Additionally, DBS offers promising outcomes for the treatment of other neurological conditions ranging from bipolar disorder [45] and Alzheimer's disease [58] to addiction [65], cerebral palsy [69, 107], and hyperphagic obesity [103] (Table 1). However, the underlying therapeutic mechanisms of DBS remain unknown despite years of research and successful clinical application.

The predominant hypothesis suggests that DBS modulates pathological activity via excitation of axonal fibers of passage and inhibition of local cell bodies [28, 73]. However, the integration of spatially and temporally distant signals suggests that the neural mechanisms underlying DBS efficacy may be far more complex [43]. Therefore, further advancement of DBS technology will require a greater understanding of the response to DBS on the molecular, biochemical, cellular, and circuitry levels. For example, the brain is a highly complex organ, with innumerable neural signals transmitted via distinct neurotransmitters capable of modulating neural activity across both local and global neural circuitry [40]. Thus, DBS may induce complex changes in synaptic plasticity that reorganize neural circuits and rectify neuropathological changes associated with neurological disorders. These changes could help explain the different timescales in therapeutic efficacy observed across different disorders. For example, DBS patients with Parkinson's disease who experience immediate symptomatic relief, and patients with major depression who require longer intervals before symptomatic improvement can be observed [42].

Table 1 Neurologic disorders and deep brain stimulation targets

Neurologic Disorders	Targets	References
Addiction	NAc, STN	[32]
Alzheimer's disease	NBM, fornix	[59]
Depression	Cg25, ALIC, NAc	[14, 25, 49, 63, 67, 72], Schlepper et al. (2008)
Dystonia	GPi, (STN)	[50, 57, 81, 107]
Epilepsy	ATN, (cerebellum, CN, STN, hippocampus, CM, CC, LoC, MB)	[12, 27]
Essential Tremor	Vim, (STN)	[8, 16, 54, 61, 84, 101, 117]
Neuropathic Pain	PAG, VPL/VPM	[86]
Hyperphagic obesity	VMH, LH	[38, 112]
Obsessive-compulsive disorder	VC/VS, (ALIC, NAc, STN, ITP)	[21, 36, 66]
Parkinson's disease	GPi, STN, (PPN)	Deep Brain Stimulation for Parkinson's Disease Study Group (2001), Schuepbach et al. (2013), [22, 23, 29, 76, 78, 98, 111, 113]
Tourette syndrome	CM thalamus, GPi, ALIC, NAc	[109]

List of neurologic disorders and deep brain stimulation targets. Targets listed in parentheses are non-validated potential targets. Anterior limb of the internal capsule: ALIC, Anterior thalamic nucleus: ATN, Cingulate area 25 or subgenus cingulate: Cg25, Centromedian nucleus of the thalamus: CM, Caudate nucleus: CN, Globus pallidus internus: GPi, Interior thalamic peduncle: ITP, Lateral hypothalamus: LH. Locus coeruleus: LoC, Mammillary bodies: MB, Nucleus accumbens: NAc, Nucleus basalis of Meynert: NBM, Periaqueductal gray: PAG, Pedunculopontine nucleus: PPN, Subthalamic nucleus; STN, Ventral capsule/ventral striatum: VC/VS, Ventral intermediate uncles of the thalamus: Vim, Ventral intermediate nucleus of the thalamus: VMH, Ventral posterolateral thalamus: VPL, Ventral posteromedial thalamus: VPM, Ventral segmental area: VTA

2 Monitoring of Neural Activity

Numerous techniques are being utilized in clinical and preclinical studies to unravel the mechanisms of action of DBS and assist in target selection as well as optimization of stimulation parameters [35, 42]. These techniques allow examination of the neural responses to DBS at local and network-levels with high spatio-temporal resolutions. The most commonly utilized techniques include electrophysiological measurement of compound neural activity, electrochemical measurement of neurotransmitter signaling, and functional imaging techniques.

2.1 Electrophysiological Monitoring

Electrophysiological analysis has played an instrumental role in unraveling the function of the central nervous system (CNS) since the early 1950s, when Hodgkin and Huxley demonstrated the electrical nature of the action potential [44]. Since then, electrophysiological analysis techniques have evolved to enable the analysis of a broad range of neurological activity, from patch-clamp techniques that allowed the study of single ion channels, to single-unit recordings and global field potentials via multiunit recording arrays [75]. This technological diversity permits comprehensive evaluation of neurological activity from the subcellular to circuitry levels [35]. For example, electrophysiological techniques have been utilized to investigate the physiological mechanisms underlying DBS efficacy in the treatment of Tourette syndrome [47]. In this study, local field potential (LFP) recording electrodes were implanted into the thalamus, globus pallidus pars internal (GPi), or nucleus accumbens to analyze neural activity before, during, and after stimulation. Results from this study suggest that the pathophysiology of Tourette syndrome is related to dysfunctional synaptic transmission within deep brain nuclei, producing oscillations of inappropriate frequency and amplitude, and preventing the effective inhibition of stereotypical behaviors and tics such as blinking, head jerking, sniffing, throat clearing, and other vocalizations [2]. Similarly, excitatory postsynaptic potentials (EPSPs) evoked by high and low frequency stimulation of neurons within the subthalamic nucleus (STN) of 6-hydroxydopamine (6-OHDA)-lesioned rats revealed that high frequency stimulation produced significant EPSP depression in dopamine-depleted rats [115]. Similarly, low frequency stimulation resulted in EPSP augmentation in dopamine-intact rats [115].

2.2 Neurochemical Monitoring

Preclinical studies have demonstrated that neurotransmitter release is evoked by high frequency stimulation, and thus may be associated with the effects of DBS [60]. Fixed potential amperometry is a technique for measuring neurotransmitters and other analytes such as Glutamate, and involves the application of a constant voltage through a carbon fiber microelectrode implanted within the target of interest (Gale et al. 2013; Tye et al. 2013). Carbon fiber microelectrodes are coated with specific enzymes that react with non-electroactive analytes of interest, resulting in electroactive products that can be electrically measured [53]. The signals detected are caused by oxidative reactions between the applied voltage and the molecules of analyte within the extracellular space (Van Gompel et al. 2010). Unfortunately, the continuous enzyme delivery required to detect the neurotransmitter of interest makes this technique impractical for chronic in vivo detection of neurochemicals (Jacobs et al. 2010). Fast scan cyclic voltammetry (FSCV) is an alternative electroanalytical technique capable of real-time detection of electroactive neurotransmitters, hormones, and other metabolites [34, 51]. Previous studies have demonstrated that FSCV can effectively detect serotonin, norepinephrine, epinephrine, dopamine, and adenosine, as well as changes in oxygen and pH [41, 88, 106]. FSCV relies on the delivery of rapid voltage oscillations to allow oxidation and reduction of electrically active compounds of interest, resulting in the generation of unique electrical voltage versus current signatures specific for each analyte [88, 92]. By taking advantage of FSCV and a Wireless Instantaneous Neurotransmitter Concentration Sensing (WINCS) system designed to wirelessly measure neurochemical responses during DBS, Chang and colleagues showed that the immediate symptomatic relief induced by implantation of the DBS lead, also known as the microthalamotomy effect (Tasker 1998) was accompanied by a neurochemical signature resembling that of adenosine release [18, 19] (Fig. 1). Previous preclinical animal studies have demonstrated that administration of

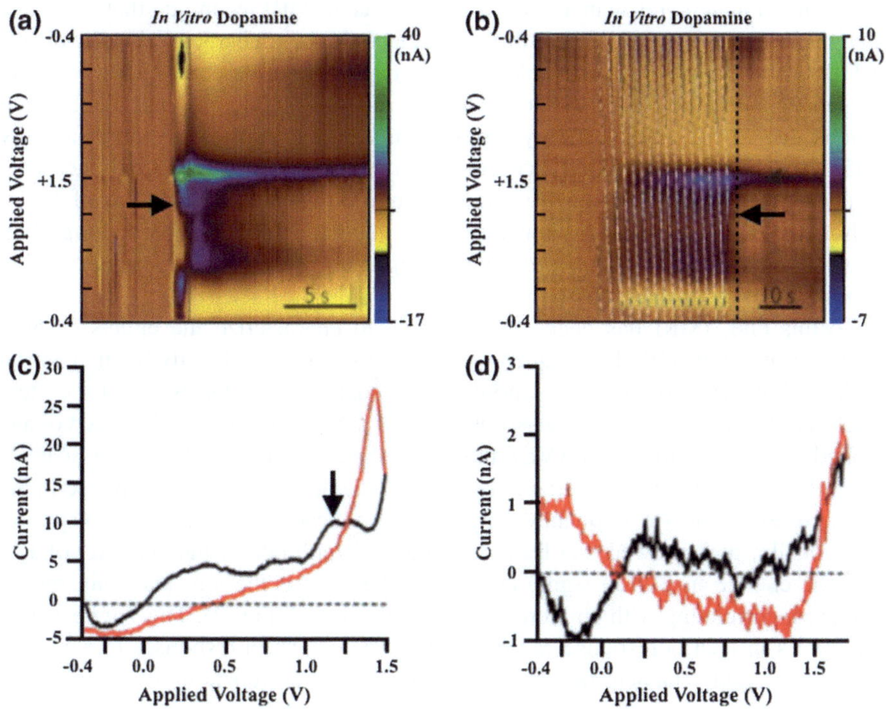

Fig. 1 Fast scan cyclic voltammetry recording during DBS electrode implantation into the VIM of the thalamus. **a** Pseudocolor plot collected from an awake patient. The plot depicts oxidation currents immediately on DBS electrode insertion. Black arrow indicates a second oxidation current peak. **b** Pseudocolor plot collected during electrical stimulation (130 Hz, 60 μsec pulse width, 2 V) through DBS electrode (in 4 out of 7 awake patients; no significant oxidation currents were not observed in the other 3 patients). The *plot* indicates oxidation current at a switching potential (1.5 V). Black arrow points to a much smaller second oxidation current. **c**, **d** Cyclic voltammograms (current versus voltage) at *black dotted lines* from A and B. Modified from [18]

adenosine A_1 agonists alleviates tremor symptoms in mice models of harmaline-induced essential tremor [7]. Therefore, it is possible that enhancement of adenosine signaling induced by the microthalamotomy effect, or by the administration of adenosine agonists, inhibits the maladaptive excitatory signaling producing uncontrolled movement in patients with essential tremor. However, the CNS contains multiple receptors for adenosine that operate through a variety of G-proteins, including G_s, G_q, and G_i. Variable effects in synaptic transmission are observed based on their localization on presynaptic, postsynaptic, or astrocytic membranes and the subsequent downstream signaling from the G-protein coupled receptor [79].

2.3 Functional Imaging

Functional magnetic resonance imaging (fMRI) is an MRI technique that measures changes in blood flow [62]. The utility of this technique for characterizing the effects of DBS is based upon the principle that the magnitude of cerebral blood flow and oxygen consumption is proportional to the relative activity of individual brain regions due to differential energy requirements during periods of neural activity [105]. Specifically, the blood-oxygen-level dependent (BOLD) signal measures changes in the magnetization of hemoglobin following deoxygenation of blood within the CNS [77]. This enables the generation of an oxyhemoglobin/deoxyhemoglobin heat map based upon changes in oxygen consumption resulting from modulation of neural activity [94].

To this end, fMRI has been recently used to characterize the effects of high (130 Hz) and low (10 Hz) frequency stimulation on neural activity in a swine model of DBS [82]. In this study, modulation of activity in the sensorimotor cortex, basal ganglia, and cerebellum was observed as a function of the stimulus voltage applied to the ventrolateral (VL) thalamus. Specifically, Paek and colleagues showed that high frequency stimulation produced a negative BOLD response in the motor cortex, while low frequency stimulation produced a positive BOLD response. Additionally, they showed that by increasing the amplitude of the applied voltage, both the change in BOLD signal as well as the size of the affected brain region increased, correlating with an increase of neural activity (Fig. 2). This suggests that differences in BOLD response can be used to analyze brain responses to electrical stimulation and characterize these responses as a function of stimulation parameters.

In addition to characterization of the neural response to DBS, functional imaging may be also effectively utilized to facilitate identification of optimal DBS targets [87]. In this context, fMRI and positron emission tomography (PET) have already demonstrated clinical utility in the treatment of treatment-resistant depression. For

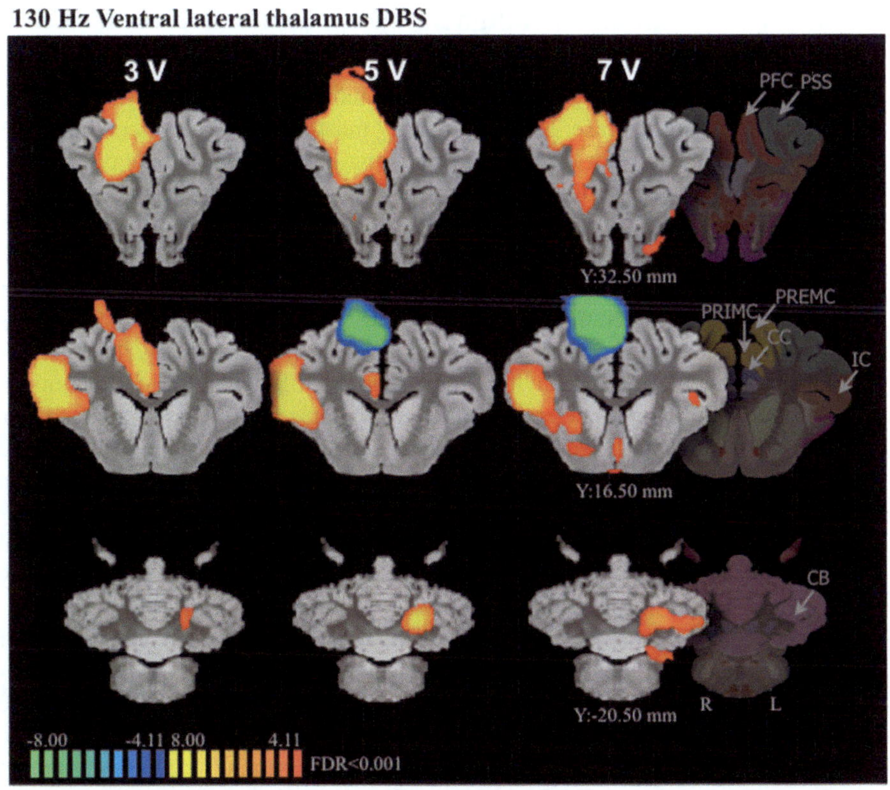

Fig. 2 Ventral lateral thalamus stimulation with different voltage intensities (3, 5, and 7 V). All voltages evoked increased BOLD signals in the ipsilateral prefrontal, primary somatosensory, insular cortices, and contralateral cerebellum. Also all voltages evoked decreased BOLD signal in the ipsilateral primary motor and premotor cortices. Abbreviations: BOLD, blood-oxygen-level dependent; CB, cerebellum; CC, cingulate cortex; DBS, deep brain stimulation; FDR, false discovery rate; IC, insular cortex; PFC, prefrontal cortex; PRIMC, primary motor cortex; PREMC, premotor cortex; V, voltage. Modified from [82]

example, an overactive subgenual cingulate cortex (Brodmann area 25) observed with PET has been shown in patients with acute sadness [71]. This overactivity has subsequently decreased in clinical responders following antidepressant treatments [25, 70, 71]. In 2005, Mayberg and colleagues used PET to show that chronic stimulation of the cingulate cortex in patients with treatment-resistant depression normalized metabolic hyperactivity and produced clinical benefits [72]. In that study, remission was accomplished in four out of six patients, with decreases in cerebral blood flow to subgenual cingulate, orbital frontal cortex, hypothalamus, anterior insula, and medial frontal cortex. At the same time, their results showed increases in cerebral blood flow within the dorsolateral prefrontal, dorsal anterior, posterior cingulate, premotor, and parietal regions.

3 Novel Stimulation Paradigms

While DBS has been used effectively to treat multiple disorders such as essential tremor and Parkinson disease, DBS technology must be further developed in order to improve patient care. For example, the existing DBS paradigm requires patients to return to the clinic for periodic adjustment of their stimulation parameters as their disease progresses [23, 35, 72]. Switching the existing DBS paradigm from an open-loop strategy where stimulation parameters are fixed, to an adaptive paradigm that relies on biological feedback to adjust stimulation parameters will be crucial for developing the next generation of DBS systems. Thus, closed-loop DBS systems equipped with electrophysiological, biochemical, and inertial sensors that monitor the molecular, cellular, and behavioral responses to DBS may allow for automated titration of stimulation parameters for sustained therapeutic benefits in the face of a changing environment. Developing a greater understanding of the cellular and molecular mechanisms of DBS by leveraging functional imaging in conjunction with neurochemical and electrophysiological techniques may also assist in optimizing DBS targets and stimulation parameters for specific disorders and individual patients [35]. Ultimately, adaptive, closed-loop DBS systems will extend battery life, reduce required hospital visits and associated healthcare costs [30].

Several key factors need to be investigated before automated adjustment of stimulation parameters can be clinically implemented. First, the complex relationships between clinical behavior and neural activity need to be elucidated. The advancement of electrochemical, electrophysiological, and functional imaging techniques from preclinical to clinical settings will be essential for the development of next-generation smart DBS systems. For example, optimal locations for recording of neural activity should be identified for specific disorders and specific patients. Furthermore, neurotransmitters critical for pathological activity and therapeutic response will need to be identified (Fitzgerald 2014). Additionally, specific neurotransmitter concentrations and their role in therapeutic efficacy will need to be elucidated. Second, the type of sensors that will be used to monitor the environment in order to effectively adjust stimulation parameters must be identified. The material of these sensors must be MRI safe and biocompatible for chronic implantation. For example, efforts are underway to develop electrochemical-sensing techniques capable of extending electrode longevity by renewing the electrochemically active surface following adsorption of chemical species [102]. This is paramount, as the carbon fiber microelectrodes typically used for FSCV are subject to electrode fouling due to the charge imbalance of the waveforms required for FSCV and are thus not suitable for chronic measurements [11]. Recent efforts to overcome this limitation have been focused on the use of boron-doped diamond microelectrodes that can be used over longer periods of time [15, 17, 39, 68, 83, 85, 93, 97, 99, 114, 116, 118]. Third, correlation of multimodal electrophysiological and neurochemical activity may provide new insight into the cellular and molecular mechanisms of therapeutic neuromodulation. Therefore, efforts should be directed toward developing smart DBS controllers that rely on the relationships between neural activity and the clinical effects of DBS to replace the trial-and-error process currently used for clinical DBS programming [35, 91, 100]. The versatility and adaptability of such controllers will allow optimization of DBS therapies to individual patients and symptoms. In turn, this will likely improve clinical outcomes, reduce the time and frequency of patient visits, and decrease overall health care costs.

4 Conclusion

Despite its clinical efficacy, limitations in existing DBS technology make optimization of therapeutic benefits a difficult and expensive endeavor. However, combination of multimodal electrophysiological and neurochemical sensing with functional imaging techniques may provide new insight into the cellular and molecular mechanisms of therapeutic DBS. By focusing on these techniques that further the efforts in understanding the underlying therapeutic mechanisms of DBS, we may be able to tailor application of DBS to individual patients and symptoms. Furthermore, the development of closed-loop DBS strategies will likely lead to the improvement of therapeutic outcomes.

References

1. Abelson, J.L., Curtis, G.C., Sagher, O., Albucher, R.C., Harrigan, M., Taylor, S.F., Martis, B., Giordani, B.: Deep brain stimulation for refractory obsessive-compulsive disorder. Biol. Psychiatr. **57**(5), 510–516 (2005). doi:10.1016/j.biopsych.2004.11.042
2. Albin, R.L., Mink, J.W.: Recent advances in Tourette syndrome research. Trends Neurosci. **29**(3), 175–182 (2006). doi:10.1016/j.tins.2006.01.001
3. Amorim, B.O., Covolan, L., Ferreira, E., Brito, J.G., Nunes, D.P., de Morais, D.G., Nobrega, J.N., Rodrigues, A.M., deAlmeida, A.C., Hamani, C.: Deep brain stimulation induces antiapoptotic and anti-inflammatory effects in epileptic rats. J. neuroinflamm **12**, 162 (2015). doi:10.1186/s12974-015-0384-7
4. Aziz, T.Z., Peggs, D., Sambrook, M.A., Crossman, A.R.: Lesion of the subthalamic nucleus for the alleviation of 1-methyl-4-phenyl-1,2,3,6-tetrahydropyridine (MPTP)-induced parkinsonism in the Primate. Mov. Disord. **6**(4), 288–292 (1991). doi:10.1002/Mds.870060404
5. Barbas, H., Saha, S., Rempel-Clower, N., Ghashghaei, T.: Serial pathways from primate prefrontal cortex to autonomic areas may influence emotional expression. BMC neurosci. **4**, 25 (2003). doi:10.1186/1471-2202-4-25
6. Bejjani, B.P., Arnulf, I., Vidailhet, M., Pidoux, B., Damier, P., Papadopoulos, S., Bonnet, A. M., Cornu, P., Dormont, D., Agid, Y.: Irregular jerky tremor, myoclonus, and thalamus: a study using low-frequency stimulation. Mov. Disord. **15**(5), 919–924 (2000). doi:10.1002/1531-8257(200009)15:5<919:Aid-Mds1024>3.0.Co;2-0
7. Bekar, L., Libionka, W., Tian, G.F., Xu, Q., Torres, A., Wang, X., Lovatt, D., Williams, E., Takano, T., Schnermann, J., Bakos, R., Nedergaard, M.: Adenosine is crucial for deep brain

stimulation-mediated attenuation of tremor. Nat. Med. **14**(1), 75–80 (2008). doi:10.1038/nm1693
8. Benabid, A.L., Pollak, P., Gao, D., Hoffmann, D., Limousin, P., Gay, E., Payen, I., Benazzouz, A.: Chronic electrical stimulation of the ventralis intermedius nucleus of the thalamus as a treatment of movement disorders. J. Neurosurg. **84**(2), 203–214 (1996). doi:10.3171/jns.1996.84.2.0203
9. Benabid, A.L., Pollak, P., Gervason, C., Hoffmann, D., Gao, D.M., Hommel, M., Perret, J. E., de Rougemont, J.: Long-term suppression of tremor by chronic stimulation of the ventral intermediate thalamic nucleus. Lancet **337**(8738), 403–406 (1991)
10. Benazzouz, A., Gross, C., Feger, J., Boraud, T., Bioulac, B.: Reversal of rigidity and improvement in motor-performance by subthalamic high-frequency Stimulation in MPTP-treated monkeys. Eur. J. Neurosci. **5**(4), 382–389 (1993). doi:10.1111/J.1460-9568.1993.Tb00505.X
11. Bennet, K.E., Tomshine, J.R., Min, H.K., Manciu, F.S., Marsh, M.P., Paek, S.B., Settell, M. L., Nicolai, E.N., Blaha, C.D., Kouzani, A.Z., Chang, S.Y., Lee, K.H.: A diamond-based electrode for detection of neurochemicals in the human brain. Front. Hum. Neurosci. **10**, 102 (2016). doi:10.3389/fnhum.2016.00102
12. Bergey, G.K., Morrell, M.J., Mizrahi, E.M., Goldman, A., King-Stephens, D., Nair, D., Srinivasan, S., Jobst, B., Gross, R.E., Barkley, G., Salanova, V., Olejniczak, P., Cole, A., Cash, S.S., Noe, K., Wharen, R., Worrell, G., Murro, A.M., Edwards, J., Duchowny, M., Spencer, D., Smith, M., Geller, E., Gwinn, R., Skidmore, C., Eisenschenk, S., Berg, M., Heck, C., Van Ness, P., Fountain, N., Rutecki, P., Massey, A., O'Donovan, C., Labar, D., Duckrow, R.B., Hirsch, L.J., Courtney, T., Sun, F.T., Seale, C.G.: Long-term treatment with responsive brain stimulation in adults with refractory partial seizures. Neurology **84**, 810–817 (2015)
13. Bergman, H., Wichmann, T., Delong, M.R.: Reversal of experimental parkinsonism by lesions of the subthalamic nucleus. Science **249**(4975), 1436–1438 (1990). doi:10.1126/Science.2402638
14. Bewernick, B.H., Kayser, S., Sturm, V., Schlaepfer, T.E.: Long-term effects of nucleus accumbens deep brain stimulation in treatment-resistant depression: evidence for sustained efficacy. Neuropsychopharmacology **37**, 1975–1985 (2012)
15. Bitziou, E., O'Hare, D., Patel, B.A.: Simultaneous detection of pH changes and histamine release from oxyntic glands in isolated stomach. Anal. Chem. **80**(22), 8733–8740 (2008). doi:10.1021/ac801413b
16. Blomstedt, P., Sandvik, U., Tisch, S.: Deep brain stimulation in the posterior subthalamic area in the treatment of essential tremor. Mov. Disord. **25**, 1350–1356 (2010)
17. Chan, H.Y., Aslam, D.M., Wiler, J.A., Casey, B.: A novel diamond microprobe for neuro-chemical and -electrical recording in neural prosthesis. J. Microelectromech. Syst. **18**(3), 511–521 (2009). doi:10.1109/JMEMS.2009.2015493
18. Chang, S.Y., Kim, I., Marsh, M.P., Jang, D.P., Hwang, S.C., Van Gompel, J.J., Goerss, S.J., Kimble, C.J., Bennet, K.E., Garris, P.A., Blaha, C.D., Lee, K.H.: Wireless fast-scan cyclic voltammetry to monitor adenosine in patients with essential tremor during deep brain stimulation. Mayo Clin. Proc. **87**(8), 760–765 (2012). doi:10.1016/j.mayocp.2012.05.006
19. Chang, S.Y., Shon, Y.M., Agnesi, F., Lee, K.H.: Microthalamotomy effect during deep brain stimulation: potential involvement of adenosine and glutamate efflux. In: Conference Proceedings: Annual International Conference of the IEEE Engineering in Medicine and Biology Society IEEE Engineering in Medicine and Biology Society Conference, 2009, pp. 3294–3297 (2009). doi:10.1109/IEMBS.2009.5333735
20. Constantoyannis, C., Kumar, A., Stoessl, A.J., Honey, C.R.: Tremor induced by thalamic deep brain stimulation in patients with complex regional facial pain. Mov. Disord. Off. J. Mov. Disord. Soc. **19**(8), 933–936 (2004). doi:10.1002/mds.20047
21. Denys, D., Mantione, M., Figee, M., van den Munckhof, P., Koerselman, F., Bosch, A., Schuurman, R.: Deep brain stimulation of the nucleus accumbens for treatment-refractory obsessive-compulsive disorder. Arch. Gen. Psychiatr. **67**, 1061–1068 (2010)

22. Deuschl, G., Agid, Y.: Subthalamic neurostimulation for Parkinson's disease with early fluctuations: balancing the risks and benefits. Lancet Neurol. **12**, 1025–1034 (2013)
23. Deuschl, G., Herzog, J., Kleiner-Fisman, G., Kubu, C., Lozano, A.M., Lyons, K.E., Rodriguez-Oroz, M.C., Tamma, F., Troster, A.I., Vitek, J.L., Volkmann, J., Voon, V.: Deep brain stimulation: postoperative issues. Mov. Disord. Off. J. Mov. Disord. Soc. **21**(Suppl 14), S219–S237 (2006). doi:10.1002/mds.20957
24. Diederich, N.J., Kalteis, K., Stamenkovic, M., Pieri, V., Alesch, F.: Efficient internal pallidal stimulation in Gilles de la Tourette syndrome: a case report. Mov. Disord. Off. J. Mov. Disord. Soc. **20**(11), 1496–1499 (2005). doi:10.1002/mds.20551
25. Dougherty, D.D., Rezai, A.R., Carpenter, L.L., Howland, R.H., Bhati, M.T., O'Reardon, J. P., Eskandar, E.N., Baltuch, G.H., Machado, A.D., Kondziolka, D., Cusin, C., Evans, K.C., Price, L.H., Jacobs, K., Pandya, M., Denko, T., Tyrka, A.R., Brelje, T., Deckersbach, T., Kubu, C., Malone Jr., D.A.: A randomized sham-controlled trial of deep brain stimulation of the ventral capsule/ventral striatum for chronic treatment-resistant depression. Biol. Psychiatr. **78**(4), 240–248 (2015). doi:10.1016/j.biopsych.2014.11.023
26. Faria, M.: Violence, mental illness, and the brain—a brief history of psychosurgery: part 2—from the limbic system and cingulotomy to deep brain stimulation. Surg. Neurol. Int. **4** (2013)
27. Fisher, R., Salanova, V., Witt, T., Worth, R., Henry, T., Gross, R., Oommen, K., Osorio, I., Nazzaro, J., Labar, D., Kaplitt, M., Sperling, M., Sandok, E., Neal, J., Handforth, A., Stern, J., De Salles, A., Chung, S., Shetter, A., Bergen, D., Bakay, R., Henderson, J., French, J., Baltuch, G., Rosenfeld, W., Youkilis, A., Marks, W., Garcia, P., Barbaro, N., Fountain, N., Bazil, C., Goodman, R., McKhann, G., Babu Krishnamurthy, K., Papavassiliou, S., Epstein, C., Pollard, J., Tonder, L., Grebin, J., Coffey, R., Graves, N., Group, S.S.: Electrical stimulation of the anterior nucleus of thalamus for treatment of refractory epilepsy. Epilepsia **51**(5), 899–908 (2010). doi:10.1111/j.1528-1167.2010.02536.x
28. Florence, G., Sameshima, K., Fonoff, E.T., Hamani, C.: Deep brain stimulation: more complex than the inhibition of cells and excitation of fibers. Neurosci. Rev. J. Bring. Neurobiol. Neurol. Psychiatr. (2015). doi:10.1177/1073858415591964
29. Follett, K.A., Weaver, F.M., Stern, M., Hur, K., Harris, C.L., Luo, P., Marks, W.J., Rothlind, J., Sagher, O., Moy, C., Pahwa, R., Burchiel, K., Hogarth, P., Lai, E.C., Duda, J.E., Holloway, K., Samii, A., Horn, S., Bronstein, J.M., Stoner, G., Starr, P.A., Simpson, R., Baltuch, G., De Salles, A., Huang, G.D., Reda, D.J., CSP 468 Study Group: Pallidal versus subthalamic deep-brain stimulation for parkinson's disease. New Engl. J. Med. **362**, 2077–2091 (2010)
30. Fraix, V., Houeto, J.L., Lagrange, C., Le Pen, C., Krystkowiak, P., Guehl, D., Ardouin, C., Welter, M.L., Maurel, F., Defebvre, L., Rougier, A., Benabid, A.L., Mesnage, V., Ligier, M., Blond, S., Burbaud, P., Bioulac, B., Destee, A., Cornu, P., Pollak, P., Group, S.S.: Clinical and economic results of bilateral subthalamic nucleus stimulation in parkinson's disease. J. Neurol. Neurosurg. Psychiatr. **77**(4), 443–449 (2006). doi:10.1136/jnnp.2005.077677
31. Franzini, A., Ferroli, P., Leone, M., Broggi, G.: Stimulation of the posterior hypothalamus for treatment of chronic intractable cluster headaches: first reported series. Neurosurgery **52** (5), 1095–1099; discussion 1099–1101 (2003)
32. Kuhn, J., Bührle, C.P., Lenartz, D., Sturm, V.: Deep brain stimulation in addiction due to psychoactive substance use. In: Handbook *of* Clinical Neurology, vol. 116, pp. 259–269 (2013)
33. Gabriels, L., Cosyns, P., Nuttin, B., Demeulemeester, H., Gybels, J.: Deep brain stimulation for treatment-refractory obsessive-compulsive disorder: psychopathological and neuropsychological outcome in three cases. Acta Psychiatr. Scand. **107**(4), 275–282 (2003)
34. Glanowska, K.M., Venton, B.J., Moenter, S.M.: Fast scan cyclic voltammetry as a novel method for detection of real-time gonadotropin-releasing hormone release in mouse brain

slices. J. Neurosci. Off. J. Soc. Neurosci. **32**(42), 14664–14669 (2012). doi:10.1523/JNEUROSCI.1303-12.2012

35. Grahn, P.J., Mallory, G.W., Khurram, O.U., Berry, B.M., Hachmann, J.T., Bieber, A.J., Bennet, K.E., Min, H.K., Chang, S.Y., Lee, K.H., Lujan, J.L.: A neurochemical closed-loop controller for deep brain stimulation: toward individualized smart neuromodulation therapies. Front. Neurosci. **8**, 169 (2014). doi:10.3389/fnins.2014.00169
36. Greenberg, B.D., Gabriels, L.A., Malone, D.A., Rezai, A.R., Friehs, G.M., Okun, M.S., Shapira, N.A., Foote, K.D., Cosyns, P.R., Kubu, C.S., Malloy, P.F., Salloway, S.P., Giftakis, J.E., Rise, M.T., Machado, A.G., Baker, K.B., Stypulkowski, P.H., Goodman, W.K., Rasmussen, S.A., Nuttin, B.J.: Deep brain stimulation of the ventral internal capsule/ventral striatum for obsessive-compulsive disorder: worldwide experience. Mol. Psychiatr. **15**, 64–79 (2010)
37. Grubert, C., Hurlemann, R., Bewernick, B.H., Kayser, S., Hadrysiewicz, B., Axmacher, N., Sturm, V., Schlaepfer, T.E.: Neuropsychological safety of nucleus accumbens deep brain stimulation for major depression: effects of 12-month stimulation. World J. Biol. Psychiatr. **12**(7), 516–527 (2011). doi:10.3109/15622975.2011.583940
38. Halpern, C.H., Wolf, J.A., Bale, T.L., Stunkard, A.J., Danish, S.F., Grossman, M., Jaggi, J. L., Grady, M.S., Baltuch, G.H.: Deep brain stimulation in the treatment of obesity. J. Neurosurg. **109**, 625–634 (2008)
39. Halpern, J.M., Xie, S.T., Sutton, G.P., Higashikubo, B.T., Chestek, C.A., Lu, H., Chiel, H.J., Martin, H.B.: Diamond electrodes for neurodynamic studies in Aplysia californica. Diam. Relat. Mater. **15**(2–3), 183–187 (2006). doi:10.1016/j.diamond.2005.06.039
40. Hamani, C., Temel, Y.: Deep brain stimulation for psychiatric disease: contributions and validity of animal models. Sci. Trans. Med. **4**(142), 142rv148 (2012). doi:10.1126/scitranslmed.3003722
41. Heien, M.L., Johnson, M.A., Wightman, R.M.: Resolving neurotransmitters detected by fast-scan cyclic voltammetry. Anal. Chem. **76**(19), 5697–5704 (2004). doi:10.1021/ac0491509
42. Herrington, T.M., Cheng, J.J., Eskandar, E.N.: Mechanisms of deep brain stimulation. J. Neurophysiol. **115**(1), 19–38 (2016). doi:10.1152/jn.00281.2015
43. Hess, C.W., Vaillancourt, D.E., Okun, M.S.: The temporal pattern of stimulation may be important to the mechanism of deep brain stimulation. Exp. Neurol. **247**, 296–302 (2013). doi:10.1016/j.expneurol.2013.02.001
44. Hodgkin, A.L., Huxley, A.F.: A quantitative description of membrane current and its application to conduction and excitation in nerve. J. Physiol. **117**(4), 500–544 (1952)
45. Holtzheimer, P.E., Kelley, M.E., Gross, R.E., Filkowski, M.M., Garlow, S.J., Barrocas, A., Wint, D., Craighead, M.C., Kozarsky, J., Chismar, R., Moreines, J.L., Mewes, K., Posse, P. R., Gutman, D.A., Mayberg, H.S.: Subcallosal cingulate deep brain stimulation for treatment-resistant unipolar and bipolar depression. Arch. Gen. Psychiatr. **69**(2), 150–158 (2012). doi:10.1001/archgenpsychiatry.2011.1456
46. Houeto, J.L., Karachi, C., Mallet, L., Pillon, B., Yelnik, J., Mesnage, V., Welter, M.L., Navarro, S., Pelissolo, A., Damier, P., Pidoux, B., Dormont, D., Cornu, P., Agid, Y.: Tourette's syndrome and deep brain stimulation. J. Neurol. Neurosurg. Psychiatr. **76**(7), 992–995 (2005). doi:10.1136/jnnp.2004.043273
47. Israelashvili, M., Loewenstern, Y., Bar-Gad, I.: Abnormal neuronal activity in Tourette syndrome and its modulation using deep brain stimulation. J. Neurophysiol. **114**(1), 6–20 (2015). doi:10.1152/jn.00277.2015
48. Katayama, Y., Yamamoto, T., Kobayashi, K., Kasai, M., Oshima, H., Fukaya, C.: Motor cortex stimulation for phantom limb pain: comprehensive therapy with spinal cord and thalamic stimulation. Stereotact. Funct. Neurosurg. **77**(1–4), 159–162 (2001). 64593

49. Kennedy, S.H., Giacobbe, P., Rizvi, S.J., Placenza, F.M., Nishikawa, Y., Mayberg, H.S., Lozano, A.M.: Deep brain stimulation for treatment-resistant depression: follow-up after 3 to 6 years. Am. J. Psychiatr. **168**, 502–510 (2011)
50. Kiss, Z.H., Doig-Beyaert, K., Eliasziw, M., Tsui, J., Haffenden, A., Suchowersky, O.: Functional and stereotactic section of the Canadian neurosurgical society, Canadian movement disorders group. The Canadian multicentre study of deep brain stimulation for cervical dystonia. Brain **130**, 2879–2886 (2007)
51. Kita, J.M., Wightman, R.M.: Microelectrodes for studying neurobiology. Curr. Opin. Chem. Biol. **12**(5), 491–496 (2008). doi:10.1016/j.cbpa.2008.06.035
52. Kocabicak, E., Temel, Y., Hollig, A., Falkenburger, B., Tan, S.: Current perspectives on deep brain stimulation for severe neurological and psychiatric disorders. Neuropsychiatr. Dis. Treat. **11**, 1051–1066 (2015). doi:10.2147/NDT.S46583
53. Koehne, J., Marsh, M., Boakye, A., Douglas, B., Kim, I., Chang, S., Jang, D., Bennet, K., Kimble, C., Andrews, R., Meyyappan, M.: Carbon nanofiber electrode array for electrochemical detection of dopamine using fast scan cyclic voltammetry. Analyst **136** (9), 1802–1805 (2011)
54. Koller, W., Pahwa, R., Busenbark, K., Hubble, J., Wilkinson, S., Lang, A., Tuite, P., Sime, E., Lazano, A., Hauser, R., Malapira, T., Smith, D., Tarsy, D., Miyawaki, E., Norregaard, T., Kormos, T., Olanow, C.W.: High-frequency unilateral thalamic stimulation in the treatment of essential and parkinsonian tremor. Ann. Neurol. **42**, 292–299 (1997)
55. Kumar, R., Dagher, A., Hutchison, W.D., Lang, A.E., Lozano, A.M.: Globus pallidus deep brain stimulation for generalized dystonia: clinical and PET investigation. Neurology **53**(4), 871–874 (1999)
56. Kuncel, A.M., Grill, W.M.: Selection of stimulus parameters for deep brain stimulation. Clin. Neurophysiol. Off. J. Int. Fed. Clin. Neurophysiol. **115**(11), 2431–2441 (2004). doi:10.1016/j.clinph.2004.05.031
57. Kupsch, A., Benecke, R., Müller, J., Trottenberg, T., Schneider, G.H., Poewe, W., Eisner, W., Wolters, A., Müller, J.U., Deuschl, G.: Pallidal deep-brain stimulation in primary generalized or segmental dystonia. N. Engl. J. Med. **355**, 1978–1990 (2006)
58. Laxton, A.W., Lozano, A.M.: Deep brain stimulation for the treatment of Alzheimer disease and dementias. World Neurosurg. **80** (3–4), S28 e21–28. doi:10.1016/j.wneu.2012.06.028
59. Laxton, A.W., Tang-Wai, D.F., McAndrews, M.P., Zumsteg, D., Wennberg, R., Keren, R., Wherrett, J., Naglie, G., Hamani, C., Smith, G.S., Lozano, A.M.: A phase I trial of deep brain stimulation of memory circuits in Alzheimer's disease. Ann. Neurol. **68**, 521–534 (2010)
60. Lee, K.H., Chang, S.Y., Roberts, D.W., Kim, U.: Neurotransmitter release from high-frequency stimulation of the subthalamic nucleus. J. Neurosurg. **101**(3), 511–517 (2004). doi:10.3171/jns.2004.101.3.0511
61. Limousin, P., Speelman, J.D., Gielen, F., Janssens, M.: Multicentre European study of thalamic stimulation in parkinsonian and essential tremor. J. Neurol. Neurosurg. Psychiatr. **66**, 289–296 (1999)
62. Logothetis, N.K.: The neural basis of the blood-oxygen-level-dependent functional magnetic resonance imaging signal. Philos. Trans. R. Soc. Lond. B Biol. Sci. **357**(1424), 1003–1037 (2002). doi:10.1098/rstb.2002.1114
63. Lozano, A.M., Giacobbe, P., Hamani, C., Rizvi, S.J., Kennedy, S.H., Kolivakis, T.T., Debonnel, G., Sadikot, A.F., Lam, R.W., Howard, A.K., Ilcewicz-Klimek, M., Honey, C.R., Mayberg, H.S.: A multicenter pilot study of subcallosal cingulate area deep brain stimulation for treatment-resistant depression. J. Neurosurg. **116**, 315–322 (2012)
64. Lozano, A.M., Kumar, R., Gross, R.E., Giladi, N., Hutchison, W.D., Dostrovsky, J.O., Lang, A.E.: Globus pallidus internus pallidotomy for generalized dystonia. Mov. Disord. Off. J. Mov. Disord. Soc. **12**(6), 865–870 (1997). doi:10.1002/mds.870120606

65. Luigjes, J., van den Brink, W., Feenstra, M., van den Munckhof, P., Schuurman, P.R., Schippers, R., Mazaheri, A., De Vries, T.J., Denys, D.: Deep brain stimulation in addiction: a review of potential brain targets. Mol. Psychiatr. **17**(6), 572–583 (2012). doi:10.1038/mp.2011.114
66. Mallet, L., Polosan, M., Jaafari, N., Baup, N., Welter, M.L., Fontaine, D., du Montcel, S.T., Yelnik, J., Chéreau, I., Arbus, C., Raoul, S., Aouizerate, B., Damier, P., Chabardès, S., Czernecki, V., Ardouin, C., Krebs, M.O., Bardinet, E., Chaynes, P., Burbaud, P., Cornu, P., Derost, P., Bougerol, T., Bataille, B., Mattei, V., Dormont, D., Devaux, B., Vérin, M., Houeto, J.L., Pollak, P., Benabid, A.L., Agid, Y., Krack, P., Millet, B., Pelissolo, A., STOC Study Group: Subthalamic nucleus stimulation in severe obsessive-compulsive disorder. N. Engl. J. Med. **359**, 2121–2134 (2008)
67. Malone, D.A., Dougherty, D.D., Rezai, A.R., Carpenter, L.L., Friehs, G.M., Eskandar, E.N., Rauch, S.L., Rasmussen, S.A., Machado, A.G., Kubu, C.S., Tyrka, A.R., Price, L.H., Stypulkowski, P.H., Giftakis, J.E., Rise, M.T., Malloy, P.F., Salloway, S.P., Greenberg, B. D.: Deep brain stimulation of the ventral capsule/ventral striatum for treatment-resistant depression. Biol. Psychiatr. **65**, 267–275 (2009)
68. Marcelli, G., Patel, B.A.: Understanding changes in uptake and release of serotonin from gastrointestinal tissue using a novel electroanalytical approach. Analyst **135**(9), 2340–2347 (2010). doi:10.1039/c0an00260g
69. Marks, W.A., Honeycutt, J., Acosta Jr., F., Reed, M., Bailey, L., Pomykal, A., Mercer, M.: Dystonia due to cerebral palsy responds to deep brain stimulation of the globus pallidus internus. Mov. Disord. Off. J. Mov. Disord. Soc. **26**(9), 1748–1751 (2011). doi:10.1002/mds.23723
70. Mayberg, H.S., Brannan, S.K., Tekell, J.L., Silva, J.A., Mahurin, R.K., McGinnis, S., Jerabek, P.A.: Regional metabolic effects of fluoxetine in major depression: serial changes and relationship to clinical response. Biol. Psychiatr. **48**(8), 830–843 (2000)
71. Mayberg, H.S., Liotti, M., Brannan, S.K., McGinnis, S., Mahurin, R.K., Jerabek, P.A., Silva, J.A., Tekell, J.L., Martin, C.C., Lancaster, J.L., Fox, P.T.: Reciprocal limbic-cortical function and negative mood: converging PET findings in depression and normal sadness. Am. J. Psychiatr. **156**(5), 675–682 (1999). doi:10.1176/ajp.156.5.675
72. Mayberg, H.S., Lozano, A.M., Voon, V., McNeely, H.E., Seminowicz, D., Hamani, C., Schwalb, J.M., Kennedy, S.H.: Deep brain stimulation for treatment-resistant depression. Neuron **45**(5), 651–660 (2005). doi:10.1016/j.neuron.2005.02.014
73. McIntyre, C.C., Thakor, N.V.: Uncovering the mechanisms of deep brain stimulation for Parkinson's disease through functional imaging, neural recording, and neural modeling. Crit. Rev. Biomed. Eng. **30**(4–6), 249–281 (2002)
74. Merkl, A., Schneider, G.H., Schonecker, T., Aust, S., Kuhl, K.P., Kupsch, A., Kuhn, A.A., Bajbouj, M.: Antidepressant effects after short-term and chronic stimulation of the subgenual cingulate gyrus in treatment-resistant depression. Exp. Neurol. **249**, 160–168 (2013). doi:10.1016/j.expneurol.2013.08.017
75. Neher, E., Sakmann, B., Steinbach, J.H.: The extracellular patch clamp: a method for resolving currents through individual open channels in biological membranes. Pflugers Arch. **375**(2), 219–228 (1978)
76. Odekerken, V.J., van Laar, T., Staal, M.J., Mosch, A., Hoffmann, C.F., Nijssen, P.C., Beute, G.N., van Vugt, J.P., Lenders, M.W., Contarino, M.F., Mink, M.S., Bour, L.J., van den Munckhof, P., Schmand, B.A., de Haan, R.J., Schuurman, P.R., de Bie, R.M.: Subthalamic nucleus versus globus pallidus bilateral deep brain stimulation for advanced parkinson's disease (NSTAPS study): a randomised controlled trial. Lancet Neurol. **12**, 37–44 (2013)
77. Ogawa, S., Menon, R.S., Tank, D.W., Kim, S.G., Merkle, H., Ellermann, J.M., Ugurbil, K.: Functional brain mapping by blood oxygenation level-dependent contrast magnetic resonance imaging. A comparison of signal characteristics with a biophysical model. Biophys. J. **64**(3), 803–812 (1993). doi:10.1016/S0006-3495(93)81441-3
78. Okun, M.S., Gallo, B.V., Mandybur, G., Jagid, J., Foote, K.D., Revilla, F.J., Alterman, R., Jankovic, J., Simpson, R., Junn, F., Verhagen, L., Arle, J.E., Ford, B., Goodman, R.R.,

Stewart, R.M., Horn, S., Baltuch, G.H., Kopell, B.H., Marshall, F., Peichel, D., Pahwa, R., Lyons, K.E., Tröster, A.I., Vitek, J.L., Tagliati, M.: Subthalamic deep brain stimulation with a constant-current device in Parkinson's disease: an open-label randomised controlled trial. Lancet Neurol. **11**, 140–149 (2012)

79. Olah, M.E., Stiles, G.L.: Adenosine receptor subtypes: characterization and therapeutic regulation. Ann. Rev. Pharmacol. Toxicol. **35**, 581–606 (1995). doi:10.1146/annurev.pa.35.040195.003053
80. Ongur, D., An, X., Price, J.L.: Prefrontal cortical projections to the hypothalamus in macaque monkeys. J. Comp. Neurol. **401**(4), 480–505 (1998)
81. Ostrem, J.L., Racine, C.A., Glass, G.A., Grace, J.K., Volz, M.M., Heath, S.L., Starr, P.A.: Subthalamic nucleus deep brain stimulation in primary cervical dystonia. Neurology **76**, 870–878 (2011)
82. Paek, S.B., Min, H.K., Kim, I., Knight, E.J., Baek, J.J., Bieber, A.J., Lee, K.H., Chang, S.Y.: Frequency-dependent functional neuromodulatory effects on the motor network by ventral lateral thalamic deep brain stimulation in swine. NeuroImage **105**, 181–188 (2015). doi:10.1016/j.neuroimage.2014.09.064
83. Park, J., Quaiserova-Mocko, V., Patel, B.A., Novotny, M., Liu, A., Bian, X., Galligan, J.J., Swain, G.M.: Diamond microelectrodes for in vitro electroanalytical measurements: current status and remaining challenges. Analyst **133**(1), 17–24 (2008). doi:10.1039/b710236b
84. Pahwa, R., Lyons, K.E., Wilkinson, S.B., Simpson, R.K., Ondo, W.G., Tarsy, D., Norregaard, T., Hubble, J.P., Smith, D.A., Hauser, R.A., Jankovic, J.: Long-term evaluation of deep brain stimulation of the thalamus. J. Neurosurg. **104**, 506–512 (2006)
85. Park, J., Show, Y., Quaiserova, V., Galligan, J.J., Fink, G.D., Swain, G.M.: Diamond microelectrodes for use in biological environments. J. Electroanal. Chem. **583**(1), 56–68 (2005). doi:10.1016/j.jelechem.2005.04.032
86. Pereira, E.A., Green, A.L., Aziz, T.Z.: Deep brain stimulation for pain. In: Handbook of Clinical Neurology, vol. 116, pp. 277–294 (2013)
87. Perlmutter, J.S., Mink, J.W.: Deep brain stimulation. Annu. Rev. Neurosci. **29**, 229–257 (2006). doi:10.1146/annurev.neuro.29.051605.112824
88. Perry, M., Li, Q., Kennedy, R.T.: Review of recent advances in analytical techniques for the determination of neurotransmitters. Anal. Chim. Acta **653**(1), 1–22 (2009). doi:10.1016/j.aca.2009.08.038
89. Pienaar, I.S., Lee, C.H., Elson, J.L., McGuinness, L., Gentleman, S.M., Kalaria, R.N., Dexter, D.T.: Deep-brain stimulation associates with improved microvascular integrity in the subthalamic nucleus in Parkinson's disease. Neurobiol. Dis. **74**, 392–405 (2015). doi:10.1016/j.nbd.2014.12.006
90. Priori, A., Foffani, G., Rossi, L., Marceglia, S.: Adaptive deep brain stimulation (aDBS) controlled by local field potential oscillations. Exp. Neurol. **245**, 77–86 (2013). doi:10.1016/j.expneurol.2012.09.013
91. Pulliam, C.L., Heldman, D.A., Orcutt, T.H., Mera, T.O., Giuffrida, J.P., Vitek, J.L.: Motion sensor strategies for automated optimization of deep brain stimulation in Parkinson's disease. Parkinsonism Relat. Disord. **21**(4), 378–382 (2015). doi:10.1016/j.parkreldis.2015.01.018. Epub 2015 Feb 11
92. Robinson, D.L., Venton, B.J., Heien, M.L., Wightman, R.M.: Detecting subsecond dopamine release with fast-scan cyclic voltammetry in vivo. Clin. Chem. **49**(10), 1763–1773 (2003)
93. Roham, M., Halpern, J.M., Martin, H.B., Chiel, H.J., Mohseni, P.: Diamond microelectrodes and CMOS microelectronics for wireless transmission of fast-scan cyclic voltammetry. In: Conference Proceedings: Annual International Conference of the IEEE Engineering in Medicine and Biology Society IEEE Engineering in Medicine and Biology Society Conference 2007, pp. 6044–6047 (2007). doi:10.1109/IEMBS.2007.4353726
94. Rosen, B.R., Buckner, R.L., Dale, A.M.: Event-related functional MRI: past, present, and future. Proc. Natl. Acad. Sci. U.S.A. **95**(3), 773–780 (1998)

95. Schlaepfer, T.E., Cohen, M.X., Frick, C., Kosel, M., Brodesser, D., Axmacher, N., Joe, A. Y., Kreft, M., Lenartz, D., Sturm, V.: Deep brain stimulation to reward circuitry alleviates anhedonia in refractory major depression. Neuropsycho-pharmacology **33**, 368–377 (2008)
96. Sellal, F., Hirsch, E., Barth, P., Blond, S., Marescaux, C.: A case of symptomatic hemidystonia improved by ventroposterolateral thalamic electrostimulation. Mov. Disord. Off. J. Mov. Disord. Soc. **8**(4), 515–518 (1993). doi:10.1002/mds.870080418
97. Singh, Y.S., Sawarynski, L.E., Michael, H.M., Ferrell, R.E., Murphey-Corb, M.A., Swain, G.M., Patel, B.A., Andrews, A.M.: Boron-doped diamond microelectrodes reveal reduced serotonin uptake rates in lymphocytes from adult rhesus monkeys carrying the short allele of the 5-HTTLPR. ACS Chem. Neurosci. **1**(1), 49–64 (2010). doi:10.1021/cn900012y
98. Schüpbach, W.M., Maltête, D., Houeto, J.L., du Montcel, S.T., Mallet, L., Welter, M.L., Gargiulo, M., Béhar, C., Bonnet, A.M., Czernecki, V., Pidoux, B., Navarro, S., Dormont, D., Cornu, P., Agid, Y.: Neurosurgery at an earlier stage of parkinson disease: a randomized, controlled trial. Neurology **68**, 267–271 (2007)
99. Suzuki, A., Ivandini, T.A., Yoshimi, K., Fujishima, A., Oyama, G., Nakazato, T., Hattori, N., Kitazawa, S., Einaga, Y.: Fabrication, characterization, and application of boron-doped diamond microelectrodes for in vivo dopamine detection. Anal. Chem. **79**(22), 8608–8615 (2007). doi:10.1021/ac071519h
100. Swann, N.C., de Hemptinne, C., Miocinovic, S., Qasim, S., Wang, S.S., Ziman, N., Ostrem, J.L., San Luciano, M., Galifianakis, N.B., Starr, P.A.: Gamma oscillations in the hyperkinetic state detected with chronic human brain recordings in parkinson's disease. J. Neurosci. **36**(24), 6445–6458 (2016). doi:10.1523/JNEUROSCI.1128-16.2016
101. Sydow, O., Thobois, S., Alesch, F., Speelman, J.D.: Multicentre European study of thalamic stimulation in essential tremor: a six year follow up. J. Neurol. Neurosurg. Psychiatr. **74**, 1387–1391 (2003)
102. Takmakov, P., Zachek, M.K., Keithley, R.B., Walsh, P.L., Donley, C., McCarty, G.S., Wightman, R.M.: Carbon microelectrodes with a renewable surface. Anal. Chem. **82**(5), 2020–2028 (2010). doi:10.1021/ac902753x
103. Tekriwal, A., Baltuch, G.: deep brain stimulation: expanding applications. Neurol Med-Chir. **55**(12), 861–877 (2015). doi:10.2176/nmc.ra.2015-0172
104. Temel, Y., Visser-Vandewalle, V.: Surgery in tourette syndrome. Mov. Disord. Off. J. Mov. Disord. Soc. **19**(1), 3–14 (2004). doi:10.1002/mds.10649
105. Uludag, K., Dubowitz, D.J., Yoder, E.J., Restom, K., Liu, T.T., Buxton, R.B.: Coupling of cerebral blood flow and oxygen consumption during physiological activation and deactivation measured with fMRI. NeuroImage **23**(1), 148–155 (2004). doi:10.1016/j.neuroimage.2004.05.013
106. Venton, B.J., Michael, D.J., Wightman, R.M.: Correlation of local changes in extracellular oxygen and pH that accompany dopaminergic terminal activity in the rat caudate-putamen. J. Neurochem. **84**(2), 373–381 (2003)
107. Vidailhet, M., Yelnik, J., Lagrange, C., Fraix, V., Grabli, D., Thobois, S., Burbaud, P., Welter, M.L., Xie-Brustolin, J., Braga, M.C., Ardouin, C., Czernecki, V., Klinger, H., Chabardes, S., Seigneuret, E., Mertens, P., Cuny, E., Navarro, S., Cornu, P., Benabid, A.L., Le Bas, J.F., Dormont, D., Hermier, M., Dujardin, K., Blond, S., Krystkowiak, P., Destee, A., Bardinet, E., Agid, Y., Krack, P., Broussolle, E., Pollak, P., French, S.-S.G.: Bilateral pallidal deep brain stimulation for the treatment of patients with dystonia-choreoathetosis cerebral palsy: a prospective pilot study. Lancet Neurol. **8**(8), 709–717 (2009). doi:10.1016/S1474-4422(09)70151-6
108. Visser-Vandewalle, V., Temel, Y., Boon, P., Vreeling, F., Colle, H., Hoogland, G., Groenewegen, H.J., van der Linden, C.: Chronic bilateral thalamic stimulation: a new therapeutic approach in intractable tourette syndrome—report of three cases. J. Neurosurg. **99**(6), 1094–1100 (2003). doi:10.3171/Jns.2003.99.6.1094
109. Viswanathan, A., Jimenez-Shahed, J., Baizabal Carvallo, J.F., Jankovic, J.: Deep brain stimulation for tourette syndrome: target selection. Stereotact Funct. Neurosurg. **90**, 213–224 (2012)

110. Walter, B.L., Vitek, J.L.: Surgical treatment for parkinson's disease. Lancet Neurol. **3**(12), 719–728 (2004). doi:10.1016/S1474-4422(04)00934-2
111. Weaver, F.M., Follett, K., Stern, M., Hur, K., Harris, C., Marks, W.J., Rothlind, J., Sagher, O., Reda, D., Moy, C.S., Pahwa, R., Burchiel, K., Hogarth, P., Lai, E.C., Duda, J.E., Holloway, K., Samii, A., Horn, S., Bronstein, J., Stoner, G., Heemskerk, J., Huang, G.D.: CSP 468 Study Group. Bilateral deep brain stimulation vs. best medical therapy for patients with advanced Parkinson disease: a randomized controlled trial. JAMA **301**, 63–73 (2009)
112. Whiting, D.M., Tomycz, N.D., Bailes, J., de Jonge, L., Lecoultre, V., Wilent, B., Alcindor, D., Prostko, E.R., Cheng, B.C., Angle, C., Cantella, D., Whiting, B.B., Mizes, J.S., Finnis, K.W., Ravussin, E., Oh, M.Y.: Lateral hypothalamic area deep brain stimulation for refractory obesity: a pilot study with preliminary data on safety, body weight, and energy metabolism. J. Neurosurg. **119**, 56–63 (2013)
113. Williams, A., Gill, S., Varma, T., Jenkinson, C., Quinn, N., Mitchell, R., Scott, R., Ives, N., Rick, C., Daniels, J., Patel, S., Wheatley, K.: PD SURG Collaborative Group. Deep brain stimulation plus best medical therapy versus best medical therapy alone for advanced Parkinson's disease (PD SURG trial): a randomised, open-label trial. Lancet Neurol. **9**, 581–591 (2010)
114. Xie, S.T., Shafer, G., Wilson, C.G., Martin, H.B.: In vitro adenosine detection with a diamond-based sensor. Diam. Relat. Mater. **15**(2–3), 225–228 (2006). doi:10.1016/j.diamond.2005.08.018
115. Yamawaki, N., Magill, P.J., Woodhall, G.L., Hall, S.D., Stanford, I.M.: Frequency selectivity and dopamine-dependence of plasticity at glutamatergic synapses in the subthalamic nucleus. Neuroscience **203**, 1–11 (2012). doi:10.1016/j.neuroscience.2011.12.027
116. Yoshimi, K., Naya, Y., Mitani, N., Kato, T., Inoue, M., Natori, S., Takahashi, T., Weitemier, A., Nishikawa, N., McHugh, T., Einaga, Y., Kitazawa, S.: Phasic reward responses in the monkey striatum as detected by voltammetry with diamond microelectrodes. Neurosci. Res. **71**(1), 49–62 (2011). doi:10.1016/j.neures.2011.05.013
117. Zhang, K., Bhatia, S., Oh, M.Y., Cohen, D., Angle, C., Whiting, D.: Long-term results of thalamic deep brain stimulation for essential tremor. J. Neurosurg. **112**, 1271–1276 (2010)
118. Zhao, H., Bian, X.C., Galligan, J.J., Swain, G.M.: Electrochemical measurements of serotonin (5-HT) release from the guinea pig mucosa using continuous amperometry with a boron-doped diamond microelectrode. Diam. Relat. Mater. **19**(2–3), 182–185 (2010). doi:10.1016/j.diamond.2009.10.004

Mechanism of Docosahexaenoic Acid in the Enhancement of Neuronal Signalling

Md Ahsan Ul Bari, Julie Gaburro, Agnes Michalczyk, M. Leigh Ackland, Catherine Williams and Asim Bhatti

Abstract Microelectrode array (MEA) has attracted paramount attention from neuroscientific community to explore and understand the working principle of nervous systems and the effect of drugs on the behaviour of neurons. In this work, we attempt to explore the effect of docosahexaenoic acid (DHA) on the overall neuronal spike activity as well as the spontaneous activity patterns of primary cortical neurons employing MEA technology. Neocortex neurons of C57BL/6 mice were cultured on MEA for 2 weeks until maturation and then treated with 10 μg/ml DHA for 48 h. Our results demonstrated that DHA supplementation enhanced the overall spike activity (454.35 spikes/s) of the neurons compared to the non-treated control (297.01 spikes/s). This is a preliminary study to explore the changes in the electrophysiological properties of neurons in response to DHA. Our results indicated the potential use of DHA in improving neuronal signalling indicating it could be helpful in improving the diseased condition of neuronal disorders particularly in Alzheimer's disease (AD).

Keywords Docosahexaenoic acid (DHA) · Microelectrode array (MEA) · Alzheimer's disease (AD) · Neuronal spike · Neuronal signalling

---

Md Ahsan Ul Bari (✉) · A. Michalczyk · M. Leigh Ackland
Centre for Cellular and Molecular Biology (CCMB), Deakin University, Victoria, Australia
e-mail: bamd@deakin.edu.au

J. Gaburro · A. Bhatti
Institute for Intelligent Systems Research and Innovation (IISRI), Deakin University, Victoria, Australia

J. Gaburro · C. Williams
Australian Animal Health Laboratory (AAHL), CSIRO, Victoria, Australia

A. Bhatti et al. (eds.), *Emerging Trends in Neuro Engineering and Neural Computation*, Series in BioEngineering,
DOI 10.1007/978-981-10-3957-7_5

1 Introduction

Gradual decline in memory and cognitive function is considered to be normal and is one of the most prominent health concerns for older individuals worldwide. Recently, a positive effect of docosahexaenoic acid (DHA) in improving cognitive function has been reported [1]. Deficiency of DHA is associated with many diseases including heart, kidney, rheumatoid arthritis, neurodegenerative disorders and cancer [2, 3]. Neuroimaging and post-mortem histological studies show that adult patients with recurrent affective disorders have neuronal atrophy in the pre-frontal cortex, a region of the brain that is susceptible to degenerative changes induced by nutritional DHA deficiency [4]. A link between DHA and Alzheimer's disease (AD) has also been established highlighting diseased patients had low levels of DHA [5]. Studies on animals fed on DHA-deficient diets displayed learning and memory deficits [6] and an increased depressive and aggressive behaviour [7].

In recent years, Alzheimer's disease (AD) has become a great concern amongst the highest health risks worldwide. The pathogenesis of AD includes synaptic loss [8, 9] and synaptic dysfunction. In this state, communication through neurotransmitters is disrupted, hindering neuronal signal processing [10, 11]. DHA plays an effective role in synaptic function and cognitive abilities by providing plasma membrane fluidity at synaptic regions [12], which in turn improves neuronal signals flow.

1.1 Docosahexaenoic Acid (DHA)

DHA is an essential polyunsaturated fatty acid (PUFA) [13] having 22 carbons with 6 double bonds (22:6n-3) and is a member of omega-3 fatty acid family [14, 15]. It is a primary component of membrane phospholipids in the central nervous system (CNS) [3], particularly in the brain, retina [16–18], growth cones [19] and sperm [20]. DHA constitutes approximately 30–40% of the phospholipids of the grey matter of cerebral cortex and photoreceptor cells in the retina. DHA is most abundant among subcellular fractions of brain tissue [12, 21]. In mammals brain size and number of brain cells vary depending on species but the DHA content of the brain cells remains relatively unique between species [22]. Metabolically active areas of the brain including the cerebral cortex, mitochondria, synaptosomes and synaptic vesicles [15, 23] are highly enriched in DHA content [24]. It has a major role in the development and function of the CNS. It is also associated with the formation of synapses between neurons in the nervous system, enhancement of memory, neuroprotection [25] and vision [26].

There have been several negative effects of deficiency of DHA documented suggesting DHA has a role in the development of brain hippocampus [27]. Hippocampus is essential for learning and memory tasks. Deficiency of DHA impairs the growth and development of hippocampus thus impairing learning ability.

Reduced DHA content also increase the chance of Hypertension, which is a major risk factor for cardiovascular and cerebrovascular disease as there is not enough DHA to lower the blood pressure [28].

DHA is incorporated into phospholipids in neuronal membranes, which can influence chemical and physical properties of membrane. It also helps regulating cell signalling that involved in neuronal survival, proliferation and differentiation [29]. Cascade of mechanism involved in this process. DHA promotes development of neurite [30] which can lead to a further anti-apoptotic properties through AKT signalling pathway [31]. This novel anti-apoptotic property of DHA would be of prime interest of in preventing diseases caused by neuronal apoptosis that subsequently leads to neurodegeneration.

1.2 Experimental Evidences Revealing the Beneficial Effects of DHA on Neuron

To reveal the effects of DHA on neurons or neuronal cells, several scientific methods have been put into practice. One of the prominent researchers in the field of DHA, Hee-Yong Kim and associates in 2000 had demonstrated anti-apoptotic feature of DHA using some basic molecular techniques [32]. PC12/Neuro 2A cells were pre-enriched with DHA (22:6n-3) before serum starvation showed decreased apoptotic cell death. DHA showed reduced DNA fragmentation and caspase-3 activity. In 2001 another study was conducted by Hee-Yong Kim et al. revealing the protective effect of DHA along with other polyunsaturated fatty acids such as arachidonic acid (20:4n-6; AA) against apoptosis induced by serum starvation. According to them protection by Arachidonic acid (AA) is due to direct involvement as free fatty acid serum starvation. On the other hand, DHA enrichment increases the phosphatidylserine accumulation by pre-treatment of the cells with DHA for at least 24 h prior serum starvation. Decreased DNA fragmentation and down regulation of caspase-3 activity was observed as a positive effect of DHA [33]. Akbar and HY Kim in 2002 showed anti-apoptotic effect of DHA towards staurosporine-induced apoptosis in neuro-2A cells. The conclusion of their study was that anti-apoptotic effect of DHA enrichment is mediated through the PI3-kinase/Akt pathway, which was enhanced by DHA-promoted phosphatidylserine (PS) accumulation on neuronal membrane [34].

Another effort made by Cao et al. in 2005 to investigate the effects of DHA on survival and neurite outgrowth in primary cultures of rat cortical neurons. A dose-dependant increase and decrease in cell viability were prominent in their study. Neurite outgrowth and a greater growth-associated protein-43 (GAP-43) immunoactivity were reported in DHA treatment. This indicates DHA-stimulating neuron-specific protein synthesis could be a result of neurite outgrowth [35]. Neuroprotective function of DHA through reduction in cellular zinc levels in opposing apoptosis was hypothesized by Suphioglu et al. in 2010, by reducing ZnT3 expression and zinc uptake in human neuroblastoma cell line M17 [36].

1.3 Biological Evidences Towards Beneficial Effects of DHA on Neuron

DHA has beneficial effects on brain only when used in appropriate concentration. High concentrations of DHA are potentially damaging to the neurons [34]. Hence, an appropriate concentration of DHA was determined to induce optimum cell growth and development of neurons. As shown in (Fig. 1a), 10 µg/ml of DHA enhanced cell growth and survival in NT-2 cells. Further increase in DHA concentrations triggered continuous decrease in cell viability. Cells without DHA supplementation grow normally without any significant physiological changes (Fig. 1b). On the other hand, DHA-treated cells promoted neurite outgrowth with visible increase in neurite length without affecting the neurite number (Fig. 1c), which can enhance neuronal signalling.

To show the anti-apoptotic effect of DHA, neuronal committed NT2 cells were exposed to apoptotic agents such as staurosporine (STS) and hydrogen peroxide (H_2O_2). STS is a protein kinase inhibitor that has a potential to induce apoptosis in several neuronal and non-neuronal cell types [37, 38]. Hydrogen peroxide has also been shown to induce neuronal cell death in undifferentiated and differentiated neuronal cell lines through the induction of generation of reactive oxygen species (ROS) [39]. The effects of DHA on cell viability were assessed after 48 h of co-treatment with STS and H_2O_2, respectively. This was a different approach of DHA enrichment from previous experiments, which used pre-enrichment of DHA to the cells [34]. In our study we used co-treatment approach. NT2 cells were exposed to DHA presence of STS or H_2O_2 for 48 h. The unpublished data showed NT2 cells enriched with DHA indicated significant increase in cell viability up to 80% compared to the cells only treated with STS (Fig. 2a) and H_2O_2 (data not shown). Western blot analysis was carried out to detect pro-apoptotic marker

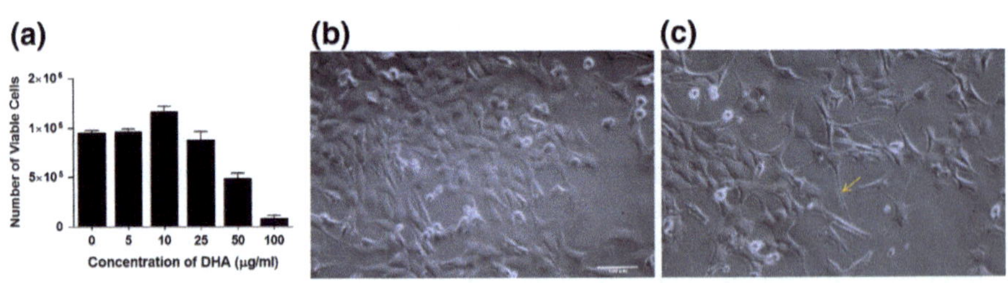

Fig. 1 Effect of DHA in cell survival and neurite outgrowth on NT-2 cell line. **a** Optimization of DHA concentration for improved cell survival, **b** normal cell growth without DHA treatment, **c** visible neurite outgrowth after DHA treatment (*yellow arrow*)

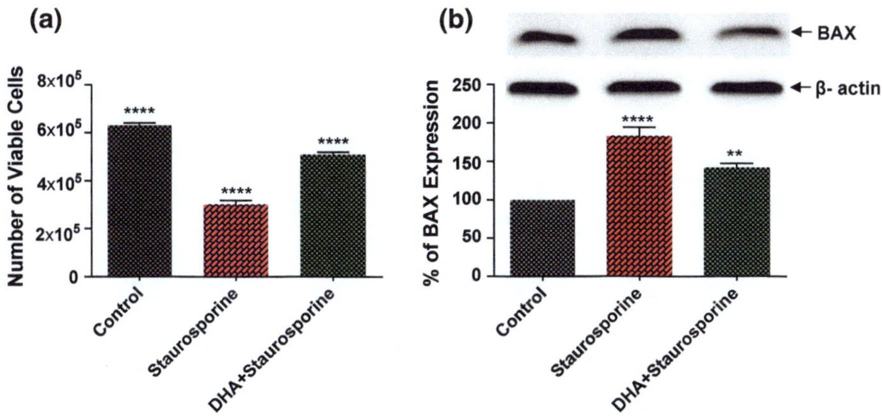

Fig. 2 Graphs showing improved cell survival by DHA treatment in NT2 cells against induced cell death by staurosporine. **a** observed increase in cell viability, **b** reduced BAX expression indicating protection against apoptotic cell death

protein BAX. The expression of BAX was increased by more than 50% compared to control when cells were co-treated with STS (Fig. 2b) and H2O2. When cells were co-treated with DHA, DHA-PLs and STS, BAX expression was decreased by 20–40% in DHA-treated cells relative to STS (Fig. 2b). Reduction of BAX expression by DHA treatment is a clear sign of anti-apoptotic feature of DHA.

Enhanced neurite growth (Fig. 3b upper panel) compared to non-treated control (Fig. 3a upper panel) can be observed as shown in Fig. 3 [35]. Enhanced neurite growth and neuronal connections are visible in DHA-treated cortical neurons that lack in untreated control. Similar condition is applied to the rat cortical neurons as well. Anti-βIII tubulin mouse antibody was used to depict the growth of neurite. In this study it was used to expose if the DHA treatment actually promotes neurite growth or not. In Fig. 3 it was clearly visible that DHA treatment had significantly increase the growth of neurite (Fig. 3b) compared to untreated control (Fig. 3a). Detailed experimental procedures are explained in methods section.

Although significant physical changes were observed in neuron from DHA treatment, there was nothing we can depict what exactly causing the physical changes or how this physical changes stimulate neuronal activity in the brain. To dig deeper into the effect of DHA on neuronal electrophysical properties we moved to a modern extracellular stimulation and recording tool named microelectrode arrays (MEAs).

Action potentials or "spikes" are the common computational currency of the brain. Numerous ways have been developed to identify the activity of individual neurons in response to external stimuli and to our perception of those stimuli. However, an individual neuron's activity does not signify the cognitive and perceptual processes. It is more likely to be the network-level phenomena in which groups of neurons act in concert [40]. One of the most powerful and successful

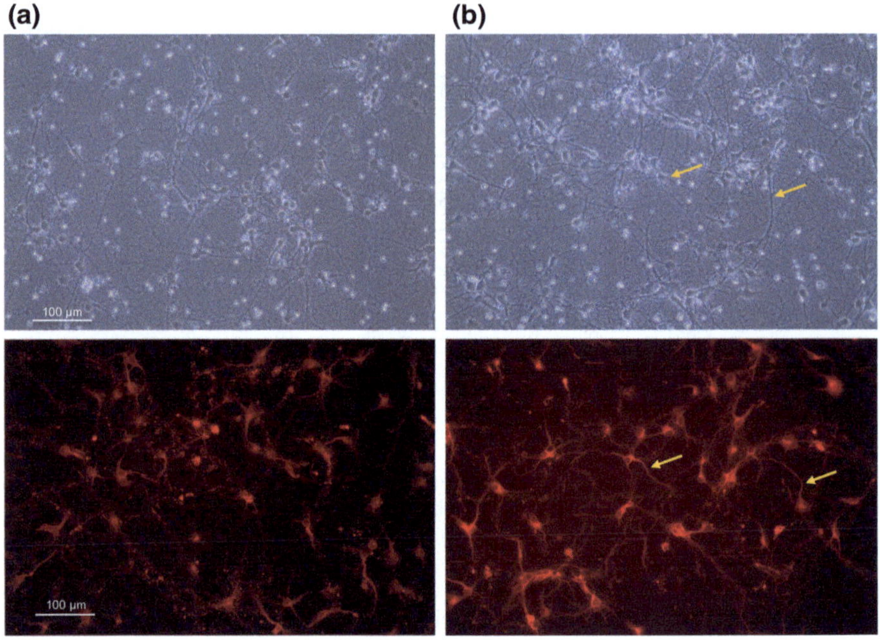

Fig. 3 Effect of DHA in neurite outgrowth on primary cortical neuron. **a** Normal cell growth without DHA treatment. **b** Visible neurite outgrowth after DHA treatment (*yellow arrows*). (*Upper panel* represents phase contrast microscopic view and *lower panel* represents fluorescent microscopic illustration of neuron treated with anti-β III tubulin antibody and conjugated with Alexa 594 fluorophore)

methods used for the recording of neural activity is extracellular recordings and stimulation by substrate-integrated microelectrode arrays (MEAs) [41]. The use of microelectrode arrays (MEAs) is constantly increasing to measure electrical activity from many excitable cells over periods of days and months [42, 43] even for a year [44]. MEA for both in vitro and in vivo is advantageous as the extracellular recording can be performed without mechanical damage to the cellular plasma membrane [43]. In this present study, we focus on elucidating beneficial effect of DHA in neuronal signalling on mouse cortical neuron and revealing the neural activity in response to DHA using MEA technology.

1.4 Microelectrode Array (MEA)

In vitro microelectrode array (MEA) technology has been widely adopted for recording signals from neurons [45, 46] to understand the fundamentals of neural network mechanisms in the brain [47]. Accurate detection and localizing signatures of neural activity waveforms are essential in understanding the functioning of neural networks [48]. The spikes of neural signals normally characterize these signatures.

The modern MEA technique has been emerged through a historical development over time [49]. Thomas et al. pioneered the first design for MEA [50] in 1972. They recorded embryonic chick heart cells on an array of 30 electrodes (2 × 15, size of 7 μm^2 and 100 μm apart). Later on, action potential from dissociated snail ganglia was recorded by Gross and his associates using MEA with an array of 36 microelectrodes (100 or 200 μm apart) [51, 52]. In 1980, a successful effort was made by Pine [53] in correlating the intracellular and extracellular events from dissociated rat superior cervical ganglion neurons by simultaneous recordings in combination of intra- and extracellular recordings by MEAs (2 × 16 array of 32 microelectrodes, 8 × 10 μm in size and 250 μm apart). This successful correlation makes the MEA recording technique a beneficial electrophysiological tool to study pharmacology [54] and network physiology. Use of high-efficiency and high-quality MEA techniques was not widely successful in the late twentieth century although the efforts made in cultured preparations of wide range of invertebrates and vertebrates [55–61].

Use of the MEA technology platform has provided us with simplified model of a neuronal network in the order of a single cortical column rather than a complete intact brain. This unique feature of MEA enables us to monitor neural activity in detail and over-defined period of time using a computer interface and record transient and steady-state behaviour of the neuronal network in response to DHA. In addition, neuron cell cultures on MEAs have been reported to survive for over a year in vitro [62], making it an ideal research model to understand the effect of different stimulations and drugs on the neuronal connectivity.

2 Materials and Methods

2.1 Cell Line and Culture Condition

NT2 cells (ATCC: CRL-1973) that have phenotypic characteristics of neuronal cells [63] were used as a model cell line to study the effects of DHA. The cells were routinely maintained in 1:1 (v/v) mixture of DMEM (Gibco, Life Technologies) and nutrient mixture F-12 (Gibco, Life Technologies) supplemented with 10% (vol/vol) foetal bovine serum (Gibco BRL, Grand Island, NY, USA), 1 mM sodium pyruvate, 1X uridine and kept at 37 °C in a tissue culture incubator with 5% CO_2 and 98% relative humidity. The exponentially growing cells were used throughout the experiments. NT2 cells were treated with 10 μg/ml DHA for 48 h [34]. The cellular responses to DHA were measured by cell viability assay using trypan blue exclusion method after 48 h treatment with different concentrations of DHA (Fig. 2a).

2.2 Acutely Dissociated Cortical Neurons and DHA Treatment

Cultures of neocortex neurons were prepared from 14–16-day gestation C57BL/6 mice. The cortices from embryonic mice brains were excised and then treated with 20X trypsin, and then manually triturated with DNase/SBTI (Sigma) that stops the action of the trypsin and allows a single-cell suspension to be prepared. The cells were then centrifuged and the resulting cell pellet was then resuspended in B27 supplemented neurobasal medium (Gibco, Life Technologies) and plated at the required density. Excess cells were then frozen down in neurobasal medium supplemented with B27, 20% foetal bovine serum and 10% DMSO, at a concentration of $10 x 10^6$ cells/ml to ensure continuous supply throughout the experiments. Cryopreserved cortical neurons of mouse embryo were stored in liquid nitrogen for long-term storage. Freshly prepared neuronal cells were diluted with pre-warmed B27-supplemented neurobasal medium (Invitrogen). For the DHA treatment cells were (4.0×10^5 cell/well) seeded on poly-D-lysine-coated 24-well culture plate. On DIV 7 cells were treated with 10 μg/ml DHA for 48 h leaving control cells aside. All treatments were done in triplicates. The picture of neurite outgrowth was captured under phase contrast microscope (Olympus, IX51). The cell density and viability was determined using a haemocytometer, allowing a stock suspension to be prepared at a density of 25,000 cells in a 30 μl drop of neurobasal medium (Gibco, life technology) and seeded to MEA's previously coated by poly-D-lysine and Laminin. At days in vitro (DIV) 7, MEAs were treated with DHA as before, after DHA treatment signals from MEA's were recorded.

2.3 Immunostaining of Cortical Neurone for Labelling Neurite

Immunostaining of cortical neuron was performed following general procedure for the immunocytochemistry (ICC) (abcam protocols) with some modification. Briefly, round coverslips rinse in ethanol were placed inside 24-well plates and coated with poly-D-lysine (100 μg/ml) for 1 h at room temperature. Coverslip wells were rinsed with sterile water three times (10 min each) allowing to dry completely and sterilized them under UV light for at least 1 h. Acutely dissociated cortical neurons were seeded at a concentration of 4×10^5 cells per well allowing them to grow in B27 supplemented neurobasal medium (Gibco, Life Technologies) for 7 days with changing the medium in two days interval. After 7 days ells were treated with DHA (10 μg/ml) and incubated for 48 h leaving untreated controls aside. Medium was aspirated and cells were rinsed in phosphate-buffered saline (PBS) three times. Plated cells were fixed in 4% paraformaldehyde in PBS (pH 7.4) for 10 min at room temperature, and then permeabilized by 0.1% Tinton-X on ice. Nonspecific labelling was blocked using 1% BSA (Sigma) and 0.2% Gelatin

(Sigma) prior to primary antibody incubation. Plates seeded with cells were incubated overnight with anti-βIII tubulin mouse monoclonal antibody (Promega, Cat#G172A) in 1% BSA. Following three times washing with PBS cells were subsequently labelled with fluorescent tagged Alexa 594 (Molecular Probe) anti-mouse secondary antibody for 1 h at room temperature. Cells were visualized under fluorescent microscope and photographed.

2.4 MEA Preparation

MEAs were pre-treated for coating with poly-ethylene-imine (PEI) and laminin (Sigma Aldrich). Laminin (0.02 mg/ml in Neurobasal medium) was applied directly to the centre of the array, in drops of 20 μl. After incubation, laminin drops were removed by vacuum aspiration just prior to plating cells. Primary neurons were allowed to rest for 30 min in the incubator for adhesion and then 1 ml of media was added in the MEA dish. Cultures were maintained in petri dishes with lids on, in an incubator with 5% CO_2, 37 °C and 65% relative humidity. The media was regularly changed after 4–5 days during 2 weeks of cells growth and maturation.

2.5 Extracellular Microelectrode Recordings

On the 15–17 DIVs MEA recording was performed at 37 °C in the standard growth medium in a cell culture incubator (37 °C, 5% CO_2, 65% relative humidity). In order to avoid contamination, culture chambers' lids were tight under biosafety cabinet before starting recording. A high band pass digital filter (200 Hz) was applied to the raw signal in order to remove electrical background noise. Recording sessions lasted for 3–5 min to record spontaneous signals from the control and treated cultures. There is a specific pathway towards the accumulation of signals from neuron through MEA system devices that is described in Fig. 2. Neuronal responses from drug treatment are transformed by different parameters across the components of the MEA toward the recorded signal typically as a form of action potential.

Simultaneous signals from all 60 electrodes were sampled at 25 kHz, visualized and stored using the standard software MC-Rack provided by Multi Channel System (Reutlingen, Germany). Later on spike and burst detection analysis was performed using specialized software NeuroSigX developed by IISRI, Deakin University, Australia. Spike's detection threshold with justification: −15 μV was set to record the response from neuron from MEA.

3 Experimental Outcomes

3.1 Spatio-Temporal Spike Analysis

To explore the effect of DHA on the electrophysiological behaviour of the neurons, spike activity behaviour was analysed using spatio-temporal analysis. Figure 4 shows the raster plot of spatio-temporal analysis where spike activity of all 60 electrodes is distributed over time. An increased activity in the DHA-treated MEA (Fig. 4b) compared to control (Fig. 4a) is clearly visible. This raster plot represents a zoomed section of the whole. Each dot in the raster plot represents electrode activity over a specific time window. Increase in the electrical signalling in between the cells will subsequently increase amount of dots in the raster plot.

3.2 Burst Activity Analysis

To further analyse the effect of DHA, quantitative spike analysis was performed employing activity histogram, as shown in Fig. 5. An increase in the spike activity can be observed from control to DHA-treated cells. Table 1 presents estimated average spike rate that increases from 297.01 spikes per second to 454.35 spikes per second for control and DHA-treated neuronal cultures, respectively.

Minimum and maximum spike rate per channel and mean spike rate per channel have also been shown as increased (Table 1) in DHA treatment compared to control. These differences are visible in Fig. 5. In control the maximum spike rate and the mean spike rate per channel were recorded as 4714 and 1485.6, respectively (Fig. 5a), where they were 14577 and 2296.7 for DHA treatment (Fig. 5b). This

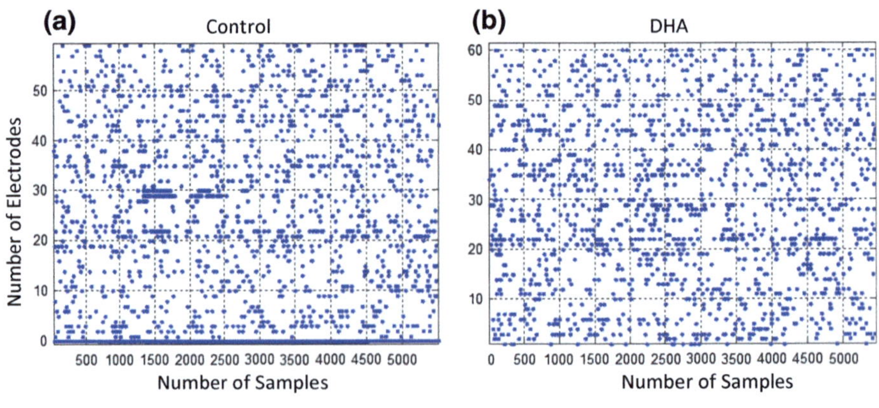

Fig. 4 Effect of DHA treatment in neuronal signalling. Neuronal signalling pattern presented in raster plot **a** in control and **b** DHA-treated cortical neuron

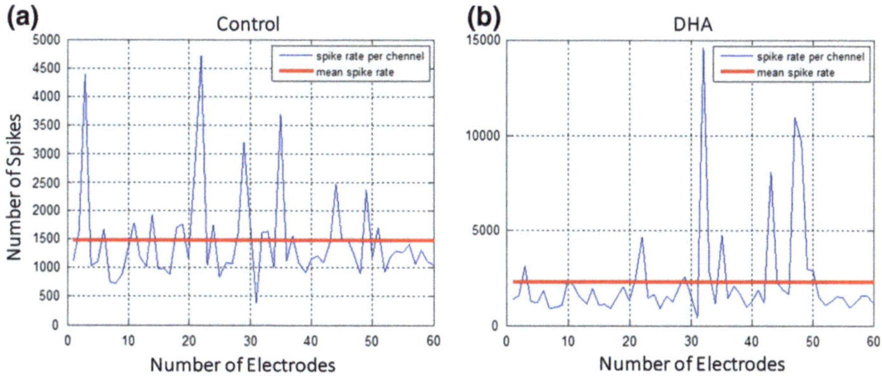

Fig. 5 Spike activity analysis of **a** control and **b** DHA-treated cortical neurons. *Blue line* highlights variation of the spike activity over 60 channels and *red line* highlights the mean spike activity over all channels

Table 1 Spike rate of the control and DHA-treated neuronal network

Criteria	Control	DHA-treated
Spike rate per second	297.01	454.35
Min spike rate per channel	393	408
Max spike rate per channel	4714	14577
Mean spike rate per channel	1485.6	2296.7

improved spike rate in DHA treatment clearly demonstrates the beneficial effect of DHA in neuronal signalling.

Furthermore, spontaneous network burst activity behaviour was analysed to further assess the effect of DHA on primary cortical neurons. Network bursts are characterized by short transients of rapid spiking activity flanked by silent periods indicating activation patterns of neurons defined by rapid action potential spiking over narrow time intervals [64, 65]. Network bursts are common feature of the central nervous system used in several in vitro preparations and are considered to be indicators of robust neural operation, communication and neuropathology [66]. Bursts were detected based on calculating the total spiking rate, which is the total number of spikes from all electrodes within 100 ms time bin. As a criterion of spotting the burst, a rapid appearance of a large number of spikes over all electrodes in a small (100 ms) time bin was used [67].

Burst activity analysis is shown in Fig. 6, where the estimated bursts are highlighted with green and red line. Green and red lines indicate the start and end of the burst activity with respect to time. A clear difference and improvement in bursting pattern from control (18 distinct burst) (Fig. 6a) to DHA-treated (20 distinct burst) (Fig. 6b) cortical neurons is apparent. By the rate of occurrence, burst patterns are classified differently. Typically, burst rates are relatively constant over time, evident with either regular or more chaotic spacing. Differently, the burst rate can be varied by more than an order of magnitude over the course of a recording

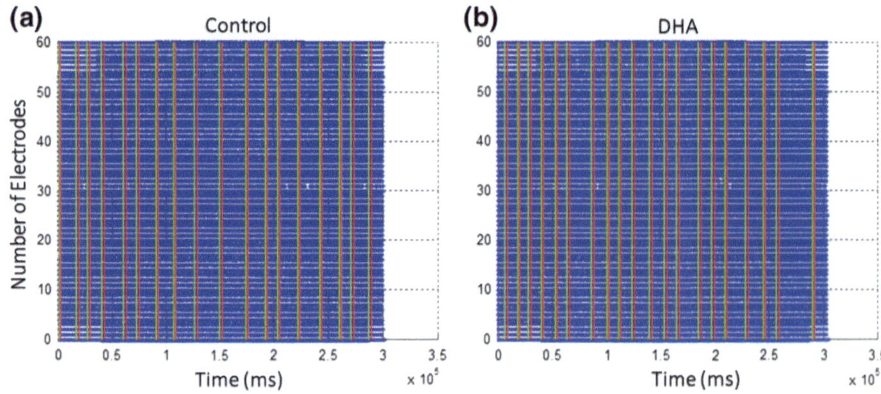

Fig. 6 Changes in spontaneous network burst activity of neuronal signalling subsequent electrode activity during data recording in **a** control and **b** in DHA-treated cortical neuron

and evident with the burst rate which is highly variable [68]. Looking into the control, it was apparent that the burst pattern was more likely to be continuous (Fig. 6a) than the pattern detected in DHA treatment (Fig. 6b) tend to be versatile. This criterion indicates the positive effect of DHA on neuron in terms of signal processing.

Bursting is a general amplifying mechanism of signals from dendritic structures, synaptic inputs and voltage-gated channels of neurons. Changes of any of these structural cores of neurons may bring about the alterations in bursting pattern in more versatile way. It is acknowledged that bursting is effective at enhancing transmitter release and promoting activity-dependent synaptic plasticity [69]. Our experimental phenomenon coincides with the statement and indicates the changes in neurite outgrowth due to DHA treatment may have been improve the bursting pattern.

3.3 *Energy Profile of the Electrodes*

Electrode activity is measured based on the total active electrodes and their intensity of activeness thought out the recorded time. These activities are represented as heat map in Fig. 7 (2D view in top panel and 3d view at the bottom). Colour-coded intensity map shows the position and intensity of a particular electrode during recording.

3D view gives us a vertical representation of the degree of intensity of a particular electrode. From our experiment it is evident that the intensity of electrodes activity for a period of time is more likely to be three times higher in DHA treatment (count of 14,000) (Fig. 7b top panel) than the untreated control

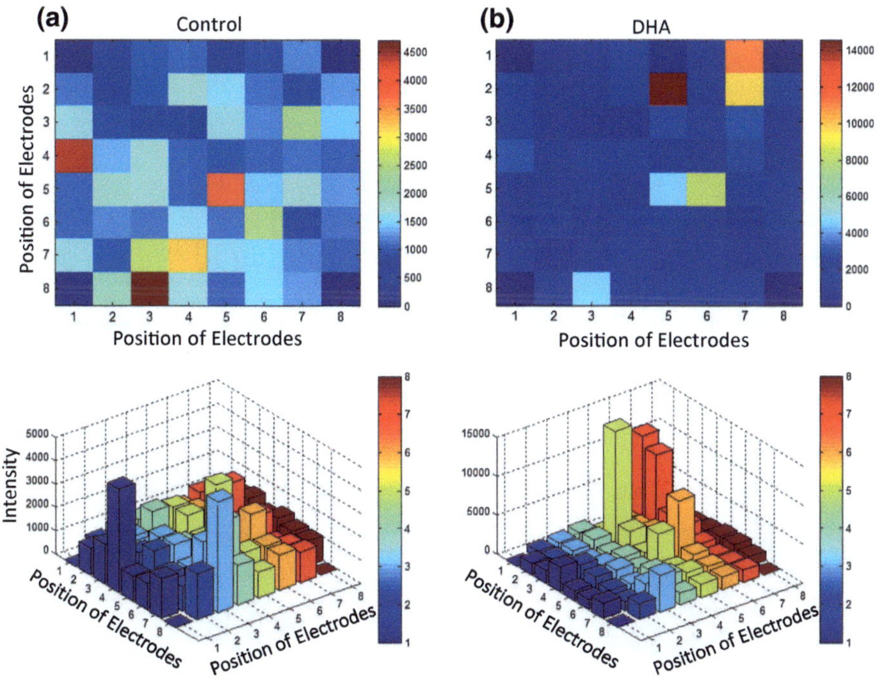

Fig. 7 Spontaneous electrode activity pattern of MEA's in **a** control and **b** DHA-treated cortical neuron on MEA

(4500 count) (Fig. 7a to panel). A fair bit of conclusion could be drawn as DHA treatment promotes the neuronal activity.

3.4 *Neural Activity Pattern Variance*

To understand the variance between neural activity patterns, dynamic time warping technique was employed. Dynamic time warping (DTW) was developed originally for the purpose of speech recognition [70]. It is a well-known technique of time series alignment algorithm. DTW is used to find an optimal alignment between two given (time-dependent) sequences of feature vectors implying certain restrictions. The sequences are warped spontaneously in a nonlinear fashion to match each other in order to find similarity variance between them [71]. DTW has been successfully applied in the fields of data mining and information retrieval and can automatically deal with time deformations and different speeds of time-dependent data. The distance between two sequences or points is measured by the function of manhattan distance. The Manhattan distance function of metric computes the distance that would travel to get from one data point to the other following a grid-like path.

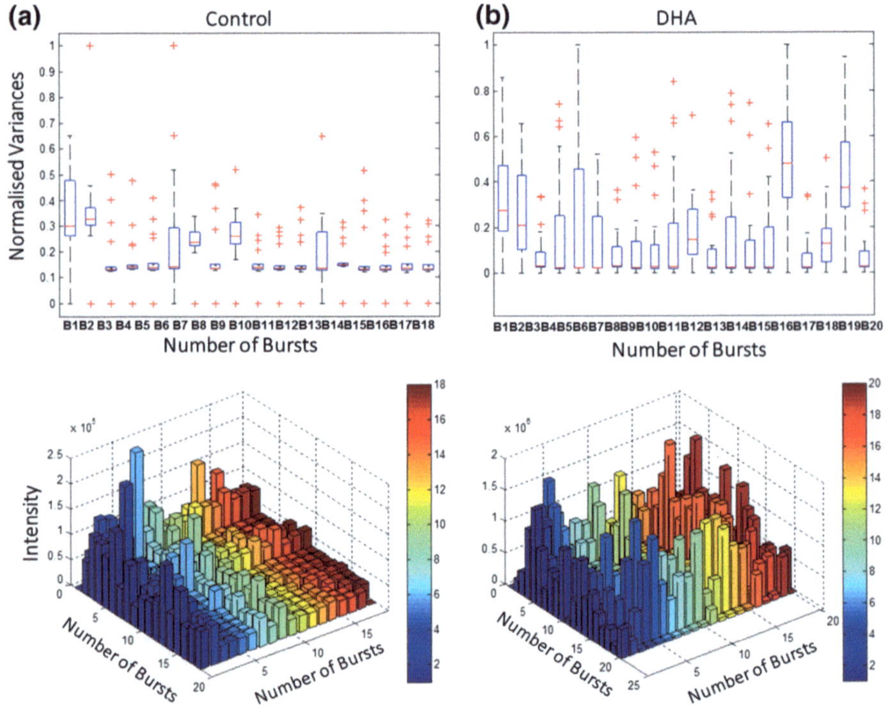

Fig. 8 Dynamic time warping (DTW) of MEA-recorded data response from **a** control (*top panel*) and graphical distribution of variance of distances (*bottom panel*) and **b** DHA-treated cortical neuron (*top panel*) and graphical distribution of variance of distances (*bottom panel*)

The distance between two items is the sum of the differences of their corresponding components.

In this experiment DTW was measured considering Windows (BW) of 100 ms and minimum electrode (ME) umber 10. It can be seen that the similarity variance calculated based on manhattan distance of detected burst has a low variance in control and were presented in a form of 2D bar graph (Fig. 8a top panel). Distances of variances were presented as 3D graph (Fig. 8a bottom panel) along with the intensity of individual burst. On the other hand DTW from DHA-treated samples were detected with high variance (Fig. 8b top panel) and the distances are highly diversified (Fig. 8b bottom panel), which may suggest the physical changes in neuron promoted by DHA treatment can bring along the variation in the signal processing as well.

3.5 Network of Signal Flow

Network flow diagram is usually used to express the connective network of processed signal during MEA recording. A complete informative idea can be gained from network flow. Network flow describes the dynamic utilization of a network as communication medium such as electrical grids anatomical connections in the brain [72] and is related to network connectivity. Following Fig. 9, organized two-dimensional network connectivity of signals from individual electrode in control (Fig. 9a top panel) is seen. Green thread-like structures represent the connective paths from one electrode to another. Bottom panel (Fig. 9a) represents the intensity of activities of the electrodes. The taller the red bars, the more active the electrodes are. On the other hand response from the DHA-treated electrodes mores dense and organized network activity were observed (Fig. 9b top and bottom panel). This improved activity in the network flow by DHA treatment suggests that supplementation of DHA may have some progressive effect on neuron that in turn can improve the signalling mechanism of central nervous system.

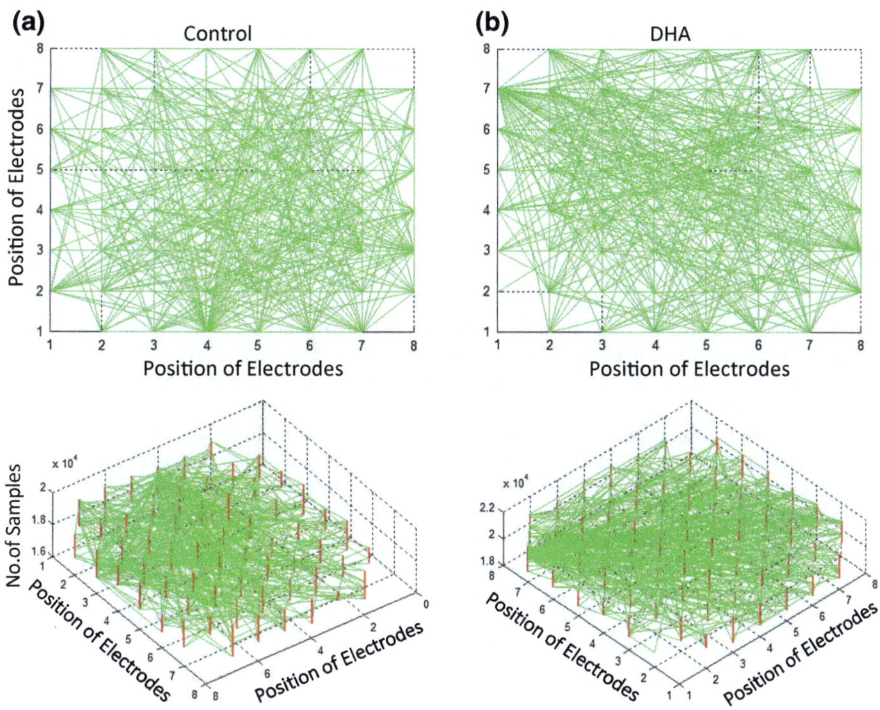

Fig. 9 Network flow of MEA-recorded data of responses from **a** control, 2D representation of connected electrodes (*top panel*) 3D representation of connected electrodes (*bottom panel*). **b** DHA-treated cortical neuron, 2D representation of connected electrodes (*top panel*) 3D representation of connected electrodes (*bottom panel*)

4 Conclusion

This study was conducted towards revealing the changes in electrophysical properties of neuron due to DHA treatment. The aim of the current study was to examine whether DHA have any effects on neuron that can enhances physical properties hence improving neuronal signalling pattern. Our study demonstrates a positive function of DHA in modulating action potential duration, affecting the information sent by one neuron and received by others and thus enhances neuronal signalling. Increased signalling may be due to increase in neurite growth, synaptogenesis and synaptic membrane fluidity promoted by DHA supplementation and increasing signal flow. These phenomena were supported by the results of spike detection, Busts analysis and the DTW function. Our experiment is pioneering the preliminary approach to highlight the significant role of DHA in neuronal signal processing. This was a short-time treatment approach for DHA. Long-term treatment with DHA for prolonged cortical neurone culture may have given us more information of DHA activity on neuron. Further experiments in more elaborative space are to be conducted to explore the vast range of neuronal signalling pattern in response to DHA supplementation.

Acknowledgements This work was jointly supported by Centre for Cellular and Molecular Biology (CCMB), Deakin University, Australia, Institute for Intelligent Systems Research and Innovation (IISRI), Deakin University, Australia and Australian Animal Health Laboratory (AAHL), CSIRO, Australia. We would also like to thank Steve Cheung's lab for providing facilities for the immunocytochemistry work for the manuscript.

References

1. Yurko-Mauro, K., et al.: Beneficial effects of docosahexaenoic acid on cognition in age-related cognitive decline. Alzheimers Dement **6**(6), 456–464 (2010)
2. Collins, F.D., et al.: Plasma lipids in human linoleic acid deficiency. Nutr. Metab. **13**(3), 150–167 (1971)
3. Connor, W.E., Neuringer, M., Lin, D.S.: Dietary effects on brain fatty acid composition: the reversibility of n-3 fatty acid deficiency and turnover of docosahexaenoic acid in the brain, erythrocytes, and plasma of rhesus monkeys. J. Lipid Res. **31**(2), 237–247 (1990)
4. McNamara, R.K.: DHA deficiency and prefrontal cortex neuropathology in recurrent affective disorders. J. Nutr. **140**(4), 864–868 (2010)
5. Tully, A.M., et al.: Low serum cholesteryl ester-docosahexaenoic acid levels in Alzheimer's disease: a case-control study. Br. J. Nutr. **89**(4), 483–489 (2003)
6. Crawford, M.A., Bazinet, R.P., Sinclair, A.J.: Fat intake and CNS functioning: ageing and disease. Ann. Nutr. Metab. **55**(1–3), 202–228 (2009)
7. DeMar Jr., J.C., et al.: One generation of n-3 polyunsaturated fatty acid deprivation increases depression and aggression test scores in rats. J. Lipid Res. **47**(1), 172–180 (2006)
8. Lukiw, W.J., et al.: A role for docosahexaenoic acid-derived neuroprotectin D1 in neural cell survival and Alzheimer disease. J. Clin. Invest. **115**(10), 2774–2783 (2005)
9. Stephan, B.C., et al.: The neuropathological profile of mild cognitive impairment (MCI): a systematic review. Mol Psychiatry **17**(11), 1056–1076 (2012)

10. Shankar, G.M., Walsh, D.M.: Alzheimer's disease: synaptic dysfunction and Abeta. Mol. Neurodegener. **4**, 48 (2009)
11. Masliah, E.: Mechanisms of synaptic dysfunction in Alzheimer's disease. Histol. Histopathol. **10**(2), 509–519 (1995)
12. Gomez-Pinilla, F.: Brain foods: the effects of nutrients on brain function. Nat. Rev. Neurosci. **9**(7), 568–578 (2008)
13. Connor, K.M., et al.: Increased dietary intake of omega-3-polyunsaturated fatty acids reduces pathological retinal angiogenesis. Nat. Med. **13**(7), 868–873 (2007)
14. Bazan, N.G.: Cell survival matters: docosahexaenoic acid signaling, neuroprotection and photoreceptors. Trends Neurosci. **29**(5), 263–271 (2006)
15. Horrocks, L.A., Farooqui, A.A.: Docosahexaenoic acid in the diet: its importance in maintenance and restoration of neural membrane function. Prostaglandins Leukot. Essent. Fat. Acids **70**(4), 361–372 (2004)
16. Jeffrey, B.G., et al.: The role of docosahexaenoic acid in retinal function. Lipids **36**(9), 859–871 (2001)
17. Kim, H.Y., Spector, A.A., Xiong, Z.M.: A synaptogenic amide N-docosahexaenoyl ethanolamide promotes hippocampal development. Prostaglandins Lipid Mediat. **96**(1–4), 114–120 (2011)
18. Su, H.M., et al.: Bioequivalence of dietary alpha-linolenic and docosahexaenoic acids as sources of docosahexaenoate accretion in brain and associated organs of neonatal baboons. Pediatr. Res. **45**(1), 87–93 (1999)
19. Martin, R.E., Bazan, N.G.: Changing fatty acid content of growth cone lipids prior to synaptogenesis. J. Neurochem. **59**(1), 318–325 (1992)
20. Poulos, A., Darin-Bennett, A., White, I.G.: The phospholipid-bound fatty acids and aldehydes of mammalian spermatozoa. Comp. Biochem. Physiol. B **46**(3), 541–549 (1973)
21. Suzuki, H., et al.: Rapid incorporation of docosahexaenoic acid from dietary sources into brain microsomal, synaptosomal and mitochondrial membranes in adult mice. Int. J. Vitam. Nutr. Res. **67**(4), 272–278 (1997)
22. Crawford, M.A., Casperd, N.M., Sinclair, A.J.: The long chain metabolites of linoleic avid linolenic acids in liver and brain in herbivores and carnivores. Comp. Biochem. Physiol. B **54**(3), 395–401 (1976)
23. Bazinet, R.P., Laye, S.: Polyunsaturated fatty acids and their metabolites in brain function and disease. Nat. Rev. Neurosci. **15**(12), 771–785 (2014)
24. Carrie, I., et al.: Specific phospholipid fatty acid composition of brain regions in mice. Effects of n-3 polyunsaturated fatty acid deficiency and phospholipid supplementation. J. Lipid Res. **41**(3), 465–472 (2000)
25. Calon, F., et al.: Docosahexaenoic acid protects from dendritic pathology in an Alzheimer's disease mouse model. Neuron **43**(5), 633–645 (2004)
26. Zhao, Y., et al.: Docosahexaenoic acid-derived neuroprotectin D1 induces neuronal survival via secretase—and PPARgamma-mediated mechanisms in Alzheimer's disease models. PLoS ONE **6**(1), e15816 (2011)
27. Yoshida, S., et al.: Synaptic vesicle ultrastructural changes in the rat hippocampus induced by a combination of alpha-linolenate deficiency and a learning task. J. Neurochem. **68**(3), 1261–1268 (1997)
28. Morin, C., et al.: Effect of docosahexaenoic acid monoacylglyceride on systemic hypertension and cardiovascular dysfunction. Am. J. Physiol. Heart Circ. Physiol. **309**(1), H93–H102 (2015)
29. Kim, H.Y., Akbar, M., Kim, Y.S.: Phosphatidylserine-dependent neuroprotective signaling promoted by docosahexaenoic acid. Prostaglandins Leukot. Essent. Fat. Acids **82**(4–6), 165–172 (2010)
30. Kim, H.Y.: Novel metabolism of docosahexaenoic acid in neural cells. J. Biol. Chem. **282** (26), 18661–18665 (2007)
31. Akbar, M., et al.: Docosahexaenoic acid: a positive modulator of Akt signaling in neuronal survival. Proc. Natl. Acad. Sci. USA **102**(31), 10858–10863 (2005)

32. Kim, H.-Y., et al.: Inhibition of neuronal apoptosis by docosahexaenoic acid (22: 6n-3) role of phosphatidylserine in antiapoptotic effect. J. Biol. Chem. **275**(45), 35215–35223 (2000)
33. Kim, H.Y., Akbar, M., Kim, K.Y.: Inhibition of neuronal apoptosis by polyunsaturated fatty acids. J. Mol. Neurosci. **16**(2–3), 223–227 (2001) discussion 279-84
34. Akbar, M., Kim, H.Y.: Protective effects of docosahexaenoic acid in staurosporine-induced apoptosis: involvement of phosphatidylinositol-3 kinase pathway. J. Neurochem. **82**(3), 655–665 (2002)
35. Cao, D., et al.: Effects of docosahexaenoic acid on the survival and neurite outgrowth of rat cortical neurons in primary cultures. J. Nutr. Biochem. **16**(9), 538–546 (2005)
36. Suphioglu, C., et al.: The omega-3 fatty acid, DHA, decreases neuronal cell death in association with altered zinc transport. FEBS Lett. **584**(3), 612–618 (2010)
37. Koh, J.Y., et al.: Staurosporine-induced neuronal apoptosis. Exp. Neurol. **135**(2), 153–159 (1995)
38. Yang, X.P., et al.: Potential protection of 2, 3, 5, 4'-tetrahydroxystilbene-2-O-beta-D-glucoside against staurosporine-induced toxicity on cultured rat hippocampus neurons. Neurosci. Lett. **576**, 79–83 (2014)
39. Chen, L., et al.: Hydrogen peroxide-induced neuronal apoptosis is associated with inhibition of protein phosphatase 2A and 5, leading to activation of MAPK pathway. Int. J. Biochem. Cell Biol. **41**(6), 1284–1295 (2009)
40. Snyder, A.C., et al.: Global network influences on local functional connectivity. Nat. Neurosci. **18**(5), 736–743 (2015)
41. Spira, M.E., Hai, A.: Multi-electrode array technologies for neuroscience and cardiology. Nat. Nanotechnol. **8**(2), 83–94 (2013)
42. Hochberg, L.R., et al.: Neuronal ensemble control of prosthetic devices by a human with tetraplegia. Nature **442**(7099), 164–171 (2006)
43. Hai, A., Shappir, J., Spira, M.E.: In-cell recordings by extracellular microelectrodes. Nat. Methods **7**(3), 200–202 (2010)
44. Hales, C.M., Rolston, J.D., Potter, S.M.: How to culture, record and stimulate neuronal networks on micro-electrode arrays (MEAs). J. Vis. Exp. (39) (2010)
45. Nam, Y., Wheeler, B.C.: In vitro microelectrode array technology and neural recordings. Crit. Rev. Biomed. Eng. **39**(1), 45–61 (2011)
46. Johnstone, A.F., et al.: Microelectrode arrays: a physiologically based neurotoxicity testing platform for the 21st century. Neurotoxicology **31**(4), 331–350 (2010)
47. Makarova, J., et al.: Parallel readout of pathway-specific inputs to laminated brain structures. Fr. Syst. Neurosci. **5**, 77 (2011)
48. Biffi, E., et al.: Development and validation of a spike detection and classification algorithm aimed at implementation on hardware devices. Comput. Intell. Neurosci. 659050 (2010)
49. Liu, M.G., et al.: Use of multi-electrode array recordings in studies of network synaptic plasticity in both time and space. Neurosci. Bull. **28**(4), 409–422 (2012)
50. Thomas Jr., C.A., et al.: A miniature microelectrode array to monitor the bioelectric activity of cultured cells. Exp. Cell Res. **74**(1), 61–66 (1972)
51. Gross, G.W., et al.: A new fixed-array multi-microelectrode system designed for long-term monitoring of extracellular single unit neuronal activity in vitro. Neurosci. Lett. **6**(2–3), 101–105 (1977)
52. Gross, G.W.: Simultaneous single unit recording in vitro with a photoetched laser deinsulated gold multimicroelectrode surface. IEEE Trans. Biomed. Eng. **26**(5), 273–279 (1979)
53. Pine, J.: Recording action potentials from cultured neurons with extracellular microcircuit electrodes. J. Neurosci. Methods **2**(1), 19–31 (1980)
54. McConnell, E.R., et al.: Evaluation of multi-well microelectrode arrays for neurotoxicity screening using a chemical training set. Neurotoxicology **33**(5), 1048–1057 (2012)
55. Israel, D.A., et al.: An array of microelectrodes to stimulate and record from cardiac cells in culture. Am. J. Physiol. **247**(4 Pt 2), H669–H674 (1984)
56. Novak, J.L., Wheeler, B.C.: Recording from the Aplysia abdominal ganglion with a planar microelectrode array. IEEE Trans. Biomed. Eng. **33**(2), 196–202 (1986)

57. Novak, J.L., Wheeler, B.C.: Multisite hippocampal slice recording and stimulation using a 32 element microelectrode array. J. Neurosci. Methods **23**(2), 149–159 (1988)
58. Regehr, W.G., et al.: Sealing cultured invertebrate neurons to embedded dish electrodes facilitates long-term stimulation and recording. J. Neurosci. Methods **30**(2), 91–106 (1989)
59. Connolly, P., et al.: An extracellular microelectrode array for monitoring electrogenic cells in culture. Biosens. Bioelectron. **5**(3), 223–234 (1990)
60. Martinoia, S., et al.: A general-purpose system for long-term recording from a microelectrode array coupled to excitable cells. J. Neurosci. Methods **48**(1–2), 115–121 (1993)
61. Nisch, W., et al.: A thin film microelectrode array for monitoring extracellular neuronal activity in vitro. Biosens. Bioelectron. **9**(9–10), 737–741 (1994)
62. Lewicki, M.S.: A review of methods for spike sorting: the detection and classification of neural action potentials. Network **9**(4), R53–R78 (1998)
63. Pleasure, S.J., Lee, V.M.: NTera 2 cells: a human cell line which displays characteristics expected of a human committed neuronal progenitor cell. J. Neurosci. Res. **35**(6), 585–602 (1993)
64. Nick, C., et al.: DrCell—a software tool for the analysis of cell signals recorded with extracellular microelectrodes. Signal Process. Int. J. **7**, 96–109 (2013)
65. Kuebler, E.S., et al.: Burst predicting neurons survive an in vitro glutamate injury model of cerebral ischemia. Sci. Rep. **5**, 17718 (2015)
66. Mack, C.M., et al.: Burst and principal components analyses of MEA data for 16 chemicals describe at least three effects classes. Neurotoxicology **40**, 75–85 (2014)
67. Pimashkin, A., et al.: Spiking signatures of spontaneous activity bursts in hippocampal cultures. Fr. Comput. Neurosci. **5**, 46 (2011)
68. Wagenaar, D.A., Pine, J., Potter, S.M.: An extremely rich repertoire of bursting patterns during the development of cortical cultures. BMC Neurosci. **7**, 11 (2006)
69. Cooper, D.C.: The significance of action potential bursting in the brain reward circuit. Neurochem. Int. **41**(5), 333–340 (2002)
70. Sakoe, H., Chiba, S.: Dynamic programming algorithm optimization for spoken word recognition. Acoust. Speech Signal Process. IEEE Trans. **26**(1), 43–49 (1978)
71. Ahmed, R., et al.: Dynamic time warping based neonatal seizure detection system. In: Engineering in Medicine and Biology Society (EMBC), 2012 Annual International Conference of the IEEE. IEEE (2012)
72. Smith, V.A., et al.: Computational inference of neural information flow networks. PLoS Comput. Biol. **2**(11), e161 (2006)

Insects Neural Model: Potential Alternate to Mammals for Electrophysiological Studies

Julie Gaburro, Saeid Nahavandi and Asim Bhatti

Abstract Microelectrode arrays are a promising tool in the study of electrophysiology of the brain in vitro. Allowing non-invasive recording of electrical signals from neuronal networks, the device is a perfect tool to help scientists to understand brain functioning at the cell level and as a whole system. Despite the large increase of MEA technology in terms of device diversity and use in neurophysiology, biological models in vitro are mostly restrained to mammals. This chapter highlights the advantages to combine insects and MEA technology for future neurophysiological studies, and introduce the possible perspectives of this research.

Keywords Microelectrode arrays · Neuronal network · Primary neurons · Insects

1 Introduction

After the discovery of Cajal that the nervous system is made of neurons [1], Edgar Adrian was able to record signals from the nervous system for the first time [2]. His breakthrough on describing neuron's function was awarded in 1932 by the Nobel Price of Physiology community. For recording electrical impulses along the nerve, Adrian used electrodes, which was followed by the development into the voltage clamp technique by Kenneth Cole and George Marmont in 1949 [3]. This was the

J. Gaburro (✉) · S. Nahavandi · A. Bhatti
Institute for Intelligent Systems Research and Innovation (IISRI), Deakin University, Waurn Ponds, Australia
e-mail: jgaburro@deakin.edu.au; Julie.gaburro@csiro.au

S. Nahavandi
e-mail: saeid.nahavandi@deakin.edu.au

A. Bhatti
e-mail: asim.bhatti@deakin.edu.au

J. Gaburro
Australian Animal Health Laboratory (AAHL), CSIRO, Geelong, Australia

A. Bhatti et al. (eds.), *Emerging Trends in Neuro Engineering and Neural Computation*, Series in BioEngineering,
DOI 10.1007/978-981-10-3957-7_6

start of electrophysiology, which can be defined as the analysis of neuronal activity by recording neuron events, either spike activity or network oscillations [4]. As this branch of physiology was developing, techniques for recording have evolved and specialized into different types.

First, the patch-clamp technique is to record and measure ion currents from a single channel. This microscale technique is powerful, and however has some limitations. The method is indeed quite tedious and requires skills and specialized tools. The main disadvantage is that it can only be applied to a single neuron, so only few cells can be used during an experimental setup [5]. Despite the use of automated patch-clamp systems [6], measuring the characteristics of different neurons, simultaneously, is limited by the dependency on the equipment such as micromanipulators.

Second, indirect measurement technique appeared to record the full brain's or area's activity with functional resonance imaging (fRMI), positron emission tomography (PET) and electroencephalography (EEG). With those methods, scientists can work at the macroscale and determine functional connectivity, from different areas of the brain or network. Yet, those techniques are not precise enough and detect neuronal signals at low spatial resolution.

An intermediary solution to those techniques is to design an electrophysiology technique working at the mesoscale. This is the case of microelectrode arrays (MEAs), which consist of microelectrodes integrated on a chip to record extracellular signals from neuronal network. Electrodes can be either made of metal or open-gate field-effect transistors (OGFETs), or oxide-semiconductor FET (OSFET). The main advantages of this technique are that it is non-invasive to the culture, it can record activity from a network of cells, at millisecond time scale for days, weeks and even months [7]. Microelectrodes record extracellular field potentials (FPs) and reflect the spike activity of neurons in time and space building a network activity, illustrating the whole neurons culture. Spikes can be sampled either via spontaneous activity or electrically stimulated. Mostly used with mammalian neurons, those devices unlocked few mechanisms of neural network functioning.

Despite all those advances, mammal brains present a major challenge to science as their central nervous system (CNS) holds 100 trillions of connections between billions of neurons. The processing of information through these intercellular communication channels is extremely complicated as well as incredibly robust. Behaviours are the response to signals from multiple neuronal networks processed in the CNS, and it is then necessary to understand how this information is treated in order to treat behaviour and brain diseases. Insects are good candidates as model of study of the CNS as their brains contain less complex neuronal networks (about 250 000 neurons for a fruit fly). First, thought as brainless, insects are endowed of complex behaviours such as orientation, body communication and social behaviours [8, 9]. Besides, both nervous systems have anatomical similarities, sharing the same embryonic origins. On the cellular level, for both nervous systems, the basic unit is the neuron, which delivers signals, thanks to similar hormones and neurotransmitters. These properties make insects a good model, since results can be extrapolated to humans.

Combining modern technology such as MEAs and adequate neural models, such as insects, will improve our understanding of the organization of the brain as a whole. This chapter reviews the similarities between insects and mammalian CNS and the lack of studies in neurophysiology using MEA technique combined with insect neurons. Second, the manuscript will present the possibilities of insect models which could be used with MEAs and their potential and perspectives into different fields of research.

2 Insects Neuronal Networks: Potential Alternate for Mammals

The study of insects has been very helpful in understanding biological mechanisms in mammals. From understanding of taste [10], pain [11] and learning [12], studies have demonstrated the usefulness of insect model system to complement and apprehend research in vertebrates. Evidence have been published that insect and mammalian nervous system have basic mechanisms in common.

2.1 Insects and Mammalian CNS, not so Different

2.1.1 Insects Are Capable of Complex Behaviours

Although their smaller size, insects possess a complete and complex nervous system, which can be compared to mammals at different levels. As vertebrates, insect's nervous system is composed of CNS with a chain of ganglia and a peripheral nervous system, which includes a stomatogastric ganglion and sensory and motor nerves. At the end of the ganglia chain, the head of the insect includes a brain which can be divided into different regions like mammals. Insect brain consists of three main regions, known as protocerebrum, deutocerebrum, and tritocerebrum and can be subdivided in different sub-regions. Insect and mammal brains consist of neurons, a common functional and structural unit. Neuron cells are connected to each other to form networks, which process information into the brain. Basic behaviours such as walking and flying for voluntary actions or breathing regulation and digestion for involuntary ones are regulated by the brain. Studies of insect brain have been reviewed by Howse, which gives an overview of electrophysiological recordings and stimulation results [13]. Insect brain shows basic patterns, but are however capable of more complex and evolved behaviours.

Socialization or cooperation between pairs is considered as elaborated behaviours. Ants and bees are able of colony organization and are social species. They are capable food location using complex communication mechanisms with pheromones: once an individual finds food, it walks to and from this source and deposits on the ground a pheromone that other individuals of the same species can detect and follow in order to find the source. This way of communication is very effective and even inspired a number of methods and techniques for optimization. The mechanism is called "ant colony optimization" and is used to design algorithms for optimization of solutions to logistical problems [14]. Yet, the best evidence that insects are endowed with complex patterns is their capability for learning and memorization. Learning and memory are, respectively, defined as the acquisition and retention of neuronal representations of new information. Recent studies showed that insects learn and use their previous learning for basic life activities including feeding, predator avoidance, aggression, social interactions and sexual behaviour [15]. As for example, bees have been showed to have the ability to be trained to recognize visual stimuli such as colours, shapes and patterns, depth and motion contrast. Those learning capacities to retain local cues are essential for bees to characterize places of interest, essentially for food sources [16].

2.1.2 Insects Have Comparable Neurophysiological Mechanisms

Looking at the evolution tree, separation between insects and vertebrates, and so mammals and humans, is quite far. Nevertheless, their CNS has many basics structures and mechanisms in common.

First, the nerve cell or neuron is the basic structural unit of the nervous system for all species. This highly specialized has several dendrites and an axon along which electric impulses are distributed to other neurons. There are some clues in the literature that stem cells generating those neurons are not too different between vertebrates and invertebrates. For instance, CNS of *Drosophila* has been used as a key model to study asymmetric division of stem cells and revealed recently the link between unregulated stem cell division and production of tumour [17]. In the same study, similarities between fundamental aspects of neural stem cell biology in *Drosophila* and the mammalian cerebral cortex have been recognized [17].

Second, neurotransmitters, which are small molecules carrying information across synapses from a nerve cell to its neighbouring cells, can be similar in both groups. It has been showed that inhibitory (GABAergic) [18] and modulatory (dopaminergic) [19] transmitters, which facilitate the regulation and release of adaptive behaviours, are the same for insects and mammals. It means that network connectivity and neuronal activity in the substructures of each CNS are mediated by common neurotransmitters. Also for both cases, CNS dysfunctions result in behavioural defects including motor abnormalities, impaired memory formation, attention deficits, affective disorders and sleep disturbances.

Multitude similarities observed suggest deep homology of insects CNS and vertebrate basal ganglia underlying the selection and maintenance of behavioural actions [20]. As the mammalian cerebral cortex is the most highly evolved region of the CNS, use of an insect model would allow resolution of neuronal communication at a high level. Although the study of insects has been very helpful in understanding biological mechanisms in mammals in terms of anatomy, physiology, pharmacology and neurophysiology of vertebrates (including mammals) and invertebrates (including insects), it has been showed that there is a gap in the application of the MEA technique with insects [21].

2.2 *Insects Neurons on MEAs: A Missing Combination in Electrophysiology*

Since their first use in the second half of the twentieth century [8, 9], MEAs have become a promising experimental platform for electrophysiological studies of neural networks. Figure 1 illustrates a general MEA system which can be applied to different types of MEAs. There is now a great diversity of MEAs used in neurophysiology, with a large field of applications (in vivo versus in vitro, or implantable array). Recent publications have described and reviewed the recent use of those MEAs in details [22–25].

Till now, numerous studies using MEAs to record primary neurons have been reported. However, most of the preparations have been made with rodent primary neurons, closest model to human. The dissection techniques and cells seeding on the microchips are well established [26] and provide good spikes and burst data analysis. Those models are now commonly used for drug testing and neurotoxicology and open a new field of research for neurodegenerative disease studies (Alzheimer [27]). Rodent primary neurons on MEAs have been tested for toxicity of a wide panel of substances, such as ethanol, metals, neuro-excitatory or inhibitory chemicals, and insecticides. The neuro-effects of those substances are detailed in Johnstone and all in their review in 2010 [28].

Mammal or vertebrate models require strict ethical authorization and are expensive and/or limited. However, most of MEA studies use rodent as source for primary neurons. For birds, only two recent studies have been found using chicken primary neurons [29]. There are disparities of type of models used on MEAs, with a clear preference for rodent models. Figure 1 illustrates the disparities over the last 3 years of studies in this domain including all types of planar MEA and in vitro studies (primary neurons, retina and brain slices). This imbalance can be explained by the availability of mouses and rats in laboratories, which are commonly used for experimental model rather than other mammal or vertebrate. Despite of their

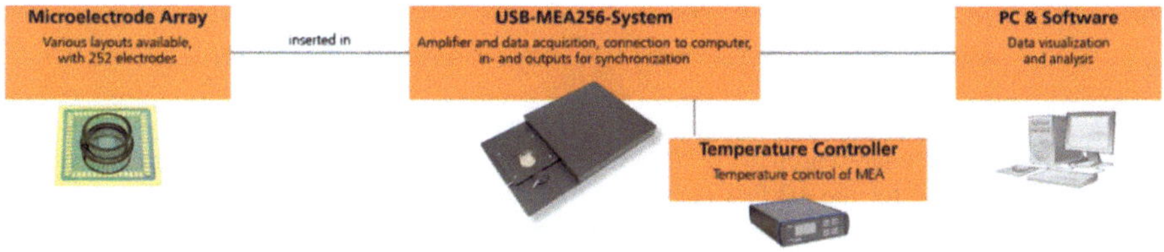

Fig. 1 MEA recording system overview

comparative simplicity, a gap is still to fill with invertebrates. Till today, the only invertebrate primary neurons cultured on MEA are from molluscs *Helix* [30–32] and one study reported with leech neurons [33]. No study using MEAs as tool for electrical recording of neurons has been done using insect's neurons so far, even for *Drosophila* a well-established model in insect neurophysiology [34].

In the next section of this chapter, we present great potential of the use of insect models with MEAs for future neurophysiological studies, still under explored today (Fig. 2).

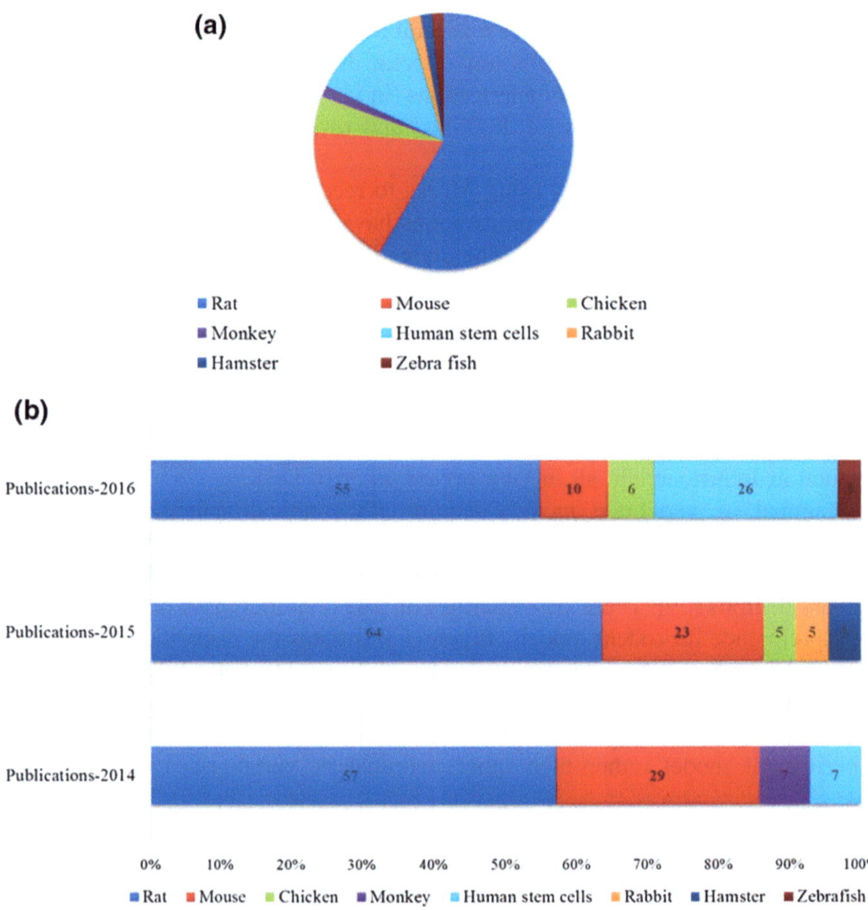

Fig. 2 The use of different models on MEAs in research studies from 2014 to 2016. Keywords used for Google Scholar search were "Microelectrode array" + neuron and in vitro biological studies were selected for the 10 first pages. **a** Proportions in percentages of the different in vitro models used with MEA technology the last three years. **b** Bar plots representing, in percentages, different models used with MEA technology per year. For 2014: N = 14, for 2015: N = 22, and for 2016: N = 31

3 Potential and Feasibility of the Insects Cultures on MEAs

As before mentioned, in entomology, insect's primary neurons have been used for neurophysiological studies, enhancing our knowledge on neurophysiological mechanisms.

Insect primary cell culture is a useful tool for studying the shape, the function and chemistry of neurons in a simplified and controlled environment [35]. It has repercussions in different areas of research such as physiology, pharmacology and cell interactions. The study of anatomy of neurons in insects is not new and spans over a century [36]. Nowadays, precise staining, powerful microscopes and 3D image analysis allow us to go further in the anatomical understanding of the brain morphology and structure. Ignell and his colleagues described in 2005, the architecture of the mosquito deutocerebrum with details of number of glomera and neurons structure in *Anopheles gambiae* and *Aedes aegypti* mosquitoes, and proved structural differences between male and female individuals [37]. Other tools have been developed to record the action potential in single-taste receptor neurons, allowing the detection of hydro-soluble compounds of single-taste receptors in insects [38].

Neuronal cultures have been developed in many different types of insects, at all development stages, reviewed by beadle in 2006 [39]. A primary culture refers to the stage of the culture after the cells are isolated from the tissue and proliferated under the appropriate conditions until they reach a stable stage and can be transferred to a new vessel with fresh growth medium to provide more room for continued growth (Life technology, 'Introduction to Cell Culture'). The first ever done with insect was designed in 1970 by Chen and Levi-Montalcini from the brains of embryonic cockroaches [40]. From there, other trials have been done with adult and embryonic stages with crickets, flies, moths and bees, all described in detailed in Beadle review. First, the studies were mostly targeted on the morphology to distinguish the different types of neurons in the insect brain [41]. Then improvement of materials and techniques allowed scientists to study neuron action potential parameters and in more detail ionic currents in insects' cultured neurons. For example, characterization of potassium channels has first been done with brain cricket primary cultures [42] and today improved cultures even provide information of potential genes involved in action potential generation in *Drosophila* [43]. Identification of GABA, glutamate, and acetylcholine receptors has been achieved for the first time on cockroach-cultured neurons respectively in 1987 [44], followed by other insect's models, such as honeybees, *Drosophila*, moths, crickets and other flies.

In *Drosophila*, the use of primary cell cultures originating from embryonic and from larval tissue has been well established and cell cultures are known since half a century [45]. Continuous neuron cell line can be established from *Drosophila* larval CNS cells and maintain their proliferation outside their endogenous environment

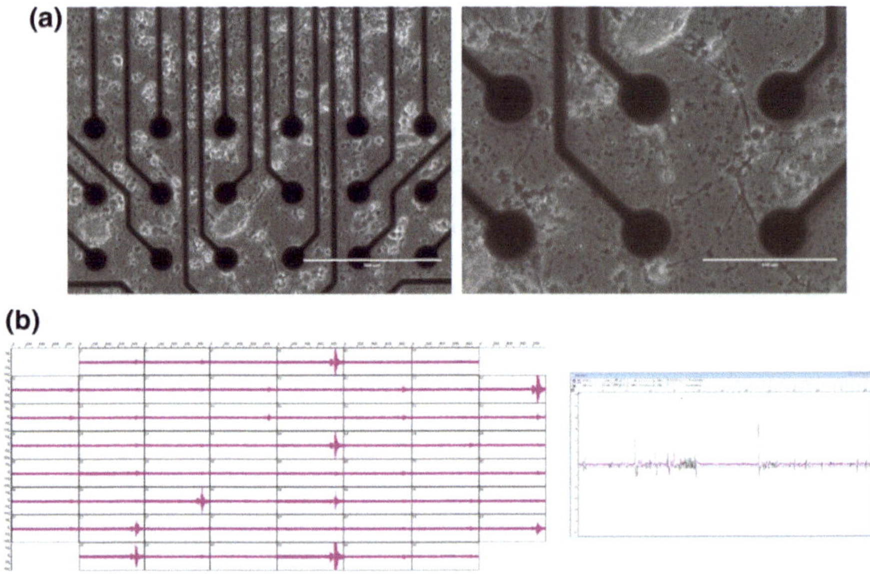

Fig. 3 Aedes aegypti primary neurons growing on MEAs (**a**). MC_Rack interface while recording spontaneous activity, all channels (*left*) and burst zooming form one channel (*right*)

[46]. Also established from *Drosophila* larval CNS, Ui and collaborators have described different stable neuron lines [47].

Figure 3 shows primary neurons from *Aedes aegypti* cultured on MEA. Trial experiments (data not published) have showed that after 7 days in Vitro, dissected and dissociated mosquito neurons are spontaneously firing on MEAs. This model is then compatible with neural network activity recording, allowing neuronal network studies of chemicals impacts on insects (pesticides), and also virus interactions on insect's nervous system.

To summarize, insect primary neurons are a well-established system used in neurophysiology. An association of those neuronal networks on MEAs would take the research of insect neurophysiology a step further.

4 Perspective and Applications

There are many reasons to believe that culturing insect cells on MEAs can open many doors to neurophysiological research. To begin with, medical research has used for more than a century, *Drosophila melanogaster* populations, to study diverse biological functions such as vision, reproduction, feeding behaviour, etc. With the emergence of neurodegenerative diseases, due to the ageing population, mutant fly models have been designed, allowing reproducible testing to be conducted more diversely and in shorter times than in rodent models. *Drosophila* is also known to be a powerful tool, thanks to its unique and successful genetic tool [48]. The highly diverse and accessible *Drosophila* mutant resources have facilitated our understanding of genetics underlying neurophysiology processes. The other advantages to use insects as model are their low cost, high-speed processes (days instead of weeks) and easy handling conditions. This is why MEA experiments could add another prospect for the application of this model to improve our global understanding of neuron communication and its implications in human and animal nervous diseases.

Neurodegenerative diseases have increased rapidly in developed countries since the end of the twentieth century, and yet no cure has been validated for Alzheimer's and Parkinson's diseases. Experiments using MEAs with mouse neurons culture [27] revealed the early steps in generation of Alzheimer's disease lesions in the hippocampus. The effectiveness of MEA technology has been demonstrated when electrophysiological properties of the sub-thalamic nucleus, an area involved in Parkinson's disease, have been identified [49]. As mentioned before, flies have been recently used as model for neurodegenerative studies [50, 51]. The use of insect brain cell cultures on MEAs could take neurodegenerative diseases research to another level in improving our understanding and the development of a cure for these neurodegenerative diseases.

Another major public health burden concerns vector-borne diseases, or diseases caused by viruses carried by a vector such as insects. Those diseases are a major issue in developing countries and represent a constant threat of emerging diseases worldwide. The best example is the case of Zika virus which has been linked in 2015 with Guillain-Barré syndrome, and 2015 microcephaly in new born babies (World Health Organization). Insects, such as mosquitoes but also flies, can be vectors of numerous pathogens, which can be responsible for diseases in humans and other mammals. Pathogens are transmitted mainly by bite allowing them to pass from animal to animal through a totally biological process. In contrast to vertebrate hosts, insect's infection by arboviruses does not show sign of infection and no or little pathology. Till now, arbovirus mechanisms and interactions with its vectors are totally unknown. Discoveries on molecular mechanisms of host interactions with pathogens could reveal a powerful tool to limit or stop arbovirus replication in insects. Combined with insect cells, MEA technology represents a good potential to understand these processes.

Finally, neurotoxicity and pharmacology research could have some advantages to use insect neurons on MEAs. Besides, enabling recordings in real time of neuronal networks culture, and so provides robust measures of network activity, MEA can have multi-wells, allowing then chemical testing on neurons at higher scale [52]. Regular or multi-well MEA systems can record physical, chemical and pharmacological perturbations after introduction of different chemicals, which are reflected in the tissue responses under different conditions. Cell or tissue responses provide important data of their impact on nervous system and those substances effects can be classified a classification [28, 53–55]. However, all studies to date have been conducted in vertebrate cell cultures, highlighting the gap for use of MEAs in insects despite their medical and ecological importance. There is a huge potential to use insect neurons on MEA systems for pesticides testing, especially with the emerging problem of insecticide resistance. This research could benefit both agriculture and medical prevention against bites.

5 Conclusion

This chapter emphasizes the importance of the use of insect neuronal networks with MEA technique. Literature has showed that insects and mammals share common and elementary molecular mechanisms. This is why studying CNS insects could be an advantage to understand the main underlying nervous processes of neural communication. Those results could be then extrapolated to mammals and even humans with supplementary tests including vertebrate models. The outputs of this research can be advantageous in diverse medical research domains, but also useful for neurotoxicity testing.

References

1. López-Muñoz, F., Boya, J., Alamo, C.: Neuron theory, the cornerstone of neuroscience, on the centenary of the Nobel Prize award to Santiago Ramón y Cajal. Brain Res. Bull. **70**(4), 391–405 (2006)
2. Adrian, E.D.: The basis of sensation (1928)
3. Hodgkin, A.L., Huxley, A.F., Katz, B.: Measurement of current-voltage relations in the membrane of the giant axon of Loligo. J. Physiol. **116**(4), 424–448 (1952)
4. Contreras, D.: Electrophysiological classes of neocortical neurons. Neural Netw. **17**(5), 633–646 (2004)
5. Wood, C., Williams, C., Waldron, G.J.: Patch clamping by numbers. Drug Discovery Today **9** (10), 434–441 (2004)
6. Kodandaramaiah, S.B., Franzesi, G.T., Chow, B.Y., Boyden, E.S., Forest, C.R.: Automated whole-cell patch-clamp electrophysiology of neurons in vivo. Nat. Methods **9**(6), 585–587 (2012)
7. Gross, G.W., Rhoades, B.K., Azzazy, H.M., Wu, M.-C.: The use of neuronal networks on multielectrode arrays as biosensors. Biosens. Bioelectron. **10**(6), 553–567 (1995)
8. Wilson, E.O.: The Insect Societies. The Insect Societies (1971)
9. von Frisch, K.: The Dance Language and Orientation of Bees. Belknap, Cambridge, Massachusetts (1967)
10. Ishimoto, H., Tanimura, T.: Molecular neurophysiology of taste in Drosophila. Cell. Mol. Life Sci. CMLS **61**(1), 10–18 (2004)
11. Manev, H., Dimitrijevic, N.: Fruit flies for anti-pain drug discovery. Life Sci. **76**(21), 2403–2407 (2005)
12. Brembs, B.: Operant conditioning in invertebrates. Curr. Opin. Neurobiol. **13**(6), 710–717 (2003)
13. Howse, P.: Brain structure and behavior in insects. Annu. Rev. Entomol. **20**(1), 359–379 (1975)
14. Dorigo, M., Stützle, T.: Ant colony optimization: overview and recent advances. Techreport, IRIDIA, Universite Libre de Bruxelles (2009)

15. Dukas, R.: Evolutionary biology of insect learning (2007)
16. Giurfa, M.: Cognitive neuroethology: dissecting non-elemental learning in a honeybee brain. Curr. Opin. Neurobiol. **13**(6), 726–735 (2003)
17. Brand, A.H., Livesey, F.J.: Neural stem cell biology in vertebrates and invertebrates: more alike than different? Neuron **70**(4), 719–729 (2011)
18. Sattelle, D.B.: GABA receptors of insects. Adv. Insect Physiol. **22**, 1–113 (1990)
19. Cayre, N.M., Buckingham, S., Yagodin, S., Sattelle, D: Cultured insect mushroom body neurons express functional receptors for acetylcholine, GABA, glutamate, octopamine, and dopamine. J. Neurophysiol. **81**(1), 1–14 (1999)
20. Strausfeld, N.J., Hirth, F.: Deep homology of arthropod central complex and vertebrate basal ganglia. Science **340**(6129), 157–161 (2013)
21. Gaburro, J., Duchemin, J.-B., Bhatti, A., Walker, P., Nahavandi, S.: Neurophysiology of insects using microelectrode arrays: current trends and future prospects. In: International Conference on Neural Information Processing, pp. 493–500. Springer (2014)
22. Obien, M.E.J., Deligkaris, K., Bullmann, T., Bakkum, D.J., Frey, U.: Revealing neuronal function through microelectrode array recordings. Front. Neurosci. **8**, 423 (2015)
23. Patil, A.C., Thakor, N.V.: Implantable neurotechnologies: a review of micro-and nanoelectrodes for neural recording. Med. Biol. Eng. Comput. **54**(1), 23–44 (2016)
24. Kang, H., Nam, Y.: In vitro neural recording by microelectrode arrays. In: Stretchable Bioelectronics for Medical Devices and Systems, pp. 275–291. Springer (2016)
25. Spira, M.E., Hai, A.: Multi-electrode array technologies for neuroscience and cardiology. Nat. Nanotechnol. **8**(2), 83–94 (2013)
26. Novellino, A., Scelfo, B., Palosaari, T., Price, A., Sobanski, T., Shafer, T.J., Johnstone, A.F., Gross, G.W., Gramowski, A., Schroeder, O.: Development of micro-electrode array based tests for neurotoxicity: assessment of interlaboratory reproducibility with neuroactive chemicals. Front. Neuroeng. **4**, 4 (2011)
27. Chong, S.-A., Benilova, I., Shaban, H., De Strooper, B., Devijver, H., Moechars, D., Eberle, W., Bartic, C., Van Leuven, F., Callewaert, G.: Synaptic dysfunction in hippocampus of transgenic mouse models of Alzheimer's disease: A multi-electrode array study. Neurobiol. Dis. **44**(3), 284–291 (2011)
28. Johnstone, A.F., Gross, G.W., Weiss, D.G., Schroeder, O.H.-U., Gramowski, A., Shafer, T.J.: Microelectrode arrays: a physiologically based neurotoxicity testing platform for the 21st century. Neurotoxicology **31**(4):331–350 (2010)
29. Kuang, S.Y., Wang, Z., Huang, T., Wei, L., Xi, T., Kindy, M., Gao, B.Z.: Prolonging life in chick forebrain-neuron culture and acquiring spontaneous spiking activity on a microelectrode array. Biotechnol. Lett. **37**(3), 499–509 (2015)
30. Claverol-Tinture, E., Ghirardi, M., Fiumara, F., Rosell, X., Cabestany, J.: Multielectrode arrays with elastomeric microstructured overlays for extracellular recordings from patterned neurons. J. Neural Eng. **2**(2), L1 (2005)
31. Massobrio, P., Giachello, C.N., Ghirardi, M., Martinoia, S.: Selective modulation of chemical and electrical synapses of Helix neuronal networks during in vitro development. BMC Neurosci. **14**(1), 1 (2013)
32. Massobrio, P., Tedesco, M., Giachello, C., Ghirardi, M., Fiumara, F., Martinoia, S.: Helix neuronal ensembles with controlled cell type composition and placement develop functional polysynaptic circuits on micro-electrode arrays. Neurosci. Lett. **467**(2), 121–126 (2009)
33. Wilson, R., Breckenridge, L., Blackshaw, S., Connolly, P., Dow, J., Curtis, A., Wilkinson, C.: Simultaneous multisite recordings and stimulation of single isolated leech neurons using planar extracellular electrode arrays. J. Neurosci. Methods **53**(1), 101–110 (1994)
34. Bellen, H.J., Tong, C., Tsuda, H.: 100 years of Drosophila research and its impact on vertebrate neuroscience: a history lesson for the future. Nat. Rev. Neurosci. **11**(7), 514–522 (2010)
35. Kreißl, S., Bicker, G.: Dissociated neurons of the pupal honeybee brain in cell culture. J. Neurocytol. **21**(8), 545–556 (1992)

36. Kenyon, F.: The brain of the bee. A preliminary contribution to the morphology of the nervous system of the Arthropoda. J. Comp. Neurol. **6**(3), 133–210 (1896)
37. Ignell, R., Dekker, T., Ghaninia, M., Hansson, B.S.: Neuronal architecture of the mosquito deutocerebrum. J. Comp. Neurol. **493**(2), 207–240 (2005)
38. Avenet, P., Lindemann, B.: Noninvasive recording of receptor cell action potentials and sustained currents from single taste buds maintained in the tongue: the response to mucosal NaCl and amiloride. J. Membr. Biol. **124**(1), 33–41 (1991)
39. Beadle, D.: Insect neuronal cultures: an experimental vehicle for studies of physiology, pharmacology and cell interactions. Invertebr. Neurosci. **6**(3), 95–103 (2006)
40. Chen, J., Levi-Montalcini, R.: Long term cultures of dissociated nerve cells from the embryonic nervous system of the cockroach Periplaneta americana. Archives italiennes de biologie **108**(4), 503 (1970)
41. Devaud, J.-M., Quenet, B., Gascuel, J., Masson, C.: A morphometric classification of pupal honeybee antennal lobe neurones in culture. Neuroreport **6**(1), 214–218 (1994)
42. Cayre, M., Buckingham, S., Strambi, A., Strambi, C., Sattelle, D.: Adult insect mushroom body neurons in primary culture: cell morphology and characterization of potassium channels. Cell Tissue Res. **291**(3), 537–547 (1998)
43. Küppers-Munther, B., Letzkus, J.J., Lüer, K., Technau, G., Schmidt, H., Prokop, A.: A new culturing strategy optimises Drosophila primary cell cultures for structural and functional analyses. Dev. Biol. **269**(2), 459–478 (2004)
44. Shimahara, T., Pichon, Y., Lees, G., Beadle, C., Beadle, D.: Gamma-aminobutyric acid receptors on cultured cockroach brain neurones. J. Exp. Biol. **131**(1), 231–244 (1987)
45. Egger, B., van Giesen, L., Moraru, M., Sprecher, S.G.: In vitro imaging of primary neural cell culture from Drosophila. Nat. Protoc. **8**(5), 958–965 (2013)
46. Moraru, M.M., Egger, B., Bao, D.B., Sprecher, S.G.: Analysis of cell identity, morphology, apoptosis and mitotic activity in a primary neural cell culture system in Drosophila. Neural Dev. **7**(1), 1 (2012)
47. Ui, K., Nishihara, S., Sakuma, M., Togashi, S., Ueda, R., Miyata, Y., Miyake, T.: Newly established cell lines from Drosophila larval CNS express neural specific characteristics. Vitro Cell. Dev. Biol. Anim. **30**(4), 209–216 (1994)
48. Venken, K.J., Simpson, J.H., Bellen, H.J.: Genetic manipulation of genes and cells in the nervous system of the fruit fly. Neuron **72**(2), 202–230 (2011)
49. Chu, J.-U., Jeong, M.J., Song, K.-I., Lee, H.-C., Kim, J., Kim, Y.-J., Choi, K., Suh, J.-K.F., Youn, I.: Spontaneous synchronized burst firing of subthalamic nucleus neurons in rat brain slices measured on multi-electrode arrays. Neurosci. Res. **72**(4):324–340 (2012)
50. Feany, M.B., Bender, W.W.: A Drosophila model of Parkinson's disease. Nature **404**(6776), 394–398 (2000)
51. Bilen, J., Bonini, N.M.: Drosophila as a model for human neurodegenerative disease. Annu. Rev. Genet. **39**, 153–171 (2005)
52. Strickland, J.D., Lefew, W.R., Crooks, J., Hall, D., Ortenzio, J.N., Dreher, K., Shafer, T.J.: In vitro screening of metal oxide nanoparticles for effects on neural function using cortical networks on microelectrode arrays. Nanotoxicology **10**(5), 619–628 (2016)
53. Mack, C.M., Lin, B.J., Turner, J.D., Johnstone, A.F., Burgoon, L.D., Shafer, T.J.: Burst and principal components analyses of MEA data for 16 chemicals describe at least three effects classes. Neurotoxicology **40**, 75–85 (2014)
54. McConnell, E.R., McClain, M.A., Ross, J., LeFew, W.R., Shafer, T.J.: Evaluation of multi-well microelectrode arrays for neurotoxicity screening using a chemical training set. Neurotoxicology **33**(5), 1048–1057 (2012)
55. Natarajan, A., Molnar, P., Sieverdes, K., Jamshidi, A., Hickman, J.: Microelectrode array recordings of cardiac action potentials as a high throughput method to evaluate pesticide toxicity. Toxicol. In Vitro **20**(3), 375–381 (2006)

Synchronization Criteria for Delay Coupled Izhikevich Neurons

Imali T. Hettiarachchi, Lakshmanan Shanmugam, Asim Bhatti and Saeid Nahavandi

Abstract In this chapter, we investigate the chaotic synchronization of two coupled Izhikevich neurons via a gap junction. In the absence of a controller, the coupled neurons will achieve complete chaotic synchronization only when the degree of connectivity or the coupling strength exceeds a critical value. This transition to synchronization with varying connectivity strengths is analysed with conditional Lyapunov exponents. Synchronization of gap junction separated, coupled Izhikevich neurons using control laws has remained non-investigated to this date. As such, in this chapter we propose a nonlinear adaptive controller, in order to obtain complete chaotic synchronization for any value of coupling strength and delay, based on the Lyapunov stability theory. Effectiveness of the proposed nonlinear controller for synchronizing delayed-coupled Izhikevich neurons are shown through numerical simulations.

Keywords Izhikevich model · Gap junction · Delay coupled · Synchronization · Nonlinear control

1 Introduction

Chaotic synchronization is one of the most interesting dynamical behaviours that can arise in a neuronal network. Important brain functions such as working memory, selective attention, sensory perception, and multisensory integration which requires

I.T. Hettiarachchi (✉) · L. Shanmugam · A. Bhatti · S. Nahavandi
Institute for Intelligent Systems Research and Innovation, Deakin University, Geelong, Australia
e-mail: imali.hettiarachchi@deakin.edu.au

L. Shanmugam
e-mail: lakshm85@gmail.com

A. Bhatti
e-mail: asim.bhatti@deakin.edu.au

S. Nahavandi
e-mail: saeid.nahavandi@deakin.edu.au

A. Bhatti et al. (eds.), *Emerging Trends in Neuro Engineering and Neural Computation*, Series in BioEngineering,
DOI 10.1007/978-981-10-3957-7_7

efficient sensory and cognitive processing are the result of synchronized firing activity of coupled neuronal networks. The information processing of the brain is a result of intensity keeping of the neuronal response during the propagation over the network, which occurs during neuronal synchronous states [1]. Further, disturbances in the synchronized network activity causing imbalance of the neurons are the cause of clinical disorders such as Parkinson's disease, schizophrenia and epilepsy [2].

Therefore, studying the synchronization and de-synchronization of neuronal spike-burst behaviours can give us insight to the information processing in the brain and even the origins of certain mental disorders. In this regard, biophysical models of single neurons have been proposed in the literature, which are useful in understanding various neuronal behaviour such as spiking, bursting, and chaos [3–7]. These models can reproduce various spiking activity observed in neural systems by tuning the parameters, and also can be used to investigate the phenomenon of synchronization of neuronal networks [8].

The Hodgkin–Huxley (HH) neuron model [3] is known to be the most comprehensive, yet the most complex spiking neuron model for simulating neurodynamics. Due to the complexity of involving biophysical parameters such as capacitance of membranes and resistance of ion channels, many alternative biologically plausible, yet computationally efficient neuron models have been proposed by researchers. Some widely used computationally efficient biophysical neuronal models are, Hindmarsh-Rose (HR) neuron model [6], the FitzHugh-Nagumo neuron model [4], the Morris–Lecar neuron model [5], and the Izhikevich neuron model [7–9].

Among such simplified models, the Izhikevich neuron model [8] is capable of reproducing almost all spiking activity observed in neural systems by tuning the parameters, with the variety of the spiking properties greater than those obtained through other models [7]. Among a wide variety of spiking and bursting neuron models proposed in the literature, we focus on the Izhikevich model of a single neuron [8].

Several studies have considered the dynamics of the Izhikevich model under different parameter settings [7, 9]. The chaotic characteristics of the Izhikevich model are examined in a few studies, where the authors examine and classify the chaotic characteristics of the single Izhikevich neuron model [10–13] and an assembly of Izhikevich neurons [11] using bifurcation diagrams, Lyapunov exponents with a saltation matrix and Poincaré section methods. Dynamics of coupled networks of Izhikevich neurons are studied in [8]. Here, a pulse coupled neural network (PCNN) model is proposed, which is capable of reproducing temporal neural activity such as spindle waves, sleep oscillations, and sustained spike synchrony.

A few studies have considered the synchronization of coupling-induced Izhikevich neurons [11, 14, 15]. In [14] two general network models of electrically coupled bursting neurons via static homotypic gap junctions (connexin-43) are studied. The degree of synchronization is discussed with respect to the strength of the gap junction connection and the population size of the network. The network models included a fully connected network (FCN) and a sparse random network (SRN), and the mean conductance of the FCN and density of connections in a SRN, characterised the strength of the gap junctions. The study shows that increasing mean conductance in a FCN and density of connections in a SRN result in increased synchronization, while the increase in population size decreases the degree of synchronization in the network. The results of the study concluded that, a large neural network could transition to a high degree of synchronization when exceeding critical parameter values.

In [15] the authors studied coupling-induced population synchronization of a population of neurons connected via excitatory AMPA synapses. This study reported, that for small coupling strengths, the population state is incoherent due to individual neurons fire spikings independently. However, when exceeding a lower threshold of the coupling strength, the coupling stimulates coherence between noise-induced spiking resulting in spike synchronization.

The studies on population synchronization of Izhikevich neurons reported in [14, 15] consider the case of bursting and regular spiking. Apart from the work presented in [11], no studies consider the case of chaotic synchronization of an assembly of Izhikevich neurons. This study also considers coupling strength dependence of signal response in the neuron assemblies consisting of strong and weak chaotic neurons.

Studies involving other neuronal models such as the HR neuron also confirms this transition to synchronization between neurons [16, 17], only occurs after a critical value of the coupling strength. In an electrically coupled HR network, sequential transitions to synchronized states are observed with the increase of coupling strength. Increasing the coupling strength from zero, first an increased incoherence and then two different transitions to synchronized states; one associated with bursts and the other with spikes occurs [16, 17]. A general conclusion of the aforementioned studies are that the coupling strength of coupled neurons with electrical or chemical excitatory synapses can enhance synchronization of the Izhikevich neuronal populations [11, 14, 15]. In this chapter, we consider chaotic synchronization of two coupled Izhikevich neurons via a gap junction and provide a control theory perspective to guarantee synchronization at any value of the connectivity strength or transmission delay.

Control law derivation of coupled neuronal models have gained vast interest among researchers during the recent years. From the control theory perspective, the goal is to achieve synchronization among coupled neurons irrespective of their coupling strength or coupling delay [17]. Many different types of controllers including nonlinear control [18], adaptive control [19], sliding mode control [20], and feedback control [21] has been proposed to synchronize chaotic neurons. Control law derivation for synchronization of gap junction separated coupled Izhikevich neurons has remained a topic yet to be discussed. In order to fill this research gap, we propose a nonlinear synchronization control strategy in this chapter.

The study is twofold. First, we derive the conditions of synchronization based on the coupling strength in the absence of control and second using a nonlinear adaptive control to guarantee chaotic synchronization. The rest of the chapter are organized as follows. Section 2 presents the Izhikevich model and its dynamic behaviour while Sect. 3 presents the coupled model and the conditions for complete transition to synchronization of the coupled neurons. An adaptive nonlinear feedback control law is derived in Sect. 4 followed by numerical simulations in Sect. 5. Section 6 concludes the chapter.

2 Methods and Material

2.1 Izhikevich Model of a Single Neuron

The single neuron model proposed by Izhikevich [7, 8] combines continuous spike-generation mechanisms and a discontinuous resetting process following the spikes. Mathematically, this is implemented as a two dimensional ordinary differential equation system given as,

$$\begin{aligned} \dot{v} &= 0.04v^2 + 5v + 140 - u + I_{ext} \\ \dot{u} &= a(bv - u) \end{aligned} \tag{1}$$

with the auxiliary after-spike resetting,

$$\text{if} \quad v \geq +30\,\text{mv} \quad \text{then} \quad \begin{cases} v \leftarrow c \\ u \leftarrow u + d \end{cases} \tag{2}$$

where v is the membrane potential of the neuron and u represents the membrane recovery variable, which accounts for the activation of K^+ ionic currents and inactivation of Na^+ ionic currents. The membrane recovery variable provides negative feedback on v. After the spike reaches its apex (+30 mv), the membrane voltage and the recovery variable are reset according to (2). I_{ext} is a direct current input. The time scale and the sensitivity of u is adjusted by the parameters a and b respectively. The Izhikevich model can emulate a wide variety of neuronal behaviour such as regular spiking, intrinsic bursting, chattering, and fast spiking [7, 8]. Figure 1 shows an example of regular spiking (RS), chattering (CH), and chaotic behaviour (CHAOS) of the Izhikevich neuron. The parameter values corresponding to each behaviour are summarized in Table 1.

In the present work, we concentrate on the chaotic behaviour of the system and chaotic synchronization of coupled Izhikevich neurons. A detailed analysis of the dynamic behaviour of the single neuron Izhikevich model can be found in [13], with reference to Lyapunov exponents and bifurcation diagrams. For the rest of the paper, we use the parameter values $a = 0.2, b = 2, c = -56, d = -16$ and $I = -99$ as in [7, 13] to simulate chaotic firing from the Izhikevich neurons.

3 Coupled Izhikevich Neuronal Model

The neurons are coupled through their synapse and this synaptic communication is twofold: electrical or chemical. Electrical synapse also referred to as gap junctions are faster synapse compared to the chemical synapses. In this paper, we consider neurons coupled through an electrical synapse and incorporate the time delay in

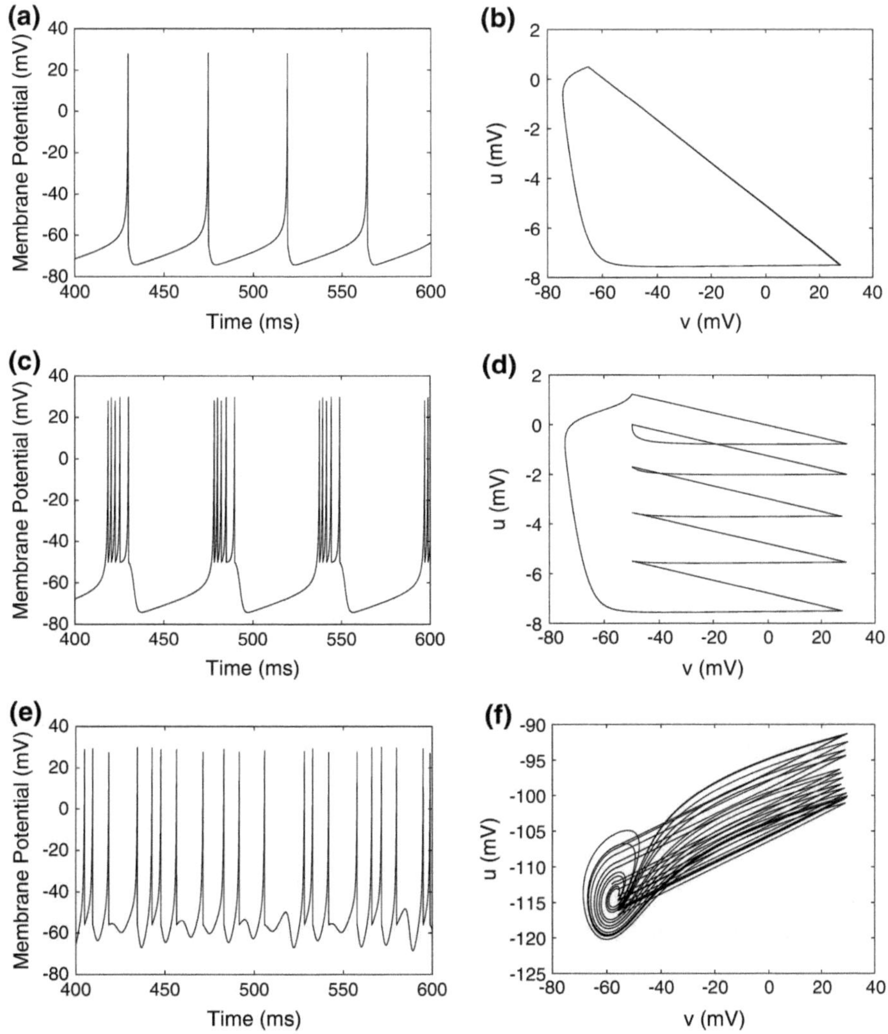

Fig. 1 Izhikevich model behaviour for different parameter combinations **a**, **b** regular spiking **c**, **d** chattering and **e**, **f** chaotic behaviour. The corresponding parameter values for each firing pattern are given in Table 1

Table 1 Parameter values for different firing patterns of the Izhikevich model

Behaviour	a	b	c	d	I
RS	0.02	0.2	−65	8	10
CH	0.02	0.2	−50	2	10
CHAOS	0.2	2	−56	−16	−99

coupling. Two identical neuron coupled Izhikevich model through gap junctions can be modelled using the Izhikevich single neuron model in (1) as,

Neuron 1,

$$\begin{aligned}\dot{v}_1 &= 0.04v_1^2 + 5v_1 + 140 - u_1 + I_{ext} - g(v_1 - v_2(t-\tau)) \\ \dot{u}_1 &= a(bv_1 - u_1)\end{aligned} \tag{3}$$

Neuron 2,

$$\begin{aligned}\dot{v}_2 &= 0.04v_2^2 + 5v_2 + 140 - u_2 + I_{ext} - g(v_2 - v_1(t-\tau)) \\ \dot{u}_2 &= a(bv_2 - u_2)\end{aligned} \tag{4}$$

with the auxiliary after-spike resetting,

$$\begin{aligned}&\text{if} \quad v_1 \geq +30\,\text{mv} \quad \text{then} \quad \begin{cases} v_1 \leftarrow c \\ u_1 \leftarrow u_1 + d \end{cases} \\ &\text{if} \quad v_2 \geq +30\,\text{mv} \quad \text{then} \quad \begin{cases} v_2 \leftarrow c \\ u_2 \leftarrow u_2 + d \end{cases}\end{aligned} \tag{5}$$

where, v_1, v_2 are the membrane voltage of the neurons and u_1, u_2 are the membrane recovery variables, g is the coupling strength and τ is the time delay during synaptic information flow. The dynamics of the membrane potentials of the coupled neurons and the effect of connectivity strengths and delay on the same will be discussed further in the Sect. 5.

3.1 *Synchrony in Coupled Izhikevich Neurons*

In this section, we investigate the chaotic synchronization of two coupled Izhikevich neurons, with the goal of exploring the effect of the coupling strength on the process of transition to synchronization. Here we assume the coupling time delay $\tau = 0$, and discuss the effect of coupling strength of synchronizing two coupled identical Izhikevich neurons.

In order to analyse the synchronization of two coupled chaotic Izhikevich neurons, we use the conditional Lyapunov exponents (CLEs), a method for chaos synchronization introduced by Pecora and Caroll [22]. To follow this method, first we define the error system of the coupled neurons as, $e_1 = v_2 - v_1$ and $e_2 = u_2 - u_1$. The linearized system that satisfies e_1, e_2 can be given as,

$$\begin{aligned} e_1 &= 0.08e_1 v_{sync} + 5e_1 - e_2 - 2ge_1 \\ e_2 &= a(be_1 - e_2) \end{aligned} \tag{6}$$

with,

$$\begin{aligned} \dot{v}_{sync} &= 0.04v_{sync}^2 + 5v_{sync} + 140 - u_{sync} + I_{ext} \\ \dot{u}_{sync} &= a(bv_{sync} - u_{sync}) \end{aligned} \tag{7}$$

with the auxiliary after-spike resetting,

$$\text{if} \quad v_{sync} \geq +30\,\text{mv} \quad \text{then} \quad \begin{cases} v_{sync} \leftarrow c \\ u_{sync} \leftarrow u_{sync} + d \end{cases} \tag{8}$$

where v_{sync} and u_{sync} are the synchronized states of the Izhikevich neuron model. At complete synchronization, i.e. at time limit $t \to \infty$, $|v_2 - v_1| = |e_1| \to 0$ and $|u_2 - u_1| = |e_2| \to 0$, reaching the synchronized state $v_2 = v_1 = v_{sync}$ and $u_2 = u_1 = u_{sync}$. Then the system in (7) gives the synchronization manifold. Lyapunov exponents of the linearized system (6), namely the CLEs determine the stability of the synchronized manifold. The stability of a synchronization manifold is attained when all the CLEs are negative [22], which implies complete synchronization of the coupled neurons. Figure 2 shows the variation of the two CLEs with varying coupling strength g between the coupled Izhikevich neurons.

As the coupling strength is increased from zero, the initially positive exponent starts to decrease. It implies that a low coupling strength makes the system incoherent. It can be observed that with increasing coupling strengths both of these exponents decrease and become negative by crossing zeros at two different cou-

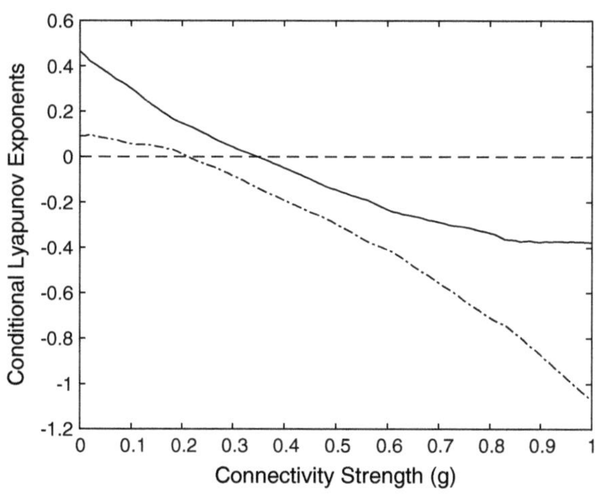

Fig. 2 Variation of the conditional Lyapunov exponents with the varying connectivity strength for two coupled chaotic Izhikevich neurons. First and the second Lyapunov exponents transition from positive to negative at $g \approx 0.212$ and $g \approx 0.348$ respectively

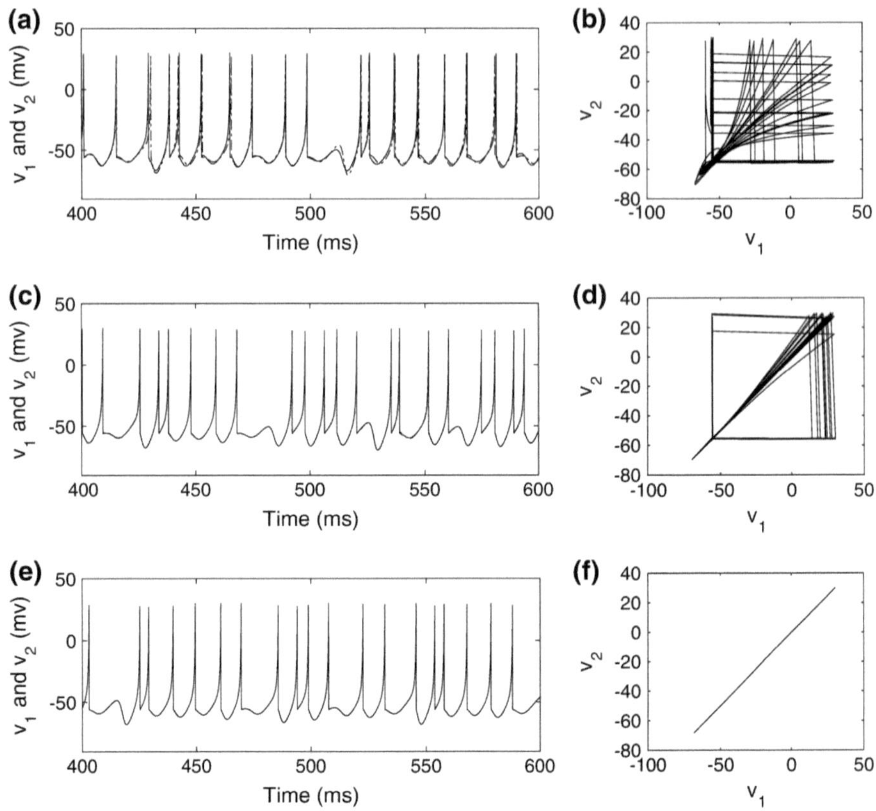

Fig. 3 **a**, **c**, **e** Time series of membrane potential of the coupled Izhikevich neurons (v_1 in *continuous line* and v_2 in *dashed line*) **b**, **d**, **f** Phase portrait of the coupled neuron membrane potentials. **a**, **b** $g = 0.1$, **c**, **d** $g = 0.22$, **e**, **f** $g = 0.35$

pling strengths, implying that the coupled Izhikevich neurons become less and less incoherent and finally completely coherent, with an increasing value of the coupling strength.

Figure 3 shows the membrane potentials of the master and slave neurons, as well as the phase portraits for varying values of connectivity strength g. The synchronization error of the system is shown in Fig. 4. It can be seen that although the second CLE crosses zero at $g = 0.212$, the other CLE is still positive. This case is shown in Figs. 3 and 4c, d with $g = 0.22$. However once both CLEs become negative for $g > 0.348$, complete synchronization occurs and it can be seen from Figs. 3 and 4e, f how the synchronization error vanishes.

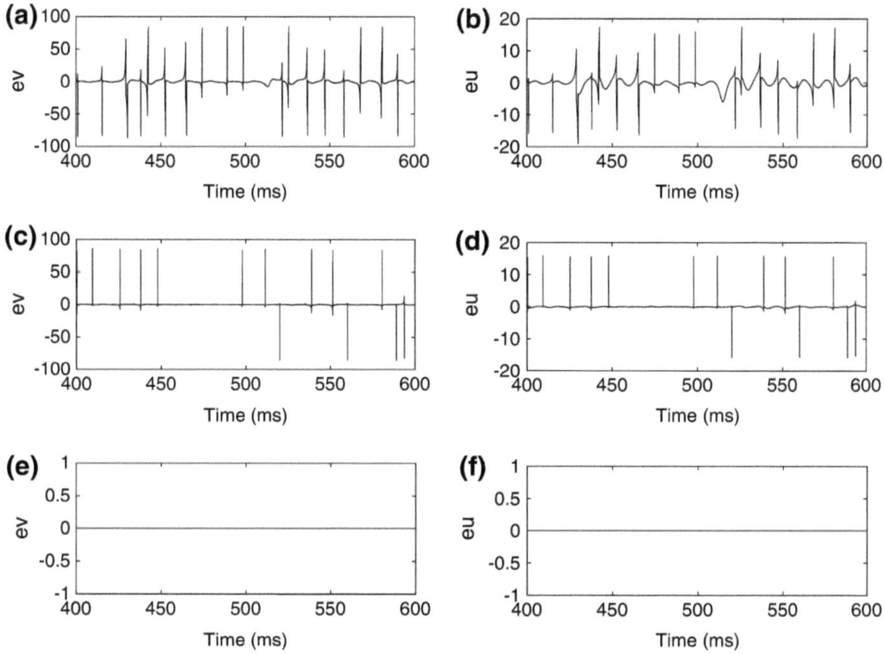

Fig. 4 Time series of the synchronization error of coupled Izhikevich neurons **a**, **c**, **e** $ev = v_1 - v_2$ **b**, **d**, **f** $eu = u_1 - u_2$. **a**, **b** $g = 0.1$, **c**, **d** $g = 0.22$, **e**, **f** $g = 0.35$

From the above discussion it is evident that in the case of two coupled identical Izhikevich neurons, complete synchronization will occur only when the connectivity strength exceeds a critical value ($g > 0.348$) in align with other population synchronization studies found in the literature [11, 14, 15]. In the next section we derive a control law which will guarantee complete synchronization between two delay coupled Izhikevich neurons at any value of their connectivity strength and delay in coupling.

4 Adaptive Synchronization via Nonlinear Feedback Control Law

The dynamical error of the two-interconnected neuronal system can be defined by $e_1 = v_2 - v_1,\ e_2 = u_2 - u_1$, as given blow,

$$\begin{bmatrix} \dot{e}_1 \\ \dot{e}_2 \end{bmatrix} = \begin{bmatrix} 0.04(v_1 + v_2)e_1 + 5e_1 - e_2 - g(e_1 - e_1(t - \tau)) + C \\ abe_1 - ae_2 \end{bmatrix} \tag{9}$$

The nonlinear control,

$$C = -0.04(v_1 + v_2)e_1 - ke_1. \tag{10}$$

The parameter update law

$$\dot{k} = \alpha e_1^2. \tag{11}$$

Theorem 1 *The asymptotic synchronization of the two neurons described in (3) and (4) can be achieved under the control law (10) and (11).*

Proof Now construct the following Lyapunov function,

$$V(t) = \frac{1}{2}e_1^2 + \frac{1}{2}e_2^2 + \frac{1}{2\alpha}(k-l)^2 + \int_{t-\tau}^{t} e_1^2(s)ds, \tag{12}$$

where $l > 0$. The derivative of the Lyapunov function in (12) with respect to (9) is,

$$\begin{aligned}
\dot{V}(t) &= e_1\dot{e}_1 + e_2\dot{e}_2 + \frac{1}{\alpha}(k-l)\dot{k} + e_1^2 - e_1^2(t-\tau) \\
&= e_1[0.04(v_1+v_2)e_1 + 5e_1 - e_2 - g(e_1 - e_1(t-\tau)) + C] \\
&\quad + e_2[abe_1 - ae_2] + \frac{1}{\alpha}(k-l)\alpha e_1^2 + e_1^2 - e_1^2(t-\tau) \\
&= e_1[6-g-l]e_1 + e_1[ab-1]e_2 + e_2[-a]e_2 + ge_1e_1(t-\tau) - e_1^2(t-\tau).
\end{aligned}$$

Moreover, we can express as matrix form

$$\dot{V}(t) = \begin{bmatrix} e_1 \\ e_2 \\ e_1(t-\tau) \end{bmatrix}^T \begin{bmatrix} 6-g-l & \frac{1}{2}(ab-1) & \frac{g}{2} \\ * & -a & 0 \\ * & * & -1 \end{bmatrix} \begin{bmatrix} e_1 \\ e_2 \\ e_1(t-\tau) \end{bmatrix} \tag{13}$$

where $*$ denote the symmetric term. We can conclude that $\dot{V}(t) < 0$ if $6 - g < l$ which implies that based on Lyapunov stability theory, the proposed error model is asymptomatically stable, i.e., two neurons are synchronized.

5 Numerical Simulations

The numerical simulations are run in MATLAB and the models are implemented using the Euler method with a step size $h = 0.01$. The parameter values of the Izhikevich neuron are chosen to be in the chaotic region of operation as, $a = 0.2, b = 2, c = -56, d = -16, I_{ext} = -99$. The connectivity strength was varied for different values $g = 0.1, 0.2,$ and 0.35, while the coupling delay was fixed at $\tau = 0.2$ ms. The initial conditions for the master model are set to $v_1(0) = -65, u_10 = b * v_1(0) = -130,$

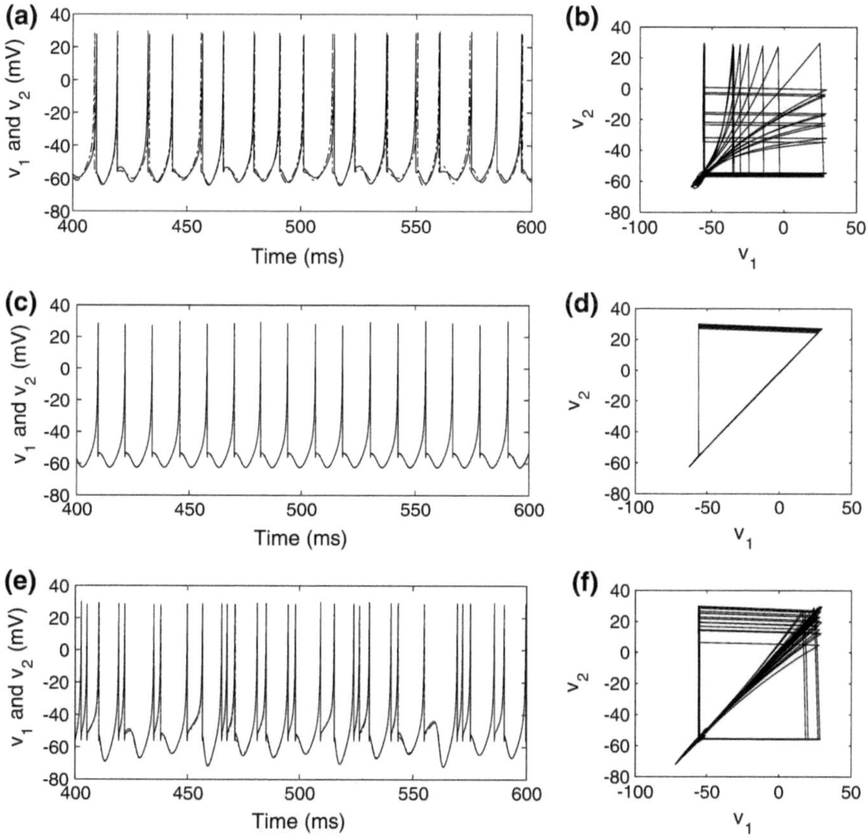

Fig. 5 **a**, **c**, **e** Membrane potential time series of the delay coupled Izhikevich neurons (v_1 in *continuous line* and v_2 in *dashed line*) **b**, **d**, **f** Phase portrait of the delay coupled neuron membrane potentials. **a**, **b** $g = 0.1$, **c**, **d** $g = 0.22$, **e**, **f** $g = 0.35$. Coupling delay $\tau = 0.2$ ms

while those of the slave neuron were set to $v_2(0) = -70, u_20 = b * v_2(0) = -140$. Figure 5 shows the membrane potentials as well as the phase portraits of the master and slave neurons for varying connectivity strength. In comparison to Fig. 3, it is evident the time delay in coupling has affected the degree of synchronization as well as the membrane potentials. This shows the effect of coupling time delay on the dynamics of the master–slave system. Also, in the absence of coupling delay the neurons show complete synchronization at $g = 0.35$ (Figs. 3 and 4e, f), however with a $\tau = 0.2$ ms coupling delay the coupled Izhikevich neurons fail to attain a complete synchronization (Fig. 5e, f). These results confirm that synchronization between two neurons cannot be attained at all settings of coupling strength and delay without control.

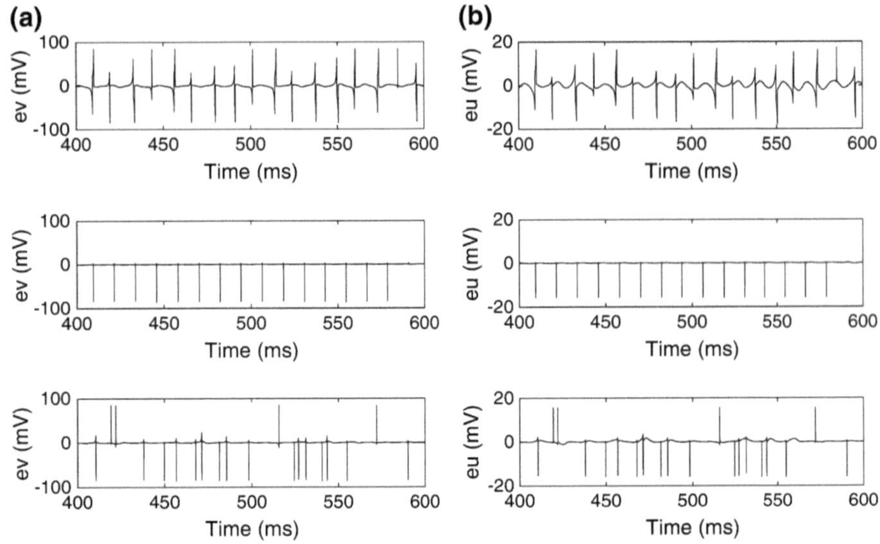

Fig. 6 Synchronization error of the delay coupled Izhikevich neurons **a** $ev = v_1 - v_2$ **b** $eu = u_1 - u_2$. *Top row* $g = 0.1$, *Middle row* $g = 0.22$, *Bottom row* $g = 0.35$. Coupling delay $\tau = 0.2$ ms

In order to investigate the effectiveness of the proposed control scheme, we apply the control law (10) and (11), for the delay coupled Izhikevich neurons with varying coupling strengths. Figure 6 shows the synchronization error of the coupled neurons in the absence of a controller. Figure 7 shows the phase portrait and the synchronization error with the control law. It can be seen that the coupled neurons achieve complete synchronization under external control. By solving the above (3) and (4) with parameter values set for chaotic behaviour, we find that in spite of the differences in the initial conditions, the two neurons synchronizes so that for $t \rightarrow \infty$, $|v_1 - v_2| \rightarrow 0$ and $|u_1 - u_2| \rightarrow 0$.

6 Conclusion

The study presented a control theory-based criterion for synchronization of two gap junction coupled chaotic Izhikevich neurons. The coupled neuron model includes coupling delay of signal transmission between both neurons. From numerical simulations, we have shown that synchronization of the coupled system cannot occur at all settings of the coupling strength and delays. Thus, in order to achieve synchronization between coupled Izhikevich neurons, independent of the coupling strength and coupling delay, we have proposed a nonlinear adaptive control scheme. We have demonstrated the effectiveness of the proposed controller through numerical

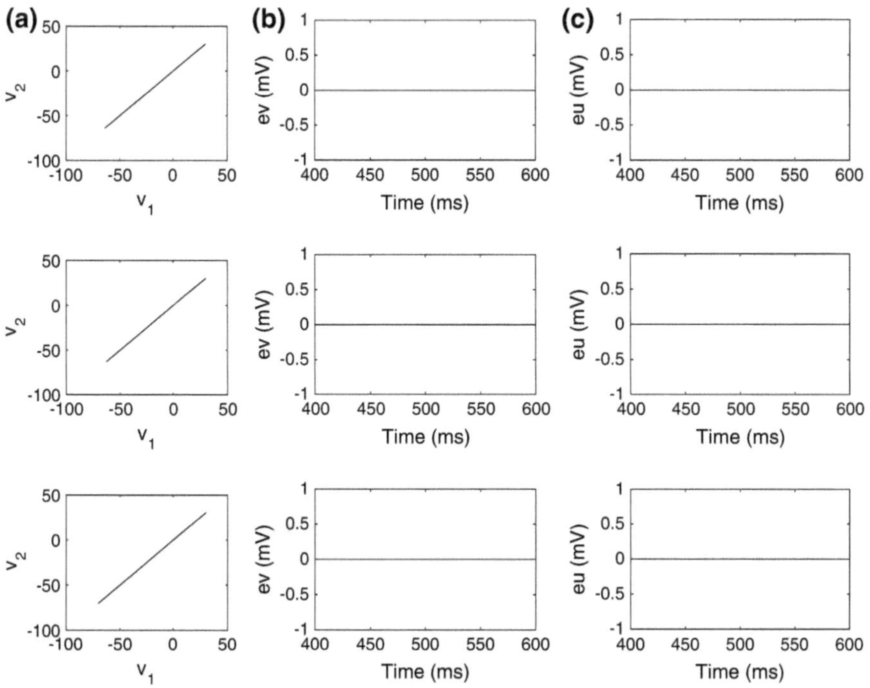

Fig. 7 Synchronization error of the delay coupled Izhikevich neurons with the nonlinear controller in (10) and (11) **a** $ev = v_1 - v_2$ **b** $eu = u_1 - u_2$. *Top row* $g = 0.1$, *Middle row* $g = 0.22$, *Bottom row* $g = 0.35$. Coupling delay $\tau = 0.2$ ms

simulations. In future, we will consider the case of non-identical and uncertainty in the model parameters of the model in network configurations consisting of two or more neurons.

References

1. Dtchetgnia Djeundam, S.R., Yamapi, R., Filatrella, G., Kofane, T.C.: Stability of the synchronized network of hindmarsh-rose neuronal models with nearest and global couplings. Commun. Nonlinear Sci. Numer. Simul. **22**, 545–563 (2015)
2. Uhlhaas, P.J., Singer, W.: Neural synchrony in brain disorders: relevance for cognitive dysfunctions and pathophysiology. Neuron **52**, 155–168 (2006)
3. Hodgkin, A.L., Huxley, A.F.: A quantitative description of membrane current and its application to conduction and excitation in nerve. J. Physiol. **117**, 500–544 (1952)
4. FitzHugh, R.: Impulses and physiological states in theoretical models of nerve membrane. Biophys. J. **1**, 445466 (1961)
5. Morris, C., Lecar, H.: Voltage oscillations in the barnacle giant muscle fiber. Biophys. J. **35**, 193–213 (1981)

6. Hindmarsh, J.L., Rose, R.M.: A model of neuronal bursting using three coupled first order differential equations. In: Proceedings of the Royal Society of London. Series B. Biological Sciences, vol. 221, pp. 87–102 (1984)
7. Izhikevich, E.M.: Which model to use for cortical spiking neurons? IEEE Trans. Neural Netw. **15**, 1063–1070 (2004)
8. Izhikevich, E.M.: Simple model of spiking neurons. IEEE Trans. Neural Netw. **14**, 1569–1572 (2003)
9. Izhikevich, E.M.: Dynamical systems in neuroscience: the geometry of excitability and bursting. In: Computational Neuroscience. MIT Press, Cambridge, Mass., London (2007)
10. Nobukawa, S., Nishimura, H., Yamanishi, T., Liu, J.: Signal response efficiency in Izhikevich neuron model. In: Proceedings of the SICE Annual Conference, pp. 1242–1247 (2011)
11. Nobukawa, S., Nishimura, H., Yamanishi, T., Liu, J.: Chaotic resonance in izhikevich neuron model and its assembly. In: 6th International Conference on Soft Computing and Intelligent Systems, and 13th International Symposium on Advanced Intelligence Systems, SCIS/ISIS 2012, pp. 49–54 (2012)
12. Nobukawa, S., Nishimura, H., Yamanishi, T., Liu, J.: Analysis of routes to chaos in izhikevich neuron model with resetting process. In: 2014 Joint 7th International Conference on Soft Computing and Intelligent Systems, SCIS 2014 and 15th International Symposium on Advanced Intelligent Systems, ISIS 2014, pp. 813–818 (2014)
13. Nobukawa, S., Nishimura, H., Yamanishi, T., Liu, J.: Analysis of chaotic resonance in Izhikevich neuron model. PLoS ONE **10** (2015)
14. Kavasseri, R., Nagarajan, R.: Synchronization in electrically coupled neural networks. Complex Syst. **16**, 369–380 (2006)
15. Kim, S., Lim, W.: Coupling-induced population synchronization in an excitatory population of subthreshold Izhikevich neurons. Cogn. Neurodyn. **7**, 495–503 (2013)
16. Dhamala, M., Jirsa, V.K., Ding, M.: Transitions to synchrony in coupled bursting neurons. Phys. Rev. Lett. **92**, 281011–281014 (2004)
17. Hettiarachchi, I.T., Lakshmanan, S., Bhatti, A., Lim, C.P., Prakash, M., Balasubramaniam, P., Nahavandi, S.: Chaotic synchronization of time-delay coupled Hindmarsh-Rose neurons via nonlinear control. Nonlinear Dyn. **86**, 1249–1262 (2016)
18. Wang, J., Deng, B., Fei, X.: Chaotic synchronization of two coupled neurons via nonlinear control in external electrical stimulation. Chaos, Solitons Fractals **27**, 1272–1278 (2006)
19. Nguyen, L.H., Hong, K.: Adaptive synchronization of two coupled chaotic Hindmarsh-Rose neurons by controlling the membrane potential of a slave neuron. Appl. Math. Model. **37**, 2460–2468 (2013)
20. Aguilar-López, R., Martónez-Guerra, R.: Synchronization of a coupled Hodgkin-Huxley neurons via high order sliding-mode feedback. Chaos, Solitons Fractals **37**, 539–546 (2008)
21. Zhang, T., Wang, J., Fei, X., Deng, B.: Synchronization of coupled Fitzhugh-Nagumo systems via mimo feedback linearization control. Chaos, Solitons Fractals **33**, 194–202 (2007)
22. Pecora, L.M., Carroll, T.L., Johnson, G.A., Mar, D.J., Heagy, J.F.: Fundamentals of synchronization in chaotic systems, concepts, and applications. Chaos **7**, 520–543 (1997)

Part II
Neural Computation

Capturing Cognition via EEG-Based Functional Brain Networks

Md. Hedayetul Islam Shovon, D. (Nanda) Nandagopal, Bernadine Cocks and Ramasamy Vijayalakshmi

Abstract The human brain is comprised of complex networks of neuronal connections, with the functioning of these networks underscoring human cognition. At any given point in time, the complexity of these networks may be greater than the entire communications network on the planet yet functional brain networks are not static; instead, they form and dissolve within milliseconds. Although much is known about the functions and actions of individual neurons in isolation, at a systems level, when billions of neurons coordinate their individual activity to create functional brain networks and thus cognition, understanding is limited. This is due in part to the system behaving completely differently to its parts; that is, emergent properties such as intelligence, emotion and cognition cannot be adequately explained from a sum-of-parts perspective; what is needed instead are powerful computational techniques to model and explore both the intricacies and dynamics of functional brain networks. Although unravelling the activity of the human brain remains circumscribed by technological and ethical constraints, complex network analysis of EEG data offers new ways to quantitatively characterize neuronal cluster patterns. This, in turn, allows the analysis of functional brain networks to understand the complex architecture of such networks. Despite the increasing attention that functional brain network analysis is gaining in computational neuroscience, the true potential of such analysis to reveal dynamic interdependencies between brain regions has yet to be realized. To address this, multi-channel EEG data has been used to examine the dynamics of such networks during cognitive activity using Information Theory based nonlinear statistical measures such as transfer entropy.

Md.H.I. Shovon (✉) · D. (Nanda) Nandagopal · B. Cocks
Cognitive Neuroengineering and Computational Neuroscience Laboratory,
School of Information Technology and Mathematical Sciences,
University of South Australia, Adelaide, Australia
e-mail: shomy004@mymail.unisa.edu.au

D. (Nanda) Nandagopal
e-mail: nanda.nandagopal@unisa.edu.au

R. Vijayalakshmi
Department of Applied Mathematics and Computational Science,
PSG College of Technology, Coimbatore, Tamil Nadu, India

A. Bhatti et al. (eds.), *Emerging Trends in Neuro Engineering and Neural Computation*, Series in BioEngineering,
DOI 10.1007/978-981-10-3957-7_8

Results across different paradigms requiring different types of cognitive effort clearly suggest that transfer entropy is a highly sensitive measure for detecting cognitive activity. Furthermore, these results demonstrate that transfer entropy has clear potential for developing cognitive metrics based on complex features such as connectivity density, clustering coefficient and weighted degree. These techniques may also have application in the clinical diagnosis of cognitive impairment as well as providing new insights into normal cognitive development and function.

Keywords Transfer entropy · Information flow · Directed functional brain network · EEG · Cognitive activity · Cognitive load

1 Introduction

The human brain is one of the most complex systems known to humankind. Weighing in at, on average, around 1.3 kg [1] the brain can be structurally divided into hemispheres (left and right) and/or lobes (occipital, frontal, parietal and temporal), with each division classically associated with different functions. For example, the left hemisphere is traditionally associated with language processing [2] while the occipital lobe is associated with visual processing [3]. Brain regions can also be categorized via relative depth; for example, the cerebral cortex, responsible for higher order cognitive processing, encompasses the surface of the human brain, while deeper structures such as the brain stem are associated primarily with non-conscious processing such as respiration [4]. In between such regions lie neural interfaces such as the thalamus [5], which mediate higher order top-down processes such as memory, with lower order, bottom-up processes such as sensory information (e.g. vision and hearing).

It is within these structural constraints that approximately 86 billion neurons [6] perform the *work* of the human brain. This is achieved via communication between neurons, both within specific neuronal groups or ensembles, and between different ensembles. Together, this communication system is comprised of over 1 trillion connections [7] or synapses, with different patterns of connections thought to represent different functional neural activity [8]. Some of these patterns can remain relatively stable over time [9], while others can form and dissolve within a matter of milliseconds [10]. Thus the first challenge for understanding the specifics of human cognition is to examine and measure neural activity with appropriate temporal sensitivity; that is, changes in patterns of neural activity must be able to be identified and then mapped with millisecond precision. Of the various technologies currently available, only electroencephalography (EEG) and magnetoencephalography (MEG) are capable of providing such measurement precision [11].

The second challenge to accurately capturing cognition is the neural property of plasticity [12]. Although gross anatomical features of the brain remain stable over time, the inner functional workings do not; rather, they adapt to the environment, both internal and external, within which the individual operates [13]. For example, new experiences are thought to be encoded through the modulation of existing neural circuits [14]; that is, *learning* occurs as a result of neural plasticity. Similarly, various types of memory processing are thought to depend on plastic responses within the brain [15]. Because of this plasticity, no two human brains are identical [16] and no single brain will necessarily behave in an identical manner at two different points in time, with a variety of endogenous and exogenous factors capable of directly influencing brain function [17–19].

As a result, human cognition cannot be considered either a static nor linear process, so measures such as Granger Causality [20] are unlikely to truly capture the intricacies and subtleties of neural function.

The final major challenge to capturing cognition is the nature of the beast itself. Human cognition is not a mere sum of parts; instead, it is an emergent property that is related to but completely different to its constituent parts [21]. To illustrate, human speech involves various sub-processes such as articulation, respiration, memory and sound processing. Although much is known about each of these sub-processes in isolation, this knowledge cannot adequately explain why a sentence such as "That is so gay" can evoke vastly different reactions depending on who, where, when, how and why the utterance is spoken [22]. Similarly, the experience of listening to and appreciating music cannot be explained simply by considering the perception of the individual musical notes [23]. As a result, considering cognitive processes such as working memory (WM) in isolation is not capturing cognition per se and it is the challenge of identifying and measuring the emergent properties of cognition that remain one of the great final frontiers for cognitive neuroengineering.

1.1 Functional Brain Networks

Although the network behaviour of the human brain has been known for some time [24], the measurement of such activity has only recently gained serious research attention. For example, the concept of small world networks, as they apply to cognitive function, was only formally introduced in 1998 [25], while the application of complex network theories such as graph theoretical analyses of cognitive function was first published in 2009 [26]. Despite this relative youth in relation to brain function such analyses have a solid, empirical support base within a variety of other domains. For example, within conservation biology, graph theoretic approaches have been applied using focal-species analysis to examine landscape or habitat connectivity [27]. From a different perspective, the World Wide Web (WWW) has been studied by numerous authors using different aspects of graph theory [28–30].

The translation from communication or social networks to brain networks, or more specifically functional brains networks (FBNs), is conceptually straight forward. A network is created by creating a graph comprised of nodes (or vertices) and edges. *Nodes*, in the case of EEG or MEG data, refer directly to electrode sites on the scalp (i.e. brain regions) whilst *edges* are the physical connections between nodes (i.e. synapses or axonal projections). As proposed by Bullmore and Sporns [26], it is these edges that represent functional associations among brain regions, thus plotting edges and edge strengths between nodes allows the creation of FBNs. Having thus constructed FBNs, it then becomes a case of applying the appropriate complex network analysis, the interpretation of which must be done within both anatomical and functional constraints.

Both linear and nonlinear statistical measures have previously been used to construct and analyse FBNs. From a linear perspective, such measures include cross correlation [31], magnitude squared coherence [32], Pearson's product moment correlation [33] and wavelet coherence [34] with all providing the degree of synchronization between signals arising from any given pair of electrodes. Such linear measures can produce linear brain network connectivity within the time domain, with stronger functional connectivity indicated by a higher correlation between the signals and vice versa, although the interpretation of any resultant FBN is restricted by the linearity of the measure. By comparison, nonlinear statistical measures provide a more realistic representation of neural activity due to the nonlinear behaviour of the brain. Granger Causality (GC) [20], Mutual Information (MI) [35] and Transfer Entropy (TE) [36] can all be used to construct FBNs with MI resulting in undirected networks, whereas GC and TE can be used to construct directed networks. As the direction of information flow within the brain is able to provide an additional layer of information on how one brain region influences another, this is clearly valuable information that MI fails to capture. It should be noted, however, that TE appears to be the most sensitive of the nonlinear measures for the construction of directed FBNs and thus the remainder of this chapter will focus on TE and its application to FBN construction and analysis.

2 Transfer Entropy

The general concept of entropy is based on the laws of thermodynamics, whereby entropy can be defined as a measure of disorganization or degradation [37]. This concept was then appropriated into Information Theory by Shannon [38] for the quantification of information with entropy defined as the average number of bits required to optimally encode independent draws of the discrete variable X following a probability distribution $p(x)$, which is known as Shannon entropy and is given in Eq. 1 [39]:

$$H(X) = -\sum_{x} p(x)\log_2 p(x) \tag{1}$$

where the sum in Eq. 1 is for all possible states of x. Shannon entropy for two variables X and Y is given in Eq. 2.

$$H(X,Y) = -\sum_{x,y} p(x,y)\log_2 p(x,y) \tag{2}$$

where the sum in Eq. 2 is for all possible states of x and y.

Assuming that the two time series of interest $X = x_t$ and $Y = y_t$ can be approximated by Markov process, Schreiber proposed a measure of causality called Transfer Entropy to compute the deviation from the following generalized Markov condition as shown in Eq. 3 [39]:

$$p\left(y_{t+1}|y_t^n, x_t^m\right) = p\left(y_{t+1}|y_t^n\right) \tag{3}$$

where $x_t^m = (x_t, \ldots, x_{t-m+1})$, $y_t^n = (y_t, \ldots, y_{t-n+1})$, while the subscript t denotes the considered state (or time step); m and n represent the orders (memory) of the Markov processes X and Y respectively. More simply, transfer entropy (TE) is an information theoretical measure which determines the direction and quantifies the information transfer between two processes [39]. Furthermore, TE estimates the amount of activity of a system which is not dependent on its own past activity but on the past activity of another system. Schreiber represented TE from X to Y as shown in Eq. 4 [40],

$$TE_{X\to Y} = \sum_{y_{t+1}, y_t^n, x_t^m} p\left(y_{t+1}, y_t^n, x_t^m\right)\log\left(\frac{p(y_{t+1}|y_t^n, x_t^m)}{p(y_{t+1}|y_t^n)}\right) \tag{4}$$

$TE_{X\to Y}$ can be regarded as the information about future observations y_{t+1} gained from the past observations of y_t^n and x_t^m minus the information about future observations y_{t+1} gained from past observations of y_t^n only. The TE measure is inherently asymmetric and based on transition probabilities, so it incorporates directional and dynamic information and can be in the range $0 \le TE_{X\to Y} < \infty$..

To improve the calculation accuracy, two additional steps can be included for the computation of TE [41, 42]. Due to the finite size and non-stationarity of EEG data, TE matrices usually contain large amounts of noise. Noise/bias can, however, be removed from the estimate of TE by subtracting the average transfer entropy from X to Y using shuffled version of X denoted by $\langle TE_{X_{shuffle}\to Y}\rangle$, over several shuffles [42]. $X_{shuffle}$ contains the same symbols as in X but those symbols are rearranged in a randomly shuffled order. Then, normalized transfer entropy is calculated from X to Y with respect to the total information in sequence Y itself. This will represent the relative amount of information transferred by X. The normalized transfer entropy (NTE) is shown in Eq. 5 as follows [41]:

$$NTE_{X\to Y} = \frac{TE_{X\to Y} - \langle TE_{X_{shuffle}\to Y}\rangle}{H(y_{t+1}|y_t)} \tag{5}$$

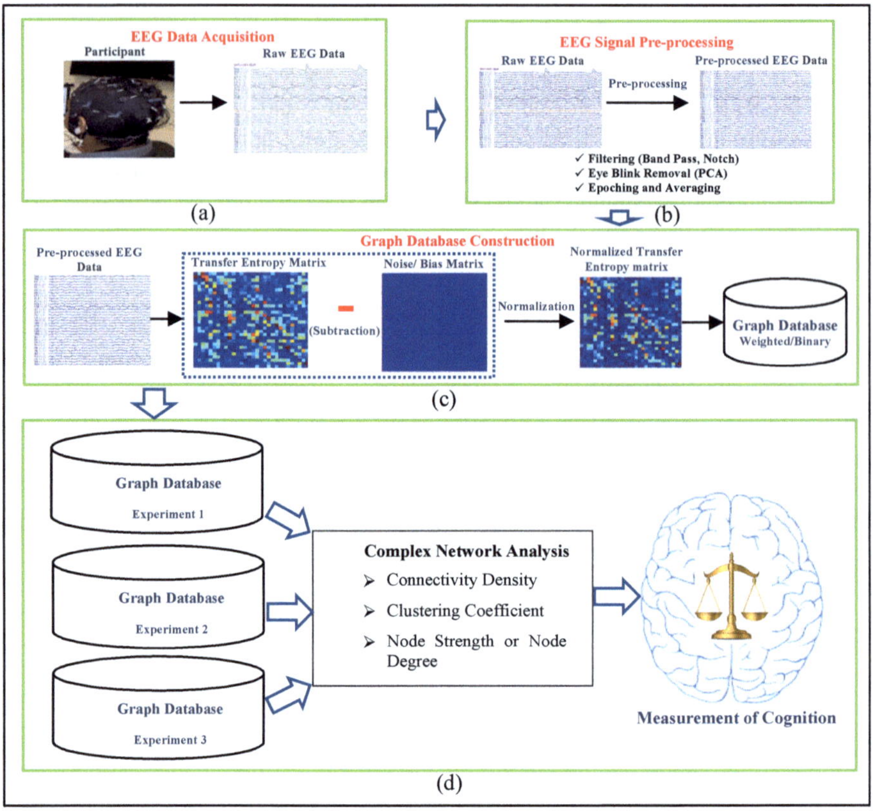

Fig. 1 Transfer entropy analysis framework. **a** EEG data acquisition, **b** EEG signal pre-processing, **c** graph database construction and **d** measurement of cognition

In Eq. 5, $H(y_{t+1}|y_t)$ represents the conditional entropy of Y at time $t+1$ given its value at time t as shown in Eq. 6.

$$H(y_{t+1}|y_t) = -\sum_{y_{t+1},y_t} p(y_{t+1},y_t)\log\frac{p(y_{t+1},y_t)}{p(y_t)} \tag{6}$$

NTE is in the range $0 \le NTE_{X\to Y} \le 1$, where 0 represents that X transfers no information to Y, and 1 represents that X transfers maximal information to Y. By computing the NTE between each pair of EEG channels FBNs can then be constructed (See Fig. 1).

3 Measuring Cognitive Activity Using Complex Network Metrics

Any FBN created via Graph Theory is a mathematical model in which EEG scalp electrodes are considered as vertices (nodes) which represent the activity of underlying neuronal populations. The connections between each pair of vertices (known as edges/links) [26, 43] are then measured using a correlation representing the weight of the edge. By then applying complex network metrics, quantitative analysis techniques can be used to characterize the different connectivity patterns of brain networks attributable to different cognitive states. Such complex network metrics include *connectivity density, clustering coefficient* and different types of *centrality measures*.

3.1 Connectivity Density

Connectivity density is the actual number of edges in the graph as a proportion of the total number of possible edges [26]. It is also referred to as the physical cost, wiring cost or connectivity cost. For a directed graph with n nodes where there are no self-connections/loops, the total number of possible connections is $n * (n-1)$.

3.2 Clustering Coefficient

The clustering coefficient for node i represents the ratio between all directed triangles actually formed by i and the number of all possible triangles that i could form. The clustering coefficient measures the cliquishness of a network and represents how well the neighbourhood of a node is connected. A high clustering coefficient is related to high local efficiency of information transfer [26] and is mostly used to measure functional segregation. A clustering coefficient for a directed network has been developed by Fagiolo [44], whereby the directed clustering coefficient, C_d of a network is the average of the clustering coefficients of all nodes as given in Eq. 7

$$C_d = \frac{1}{N}\sum_{i=1}^{N} C_i = \frac{1}{N}\sum_{i=1}^{N} \frac{\frac{1}{2}\sum_{j=1}^{N}\sum_{h=1}^{N}\left(a_{ij}+a_{ji}\right)\left(a_{ih}+a_{hi}\right)\left(a_{jh}+a_{hj}\right)}{(k_i^{out}+k_i^{in})(k_i^{out}+k_i^{in}-1)-2\sum_{j=1}^{N} a_{ij}a_{ji}} \tag{7}$$

where C_i is clustering coefficient of node i, N is the number of nodes, a_{ij} denotes the directed connection from node i to node j, and k_i^{out} and k_i^{in} are the out-degrees and in-degrees of the node i respectively.

3.3 Node Degree or Node Strength

Node strength, a centrality measure of weighted directed networks, represents the sum of all incoming and outgoing edge weights [43]. This measure not only depends on the number of incoming and outgoing edges, but also depends on the weights of the incoming and outgoing edges. For a binary directed network, node degree or total degree represents the total of incoming and outgoing edges.

3.4 Characteristic Path Length

The average shortest path length between all pairs of nodes in a network is known as the characteristic path length. The shortest path between node i and node j, $d(i,j)$ can be defined as the minimum number of nodes that it has to traverse to reach node j from node i. The shortest path length, between nodes i and j is given by

$$d_{ij} = \sum_{a_{uv} \in g i \leftrightarrow j} a_{uv}$$

where $g i \leftrightarrow j$ is the shortest path between i and j. Thus, the characteristic path length of the network [25] is given by

$$L = \frac{1}{n} \sum_{i \in N} L_i = \frac{1}{n} \sum_{i \in N} \frac{\sum_{j \in N, j \neq i} d_{ij}}{n-1}$$

where L_i is the average distance between node i and all other nodes. It is used mostly for measuring the functional integration.

3.5 Interpreting Outputs

Although the application of such complex network metrics provides accurate quantitative data, this data cannot be interpreted without considering the broader context of the experimental paradigm in which it is used, as well as the functional and anatomical constraints presented by the brain itself. For example, if an FBN is constructed that displays a uniquely left temporal lobe localisation to Brodmann Area 44 (the Inferior frontal gyrus or Pars opercularis also known as Broca's Area), it could potentially be interpreted as a language-related response [45]. If, however, the associated experimental paradigm was related to odour perception [46], then an interpretation of language processing would be erroneous. Similarly, if the same FBN had resulted in response to a visual paradigm such as a Stroop Task [47], then concluding that the FBN was vision related would likely be erroneous given the anatomical location; instead, the FBN would more likely represent internal (silent) speech generation [48]. As a result, despite the sensitivity and accuracy of using complex network metrics to detect and measure cognitive activity in the human brain, such analysis cannot always be applied post hoc; rather, the experimental paradigm needs to be planned and conducted with the demands of the analysis in mind.

4 Computational Approach to Capturing Cognition

Many statistical measures have been used to attempt to unravel the complexity of EEG data; however, the computational approach proposed here is a relatively unique blend of graph theory, information theory, statistical measures of complex networks and effective visualization of the results. These techniques serve as a lens through which the neuronal activities and interactions of complex brain networks can be explored more effectively. It specifically makes use of network theory, transfer entropy as a nonlinear statistical measure to compute pairwise correlations between EEG channels and hence construct FBNs, and the complex network metrics discussed in the previous section.

The proposed computational framework for capturing cognition is illustrated in Fig. 1.

First, multi-channel EEG data is collected during various cognitive load conditions. This data is then pre-processed to remove noise and artefacts such as eye blinks and muscle interference (see next section for more detail on data collection and pre-processing methods). This pre-processed EEG data are then used for the computation of TE matrices, where each cell of the TE matrices represents the TE value from one electrode to another. For noise removal, an average shuffled TE matrix (noise matrix) is calculated and subtracted from the original TE matrix (Fig. 1c) with the resulting noise removed normalized TE matrix called the NTE matrix. The computed NTE matrices are then subsequently used for the construction of directed FBNs and further analysis. Note also, that the constructed NTE matrices are binarized using a threshold = 0.002 for the analysis using different types of complex network metrics. Both binary and weighted directed graphs are then constructed for analysis (Fig. 1d). From this NTE computation, a graph database for each experimental paradigm, task or cognitive load state is thus established. Complex network analysis is then carried out to determine the connectivity density, clustering coefficient and node strength directly corresponding to each paradigm, task or state thereby providing quantitative measures of cognitive activity. Three examples of the practical application of this computational approach are discussed in the next section.

5 Practical Examples of Applying NTE to Measure Cognition

Three experiments [36, 49, 50] exploring the application of NTE to construct FBNs have been conducted by the Cognitive Neuroengineering and Computational Neuroscience Laboratory at the University of South Australia. In all three cases, the participation criteria, EEG signal acquisition, and EEG pre-processing were identical. A description of the three experimental paradigms is further discussed in the sections that follow.

5.1 *Participation Criteria*

Right-handed, healthy participants aged >18 years were recruited from the academic/professional staff and student populations of the University of South Australia (Mawson Lakes campus), as well as the wider Adelaide general community. All participants reported normal hearing and normal or corrected-to-normal vision; none reported any history of psychological, neurological or psychiatric disorders. Before the experiment started, experimental procedures were explained to participants and signed, informed consent was obtained.

5.2 *EEG Data Acquisition*

EEG data were acquired at a sampling rate of 1000 Hz through a 40 channel Compumedics Neuroscan Nuamps amplifier using Curry 7 software [51, 52]. Prior to data collection, each participant was fitted with an appropriate sized 32 channel Quikcap as shown in Fig. 1a. The 30 electrode sites used in the current study were based on the international 10–20 convention: FP1, FP2, F7, F3, Fz, F4, F8, FT7, FC3, FCz, FC4, FT8, T3, C3, Cz, C4, T4, TP7, CP3, CPz, CP4, TP8, T5, P3, Pz, P4, T6, O1, Oz and O2 with the average of two the earlobes (A1, A2) used as the reference. Impedance values of all electrodes were checked using Curry with no recording undertaken until all channel impedances were below 50 kΩ. All stimulus onsets and participant responses were time marked on the EEG record using Compumedics Neuroscan STIM 2 software [53].

5.3 *EEG Signal Pre-processing*

The pre-processing of EEG data was done by applying band pass filter of 1–70 Hz and a notch filter at 50 Hz as previously illustrated in Fig. 1b. To detect eye blinks, one of the typical eye blinks was selected by visual inspection and the remaining eye blinks detected using Curry 7 template matching. These eye blink artefacts were then removed using principal component analysis (PCA). Further bad blocks were removed by the Curry 7 software.

5.4 *Experimental Paradigm 1*

Participants (6 × male, 2 × female; age range 28–65 years) were first asked to sit quietly with eyes open to generate a baseline (no cognitive load) condition. Participants were then instructed to drive normally using a "Simuride" driving simulator [54] for approximately 4 min to generate a cognitive load (Drive) condition. As well a negotiating both straight and winding roads whilst maintain a virtual speed of between 50–60 km/h, participants were then also asked to respond to auditory stimuli played to them through headphones to generate a second cognitive load (DriveAdo) condition. These stimuli were comprised of a combination of digits and letters (for instance, H3, 8A) read by either a male or female voice. If the voice was male, participants then had to respond 'yes' or 'no' (by pulling left or right levers mounted behind the steering wheel) based on whether the digit was even or odd. If the voice was female, participants then had to respond based on whether the digit was even or odd.

5.5 *Experimental Paradigm 2*

Participants (same as Experiment 1) had EEG data collected during the following five conditions:

5.5.1 Baseline—Eyes Open (EOP) and Eyes Closed (EC)

To obtain baseline brain activity, participants were asked to first stare at a blue fixation star on the STIM computer monitor for 2 min. They were then asked to close their eyes and sit calmly for further 2 min, although it should be noted that only the EOP data was used as the baseline for analytical purposes.

5.5.2 Mild Cognitive Load (MiCL) 1-Back Stroop Task

To evoke a relatively mild level of cognitive activity, participants undertook a 1-back Stroop task. The standard Stroop task is one of interference. Known to evoke large effect sizes and validated as being statistically reliable [47], the task requires participants to respond to a series of visually displayed congruent and incongruent colour word stimuli. For example, a congruent stimulus would be the word *red* typed in red-coloured font; an incongruent stimulus would be the word *red* typed in yellow-coloured font. A 1-back variation on the standard Stroop task involves inserting a neutral stimulus between the presentation of the target stimulus and the cue to respond. In this experiment, a fixation cross (neutral) was inserted between the stimulus presentation (Stroop stimulus) and the cue to respond via a keyboard button press; specifically, "Y" if the font colour and word meaning were the same or "N" if they were different. This 1-back component was used to add a second layer of higher cognitive processing (working memory) to the base level of the Stroop task itself (attention).

5.5.3 Moderate Cognitive Load (MoCL) 1-Back Stroop Task with Go/No-Go Variation

To increase the cognitive activity to a moderate level, a Go/No-Go (G/NG) variation was added to the 1-back Stroop Task. As a recognition reaction time (RT) task [55], participants must respond to some stimuli (Go) whilst inhibiting their response to other stimuli (no-go). In this specific instance, participants were required to again undertake a 1-back Stroop task but with the explicit instruction to ignore any stimulus words appearing in blue font whether that stimulus was congruent. This thus placed an additional layer of load on both working memory and attention processes.

5.5.4 Heavy Cognitive Load (HCL)—Moderate Cognitive Load with Auditory Distraction

To further tax global cognitive processing, bottom-up processing was then increased through the addition of auditory distraction. Participants again performed the 1-back Stroop Task with Go/No-go Variation whilst also listening to a pre-recorded sound track constructed using dichotic listening principles [56]; that is, different sounds played to different ears at the same time (e.g. bird song on left channel, tornado siren on right channel). They were also instructed to listen for a specific stimulus (a spoken number) as part of that sound track and further told that they would be asked a question about that stimulus at the completion of the task. This thus provided personal salience, as well as encouraging attention to the task. To ensure heightened auditory monitoring throughout the task, no number was actually spoken on the soundtrack.

5.5.5 Extreme Cognitive Load (ECL)—Moderate Cognitive Load with Auditory and Visual Distraction

In the final task, participants were again asked to perform the 1-back Stroop Task with Go/No-go Variation whilst also listening to a pre-recorded sound track. Additionally, equal numbers of neutral and aversive distractor images were inserted between and with-in each visual stimulus. As per the auditory-only distraction, participants were also asked to attend to a specific stimulus type (e.g. tool) and advised that a question would be asked at the conclusion of the task; thus, participants were asked to attend to one specific distractor per modality.

5.6 *Experimental Paradigm 3*

Ten adults (7 × male, 3 × female; age range 22–59) had continuous EEG data collected during three different conditions: eyes open (baseline), visual search and Web search interaction.

5.6.1 Eyes Open/Eyes Closed (2 min each)

As per experiments 1 and 2, baseline brain activity was first obtained during both an Eyes Open (EOP) and Eyes Closed state, although only the EOP data was used for further analysis.

5.6.2 Visual Search (VS) Task

In this task, participants were asked to identify a target object from an array of distractor targets [57]. For example, as shown in Fig. 2, the top red letter is the primary target. The participants would be required to press 'y' if the target is present amongst the black distractor array, or 'n' if the target is absent.

5.6.3 Web Search Task

Participants were instructed to search for information on the Web based on three provided topic areas. They were free to use any Web browser (e.g. Internet Explorer, Google Chrome, Mozilla Firefox) and any search engine (e.g. Google, Yahoo). Participants were also free to choose the source of information. The search questions provided for the three different scenarios were:

> Scenario 1: Your employer has just told you that he is going to give you a new company car and asked you to choose one. The only restrictions are that the car must be red and be reasonably fuel efficient but it cannot be a European brand.

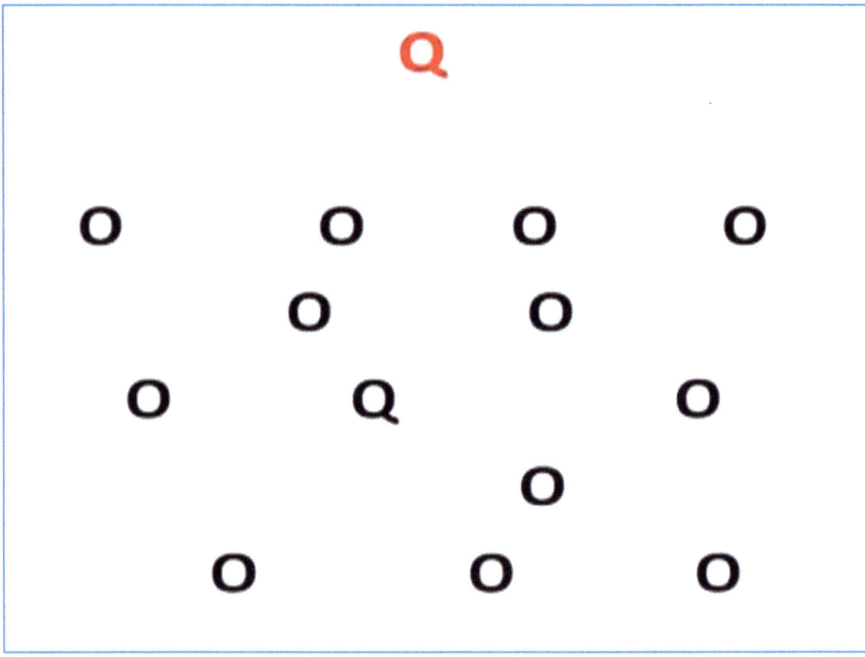

Fig. 2 An example of visual search

Scenario 2: While walking in the scrub in the Adelaide Hills you get bitten by what appears to be a tick. Should you go to the hospital Emergency Department ASAP? YES/NO and WHY?

Scenario 3: You've decided that you want to see a movie at the cinema. What movie do you decide to see, which session, which cinema and why?

In the Web search interaction, the tasks were divided into three sub-tasks: Query formulation (Q), viewing of search result List (L) and reading the each individual Content page (C).

6 Results and Discussion

6.1 Result of Experiment 1

From the continuous eyes open (EOP) and driving (Drive) data, 50 good epochs of 2 s were chosen using the back-to-back epoching process of Neuroscan Curry 7, then averaged into a single two second data epoch [12]. During audio distraction driving (DriveAdo), reaction times varied from 1 to 5 s after the stimulus onset, so for analysis purposes, extreme outliers (i.e. very fast and very slow responses) were disregarded leaving only those epochs where reaction times were 1.5–3 s. When such epochs were identified, smaller epochs were taken from 0 to 2 s from the stimulus onset. The 2 s epochs from each of the conditions (EOP, Drive and DriveAdo) were then used for the construction of binary and weighted graph database.

6.1.1 Connectivity Density

The connectivity density for all the participants is shown in Fig. 3. As indicated in the figure, connectivity density is higher in the cognitive load condition than the baseline condition, inferring more connections are established to facilitate more active information flow.

6.1.2 Clustering Coefficient

The clustering coefficient for all the participants is presented in Fig. 4. As can be seen, the, EOP condition has the smallest clustering coefficient and DriveAdo the highest. An increasing pattern of clustering coefficient across the states for all participants was also observed, presumably attributable to the increasing amount of cognitive load. This suggests that the clustering coefficient of the directed FBN

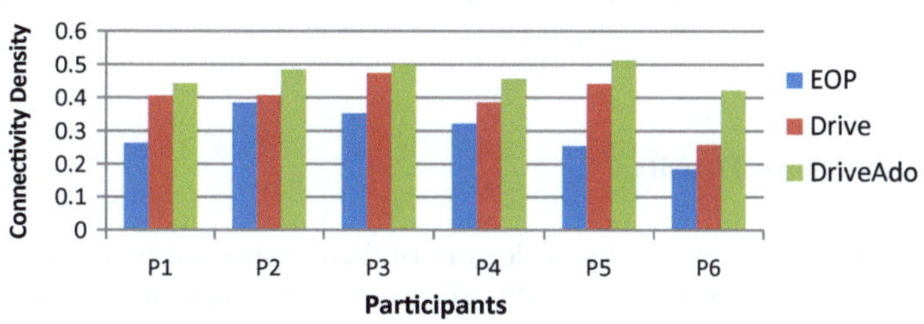

Fig. 3 Comparison of connectivity density of directed FBNs during EOP, Drive and DriveAdo

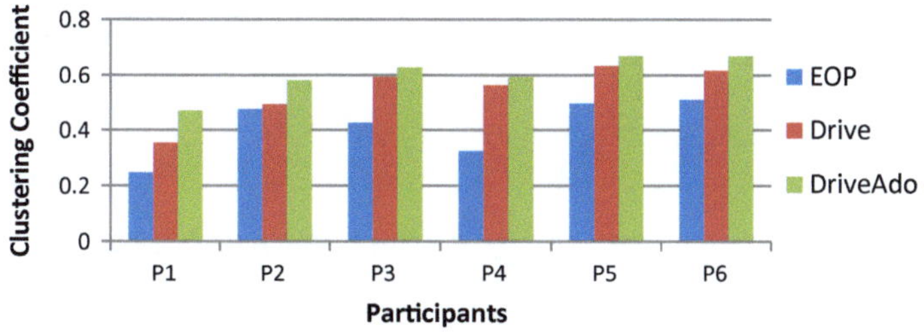

Fig. 4 Comparison of clustering coefficient of directed FBNs during EOP, Drive and DriveAdo

increases with an increase in induced cognitive load thereby further suggesting that the clustering coefficient is sensitive to cognitive load. As a result, the clustering coefficient of directed FBNs may be able to be further refined to develop a quantitative metric to measure cognitive activity with such a metric having potential application in the diagnoses of cognitive impairments.

6.1.3 Node Strength

Figure 5 represents the comparison of node strengths during different cognitive states computed using the weighted FBNs of participant P1. As shown, most of the electrodes have higher strength values during cognitive load, which indicates that each electrode sends and receives more information during cognitive load.

6.2 *Result of Experiment 2*

As per Experiment 1, extreme outliers (excessively fast and slow responses) were excluded from analysis. As a result, only those trial epochs with RTs between 1.5 and 3.5 s were used. In each cognitive load state, the selected 2 s epochs were then extracted from the stimulus onsets and averaged. For the comparison of each cognitive load state with baseline (EOP), 50 chunks of EEG data of 2 s duration were randomly selected from the EOP data, and then averaged. The 2 s epochs during EOP, MiCL, MoCL, HCL and ECL were then used for the construction of binary and weighted graph databases.

6.2.1 Connectivity Density

For all participants, the connectivity density of the constructed binary directed FBN during the different conditions (EOP, MiCL, MoCL, HCL and ECL) has been

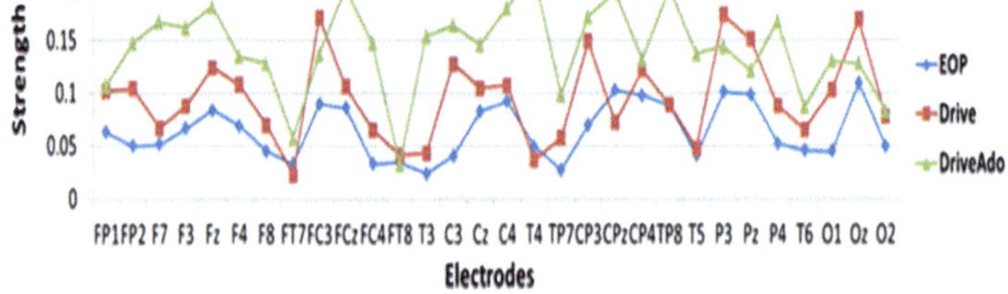

Fig. 5 Comparison of strength across electrodes during EOP, Drive and DriveAdo of participant P1

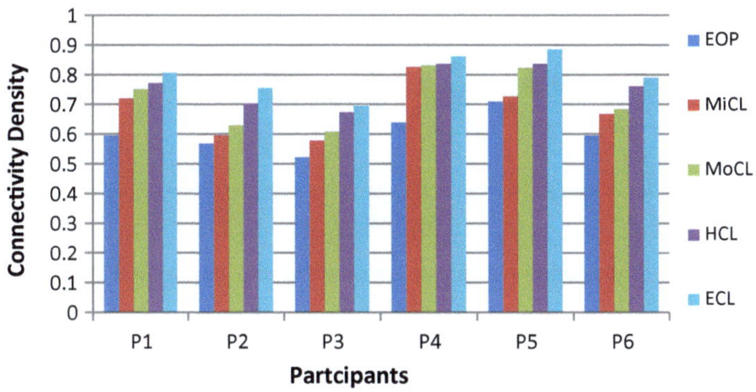

Fig. 6 Comparison of connectivity density during EOP, MiCL, MoCL, HCL and ECL

calculated and is shown in Fig. 6. Here, EOP has the least connectivity density and ECL the highest. Again, an increasing pattern of connectivity density across the states for all the participants was observed in relation to the increasing amount of cognitive load. The results demonstrated that the connectivity density of the directed FBN increases with increases in induced cognitive load; that is, connectivity density is directly proportional to the amount of cognitive load applied.

6.2.2 Clustering Coefficient

Figure 7 shows the clustering coefficient value during the different cognitive tasks (EOP, MiCL, MoCL, HCL and ECL) for all participants.

Figure 8 shows an example of the clustering coefficient value across electrodes (note, due to space restrictions, only the results of participant P1 are displayed). Of note, the clustering coefficient value increases in almost all of the electrodes during

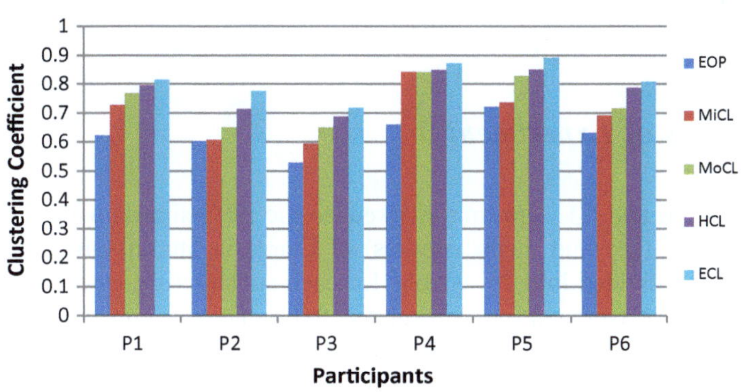

Fig. 7 Comparison of clustering coefficient of directed FBNs during EOP, MiCL, MoCL, HCL and ECL

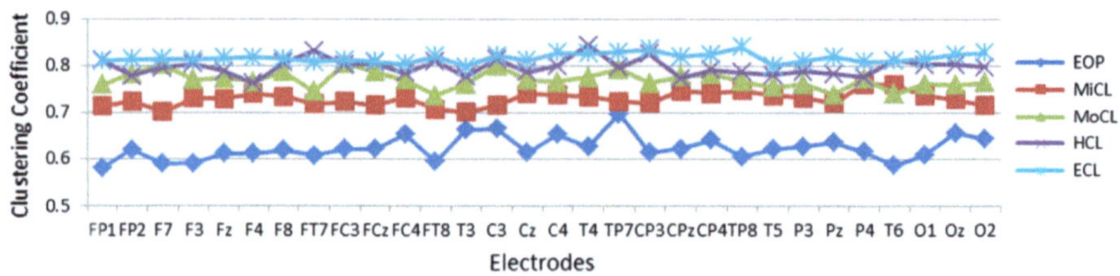

Fig. 8 Comparison of clustering coefficient across electrodes during EOP, MiCL, MoCL, HCL and ECL

cognitive load when compared to the baseline state (EOP), thus demonstrating that the information transfer among the neighbouring nodes of each electrode increases with increasing cognitive load.

Figure 9 also shows the clustering coefficient values of all the electrodes during the different cognitive tasks, this time plotted on a 3D head surface. Using the EEGLAB Headplot function, minor changes in the clustering coefficient values between the cognitive states have been visualized [58]. To effect this visualization, the colour map scale of Headplot function has been customized such that dark blue represents the minimum clustering coefficient value and the dark red the maximum clustering coefficient value. The increased clustering coefficient during increased cognitive load is demonstrated by the appearance of higher valued colour around the electrodes on the head scalp.

6.2.3 Node Strength

Node strength value across electrodes was calculated from weighted directed FBNs during the different cognitive states (EOP, MiCL, MoCL, HCL and ECL). Due to space limitations, only the results for participant P2 are shown in Fig. 10, although as expected, node strength value is higher in most of electrodes during higher cognitive load.

6.3 Results of Experiment 3

Sub-tasks (Q, L and C) were time marked on the EEG signals using the captured interaction logs (key and mouse strokes) of Camtasia Studio software [59]. EEG data were then divided into 2 s epochs for each subtask (if greater than 2 s), with those epochs then averaged for each subtask to produce one epoch of averaged data per subtask. In order to compare search features with the baseline (EOP), 50 chunks of EEG data (2 s each) were randomly selected from the EOP data and then averaged. Visual search data was also considered as baseline search activity by making 2 s epochs from the stimulus onsets, then averaging into a single 2 s

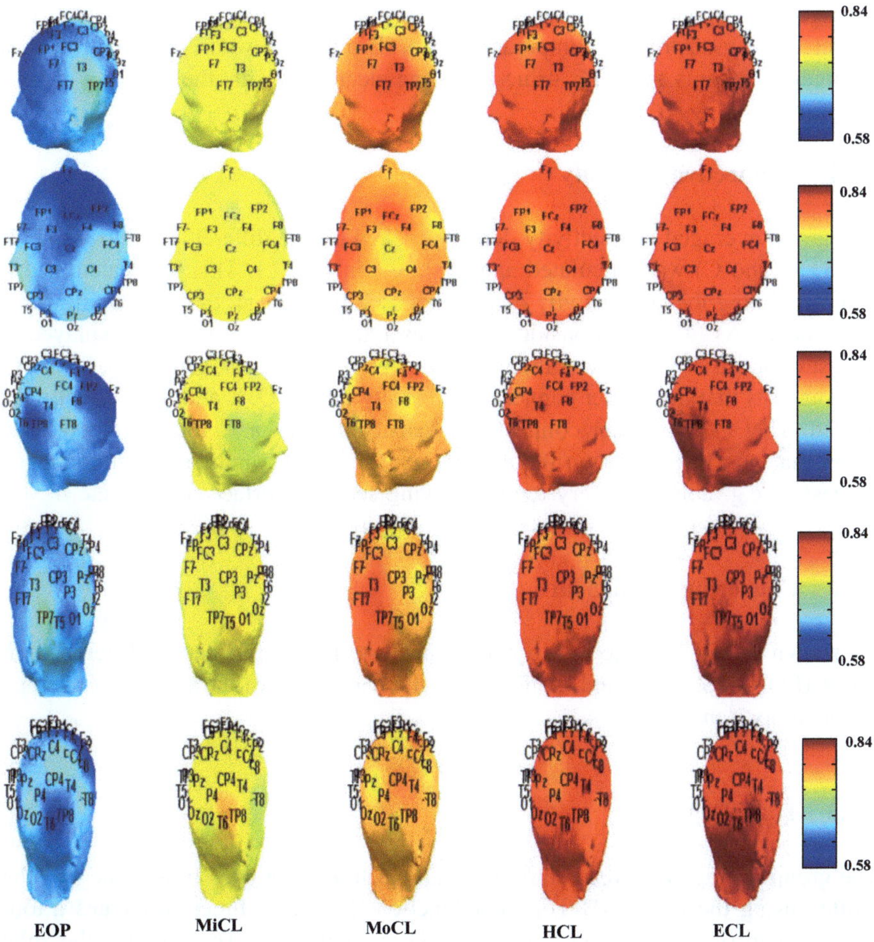

Fig. 9 Visualization of clustering coefficient value on head surface in different brain states using Headplot function

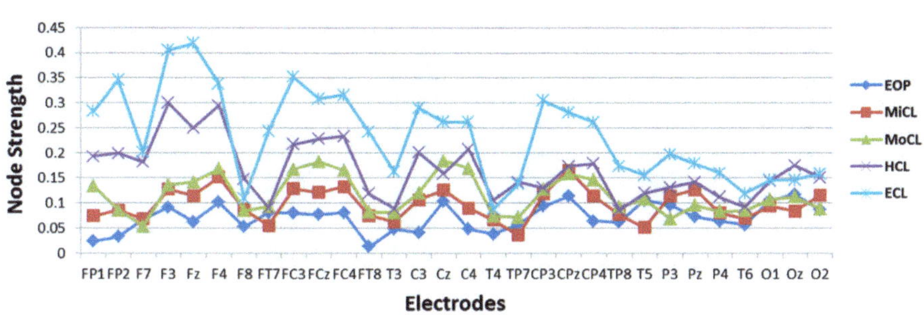

Fig. 10 Comparison of node strength across electrodes during EOP, MiCL, MoCL, HCL and ECL of participant P2

averaged data epoch. The averaged EEG data epochs relating to EOP, VS, Q, L and C were then used for the computation of NTE matrices, where each cell of these matrices represents the NTE value from one electrode to another.

6.3.1 Connectivity Density

The group averaged connectivity density evoked by the different conditions (EOP, VS, Q, L and C) was calculated and is shown in Fig. 11. This was calculated by first calculating the connectivity density for each participant, then averaging across participants. As EOP was considered the baseline cognitive state, and displayed the lowest connectivity density compared to the Web search task stages (Q, L and C), this suggests that the higher connectivity density during Q, L and, C is directly task related. That Q has higher connectivity density than L or C further supports this given that query formulation requires the execution of a number of simultaneous processes (e.g. defining query terms, viewing search interface, typing, etc.).

6.3.2 Clustering Coefficient

As shown in Fig. 12, the group averaged clustering coefficient of different brain states (EOP, VS, Q, L and C) followed the same pattern as the group averaged connectivity density.

6.3.3 Total Degree

The group averaged degree centrality of all electrodes for each brain state was then plotted using the EEGLAB Topoplot function [58]. This function creates a topographic map of a scalp data field in a 2-D circular view. To visualize the subtle variations of degree centrality values in different brain states, the colour map scale

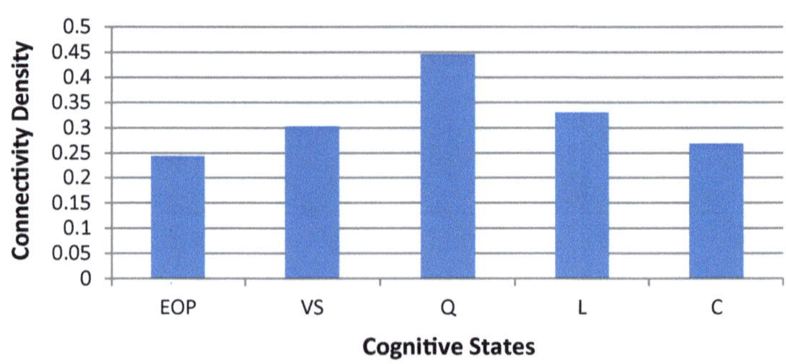

Fig. 11 Comparison of connectivity density (group average) during different brain states

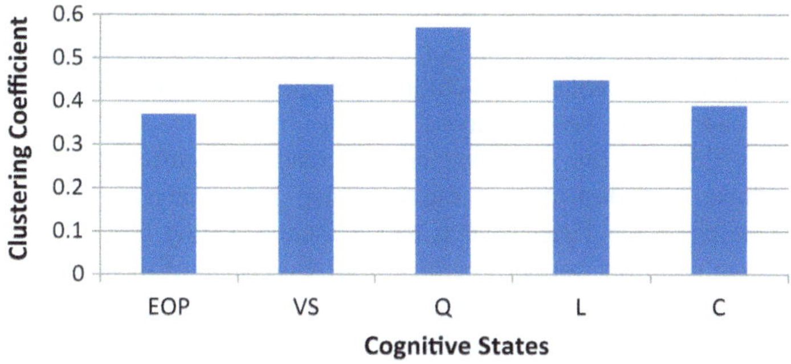

Fig. 12 Comparison of clustering coefficient (group average) during different brain states

of Topoplot was customized such that a colour map scale is used from minimum degree centrality value to maximum degree centrality value among the degree centrality value of all the electrodes of all the brain states [60]; that is, dark blue represents the minimum degree centrality value and dark red the maximum degree centrality value. As shown in the example in Fig. 13, participant P1's topoplots showed clear activity which is reflective of the degree of engagement for each task; that is, the visual search condition shows greater engagement than the eyes open condition, while the list reading condition shows greater engagement than the content reading condition. In the case of eyes open versus visual search, this is easily explainable—a visual search task requires focused, more intense visual attention than simply staring at a fixation star. Similarly, the button press response

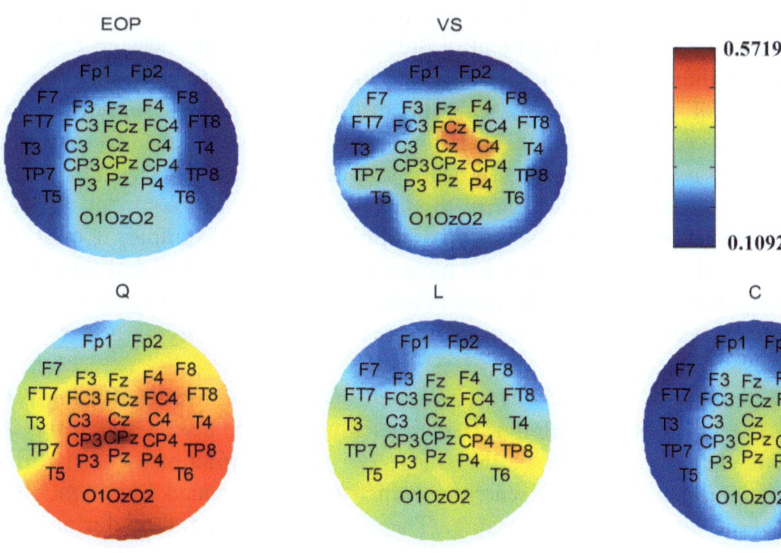

Fig. 13 Comparison of total degree across electrodes during in different brain states

requirement of the visual search task elicited higher activity in motor areas (FCz, Cz, C4) than the eyes open condition. The differences between list and content reading are also explainable, albeit conjecturally, on an attentional/focus basis; that is, content reading has been "filtered" via the list process, therefore is focused externally by this process such that content reading requires less internal focus through reduced decision-making requirements. Interestingly, the similarities in processing between the visual search and contents reading tasks suggest that the two tasks share common cortical regions in the execution of those tasks. There are, however, some obvious differences between content and list reading, with list reading eliciting higher activity at CP4 and TP8 whereas content reading exhibited higher midline activity at CPz and Pz. In the case of the query formulation task, however, the high activity is most likely a reflection of multiple processes contaminating the averaging process. Further work is currently being conducted to divide this task into smaller sub-tasks so that movement, language, decision-making and attentional processes can be further delineated during this query formation phase.

7 Conclusion

As demonstrated by the results of Experiments 1, 2 and 3, NTE shows clear promise for the construction and analysis of FBNs. In the case of Experiment 1, information theoretical NTE measures were successfully applied to construct EEG-based directed FBNs in baseline and cognitive load conditions. The overall results demonstrated that the directed FBNs constructed using NTE were sensitive to changes in cognitive load and that this sensitivity of NTE-based FBNs has the potential to assist in the development of quantitative metrics to measure cognitive activity within the human brain. Experiment 2 took NTE-based FBNs from three to five divisions of cognitive load, with the results again clearly supporting that directed FBNs constructed using NTE are highly sensitive to changes in neural activity directly related to changes in degrees of cognitive load. As well as further supporting the possible application of this approach to the development of quantitative metrics, it also suggests practical applications. For example, it might be possible to develop simple devices for measuring cognitive load in real time. Such devices could then, potentially, be used in adaptive intelligent systems as well as in safety critical systems which require intensive mental activity (e.g. air traffic control).

Finally, the results of Experiment 3 showed that FBNs constructed using NTE can be used in a variety of different research settings and contexts. In this specific instance, the key contribution of the study was the construction of such FBNs during different stages of Web searching which enabled detailed investigation of brain function during such searching. The results identified that during Web searching information transfer increases in brain networks when compared to baseline but that brain activity during different Web search task stages is not the same. Consequently, the results may have implications for examining the effects of cognitive abilities on information search behaviour/processes and search task performance/outcomes which, in turn, could potentially allow an adaptive information retrieval system to better personalize its interaction with users.

Although the results reported here are, in general, foundational in nature, they clearly show that NTE-based FBNs are sensitive to changes in cognitive load, irrespective of the actual task. This thus potentially provides new ways of developing objective, quantified scales and metrics for categorizing human cognitive activity which, in turn, may provide new ways for diagnosing and monitoring cognitive impairment. Furthermore, NTE-based FBNs may provide new avenues for the development of devices that monitor cognitive activity in real time, thereby improving safety in cognitively intense occupations such as air traffic control; for example, such a device could sound a warning when the individual approaches or passes a critical level of cognitive load. Such a device might also find use in clinical settings by allowing practitioners to monitor cognitive activity in a treatment setting; for example, gauging the degree of neural "distress" during exposure therapy [61]. Alternately, such a device might aid psycho-pharmaceutical treatment by monitoring the effects of such medications on overall cognitive function and adjusting dosages as required.

References

1. Parent, A., Carpenter, M.B.: Ch. 1. Carpenter's Human Neuroanatomy. Williams & Wilkins. ISBN 978-0-683-06752-1
2. Sperry, R.W., Gazzaniga, M.S., Bogen, J.E.: Interhemispheric relationships: the neocortical commissures; syndromes of hemisphere disconnection. Handbook of Clinical Neurology, vol. 4 (1969)
3. Chilosi, A.M., Brovedani, P., Moscatelli, M., Bonanni, P., Guerrini, R.: Neuropsychological findings in idiopathic occipital lobe epilepsies. Epilepsia **47**, 76–78 (2006)
4. Bogousslavsky, J., Khurana, R., Deruaz, J., Hornung, J., Regli, F., Janzer, R., Perret, C.: Respiratory failure and unilateral caudal brainstem infarction. Ann. Neurol. **28**, 668–673 (1990)
5. Crick, F.: Function of the thalamic reticular complex: the searchlight hypothesis. Proc. Natl. Acad. Sci. **81**, 4586–4590 (1984)
6. Azevedo, F.A., Carvalho, L.R., Grinberg, L.T., Farfel, J.M., Ferretti, R.E., Leite, R.E., Lent, R., Herculano-Houzel, S.: Equal numbers of neuronal and nonneuronal cells make the human brain an isometrically scaled-up primate brain. J. Comp. Neurol. **513**, 532–541 (2009)
7. Marois, R., Ivanoff, J.: Capacity limits of information processing in the brain. Trends Cogn. Sci. **9**, 296–305 (2005)
8. Bassett, D.S., Meyer-Lindenberg, A., Achard, S., Duke, T., Bullmore, E.: Adaptive reconfiguration of fractal small-world human brain functional networks. Proc. Natl. Acad. Sci. **103**, 19518–19523 (2006)
9. Batty, M., Taylor, M.J.: Early processing of the six basic facial emotional expressions. Cogn. Brain. Res. **17**, 613–620 (2003)
10. Sergent, C., Baillet, S., Dehaene, S.: Timing of the brain events underlying access to consciousness during the attentional blink. Nat. Neurosci. **8**, 1391–1400 (2005)

11. Nunez, P.L.: Electroencephalography (EEG). In: Ramachandran, V.S. (ed.) (In Chief) Encyclopaedia of the Human Brain, pp. 169–179
12. Huttenlocher, P.R.: Neural Plasticity. Harvard University Press (2009)
13. Kleim, J.A., Jones, T.A.: Principles of experience-dependent neural plasticity: implications for rehabilitation after brain damage. J. Speech Lang. Hear. Res. **51**, S225–S239 (2008)
14. Callan, D.E., Tajima, K., Callan, A.M., Kubo, R., Masaki, S., Akahane-Yamada, R.: Learning-induced neural plasticity associated with improved identification performance after training of a difficult second-language phonetic contrast. Neuroimage **19**, 113–124 (2003)
15. Klingberg, T.: Training and plasticity of working memory. Trends Cogn. Sci. **14**, 317–324 (2010)
16. Brierley, J.B., Beck, E.: The significance in human stereotactic brain surgery of individual variation in the diencephalon and globus pallidus. J. Neurol. Neurosurg. Psychiatry **22**, 287–298 (1959)
17. Ploughman, M.: Exercise is brain food: the effects of physical activity on cognitive function. Dev. Neurorehabil. **11**, 236–240 (2008)
18. Leigh Gibson, E., Green, M.W.: Nutritional influences on cognitive function: mechanisms of susceptibility. Nutr. Res. Rev. **15**, 169–206 (2002)
19. Janowsky, J.S., Oviatt, S.K., Orwoll, E.S.: Testosterone influences spatial cognition in older men. Behav. Neurosci. **108**, 325 (1994)
20. Friston, K., Moran, R., Seth, A.K.: Analysing connectivity with Granger causality and dynamic causal modelling. Curr. Opin. Neurobiol. **23**, 172–178 (2013)
21. Goldstein, J.: Emergence as a construct: history and issues. Emergence **1**, 49–72 (1999)
22. Cocks, B., Jamieson, G.A.: What should be the place of the normative database in speech perception research? J. Cogn. Sci. **14**, 399–417 (2013)
23. Koelsch, S., Siebel, W.A.: Towards a neural basis of music perception. Trends Cogn. Sci. **9**, 578–584 (2005)
24. Mclntosh, A., Gonzalez-Lima, F.: Structural equation modeling and its application to network analysis in functional brain imaging. Hum. Brain Mapp. **2**, 2–22 (1994)
25. Watts, D.J., Strogatz, S.H.: Collective dynamics of 'small-world' networks. Nature **393**, 440–442 (1998)
26. Bullmore, E., Sporns, O.: Complex brain networks: graph theoretical analysis of structural and functional systems. Nat. Rev. Neurosci. **10**, 186–198 (2009)
27. Bunn, A.G., Urban, D.L., Keitt, T.: Landscape connectivity: a conservation application of graph theory. J. Environ. Manage. **59**, 265–278 (2000)
28. Björneborn, L., Ingwersen, P.: Perspective of webometrics. Scientometrics **50**, 65–82 (2001)
29. Bar-Ilan, J.: Data collection methods on the Web for infometric purposes—a review and analysis. Scientometrics **50**, 7–32 (2001)
30. Almind, T.C., Ingwersen, P.: Informetric analyses on the World Wide Web: methodological approaches to 'webometrics'. J. Doc. **53**, 404–426 (1997)
31. Ts'o, D.Y., Gilbert, C.D., Wiesel, T.N.: Relationships between horizontal interactions and functional architecture in cat striate cortex as revealed by cross-correlation analysis. J. Neurosci. **6**, 1160–1170 (1986)
32. Dobie, R.A., Wilson, M.J.: Objective detection of 40 Hz auditory evoked potentials: phase coherence vs. magnitude-squared coherence. Electroencephalogr. Clin. Neurophysiol./Evoked Potentials Sect. **92**, 405–413 (1994)
33. Kowalski, C.J.: On the effects of non-normality on the distribution of the sample product-moment correlation coefficient. Appl. Stat. 1–12 (1972)
34. Sankari, Z., Adeli, H., Adeli, A.: Wavelet coherence model for diagnosis of Alzheimer disease. Clin. EEG Neurosci. **43**, 268–278 (2012)
35. Jeong, J., Gore, J.C., Peterson, B.S.: Mutual information analysis of the EEG in patients with Alzheimer's disease. Clin. Neurophysiol. **112**, 827–835 (2001)

36. Shovon, M.H.I., Nandagopal, D.N., Vijayalakshmi, R., Du, J.T., Cocks, B.: Transfer entropy and information flow patterns in functional brain networks during cognitive activity. In: Neural Information Processing, Lecture Notes in Computer Science (LNCS 8834), Part I, pp. 1–10. Springer (2014)
37. Coveney, P.V.: The second law of thermodynamics-entropy, irreversibility and dynamics. Nature **333**, 409–415 (1988)
38. Shannon, C.E.: Prediction and entropy of printed English. Bell Syst. Tech. J. **30**, 50–64 (1951)
39. Schreiber, T.: Measuring information transfer. Phys. Rev. Lett. **85**, 461 (2000)
40. Kaiser, A., Schreiber, T.: Information transfer in continuous processes. Physica D **166**, 43–62 (2002)
41. Gourévitch, B., Eggermont, J.J.: Evaluating information transfer between auditory cortical neurons. J. Neurophysiol. **97**, 2533–2543 (2007)
42. Neymotin, S.A., Jacobs, K.M., Fenton, A.A., Lytton, W.W.: Synaptic information transfer in computer models of neocortical columns. J. Comput. Neurosci. **30**, 69–84 (2011)
43. Rubinov, M., Sporns, O.: Complex network measures of brain connectivity: uses and interpretations. Neuroimage **52**, 1059–1069 (2010)
44. Fagiolo, G.: Clustering in complex directed networks. Phys. Rev. E **76**, 026107 (2007)
45. Grossman, M., Cooke, A., DeVita, C., Chen, W., Moore, P., Detre, J., Alsop, D., Gee, J.: Sentence processing strategies in healthy seniors with poor comprehension: an fMRI study. Brain Lang. **80**, 296–313 (2002)
46. Savic, I., Berglund, H.: Passive perception of odors and semantic circuits. Hum. Brain Mapp. **21**, 271–278 (2004)
47. MacLeod, C.M.: The Stroop task: the "gold standard" of attentional measures. J. Exp. Psychol. Gen. **121**, 12 (1992)
48. Friedman, L., Kenny, J.T., Wise, A.L., Wu, D., Stuve, T.A., Miller, D.A., Jesberger, J.A., Lewin, J.S.: Brain activation during silent word generation evaluated with functional MRI. Brain Lang. **64**, 231–256 (1998)
49. Shovon, M.H.I., Nandagopal, D.N., Vijayalakshmi, R., Du, J.T., Cocks, B.: Towards a cognitive metric using normalized transfer entropy. In: Proceedings of the 3rd ASE International Conference on Biomedical Computing (BioMedCom 2014), December 13–16, Cambridge, MA, USA (2015)
50. Shovon, M.H.I., Nandagopal, D.N., Du, J.T., Vijayalakshmi, R., Cocks, B.: Cognitive activity during web search. In Proceedings of the 38th International ACM SIGIR Conference on Research and Development in Information Retrieval, pp. 967–970. ACM (2015)
51. CURRY 7 EEG Acquisition and Analysis Software. Compumedics Neuroscan USA Ltd
52. Nuamps EEG Amplifier (Model 7181). Compumedics Neuroscan USA Ltd
53. STIM 2 Stimulus Delivery and Experiment Control Solution. Compumedics Neuroscan USA Ltd
54. Simuride Pro Driving Simulator Software 2010 AplusB Software Corporation
55. Rubia, K., Russell, T., Overmeyer, S., Brammer, M.J., Bullmore, E.T., Sharma, T., Simmons, A., Williams, S.C., Giampietro, V., Andrew, C.M.: Mapping motor inhibition: conjunctive brain activations across different versions of go/no-go and stop tasks. Neuroimage **13**, 250–261 (2001)
56. Wexler, B.: Dichotic presentation as a method for single hemisphere simulation studies. In: Hugdahl, K. (ed.) Handbook of Dichotic Listening: Theory, Methods and Research. John Wiley & Sons Dichotic Listening, Great Britain (1988)
57. Luck, S.J., Hillyard, S.A.: Spatial filtering during visual search: evidence from human electrophysiology. J. Exp. Psychol. Hum. Percept. Perform. **20**, 1000 (1994)
58. Delorme, A., Makeig, S.: EEGLAB: an open source toolbox for analysis of single-trial EEG dynamics including independent component analysis. J. Neurosci. Methods **134**, 9–21 (2004)
59. Camtasia Screen Recording & Video Editing. TechSmith Corporation

60. Vijayalakshmi, R., Dasari, N., Nandagopal, D., Subhiksha, R., Cocks, B., Dahal, N., Thilaga, M.: Change detection and visualization of functional brain networks using EEG data. Proc. Comput. Sci. **29**, 672–682 (2014)
61. Abramowitz, J.S., Deacon, B.J., Whiteside, S.P.: Exposure therapy for anxiety: principles and practice. Guilford Press (2012)

Modelling of Tumour-Induced Angiogenesis Influenced by Haptotaxis

Wei Chen, Li Zhang, Chengyu Liu and Alamgir Hossain

Abstract Endothelial cell (ECs) migration, influenced by both chemotaxis and haptotaxis, is a crucial step in tumour-induced angiogenesis. According to recent bioclues, we assume that some subgroups of invasive tumours generate fibronectin, leading to enhance haptotaxis. Hence, we developed a mathematical model simulating the influence of haptotaxis on angiogenesis. The simulation results show that migration of ECs can be accelerated when the invasive tumour enhances haptotaxis, which means angiogenesis can be simultaneously synthetically promoted by both chemotacxis and haptotaxis. Such results have not been reported in previous models of tumour-induced angiogenesis. According to the consensus theory, angiogenesis supplies oxygen and nutrients to facilitate tumour development. By linking the relationship of tumour invasion and angiogenesis, we propose that there is a possible mechanism of tumour-invasion and tumour-induced angiogenesis, as these promote each other and make the tumour develop faster, becoming more harmful for some subgroups of tumours. Since the proposed research on angiogenesis and the neural development systems share similar tasks of accomplishing and supporting complex networks and bear striking functional similarities, the proposed research promotes the potential connection between vascular biology and neuroscience to motivate future computational biological research.

Keywords Angiogenesis · Endothelial cell · Mathematical model · Hoptotaxis

W. Chen · L. Zhang (✉)
Computational Intelligence Research Group, Faculty of Engineering and Environment, Department of Computer Science and Digital Technologies, Northumbria University, Newcastle NE1 8ST, UK
e-mail: li.zhang@northumbria.ac.uk

C. Liu
Institute of Biomedical Engineering, Shandong University, Jinan, China

A. Hossain
Faculty of Science and Technology, Anglia Ruskin IT Research Institute, Anglia Ruskin University, Cambridge CB1 1PF, UK

A. Bhatti et al. (eds.), *Emerging Trends in Neuro Engineering and Neural Computation*, Series in BioEngineering,
DOI 10.1007/978-981-10-3957-7_9

1 Introduction

Neuro-engineering is an emerging multidisciplinary research field which has drawn uprising interests from researchers in related field. Because of the striking functional similarities between angiogenesis and the neural development systems pertaining to accomplishing and supporting complex networks, the study of the tumour-induced angiogenesis process shows great potential in inspiring future multidisciplinary neuroscience and neuro-engineering research. It also has become the focus of this research.

Specifically, angiogenesis is the process of forming new blood vessels from the existing ones, and is a crucial requirement for the growth, progression and metastasis of a tumour [1–3]. Tumour-induced angiogenesis is the phenomenon that in hypoxic microenvironment, the tumour triggers angiogenesic processes by secreting the tumour angiogenic factor (TAF), such as vascular endothelial growth factor (VEGF) [4]. The concept that tumour growth is dependent on the tumour-induced angiogenesis was first proposed by Judah Folkman in 1971 [5]. Since then, lots of studies have been involved in the tumour-induced angiogenesis.

Endothelial cells (ECs) are the most important cells involved in the angiogenesis process. Endothelial tip cells are some ECs spearheading at the nascent sprouts and guiding the newly formed vessel there. The procedure of angiogenesis is the migration of the endothelial tip cells and the proliferation of the stalk cells following the endothelial tip cells. Modelling of the migration of the endothelial tip cells is a pivotal aspect for studying tumour-induced angiogenesis process. Also, the proposed work on angiogenesis and neural systems share similar mechanisms and principles to evolve, develop, and perform biological functions [6].

2 Anderson and Chaplain's Model

2.1 Model Construction

Anderson and Chaplain's mathematical model described the relationship between the ECs density and the TAF and fibronectin concentrations. The migration rate of ECs is influenced by the random motility, chemotaxis and haptotaxis factors, which was defined as [7–9]:

$$\frac{\partial n}{\partial t} = D\nabla^2 n - \nabla \cdot (\chi(c) n \nabla c) - \nabla \cdot (\rho n \nabla f), \quad (1)$$

where n denotes the ECs density (at or near a capillary sprout tip) per unit area, c denotes the TAF concentration and f denotes the fibronectin concentration. In Eq. (1), the first term $D\nabla^2 n$ described the random motility of the ECs which follows the Ficks' diffusion law, D is a positive constant and denotes the cell diffusion rate; the second term $\nabla \cdot (\chi(c)n\nabla c)$ described the chemotactic motility characterized by the function $\chi(c) = \chi/(1+\delta c)$, which reflects the decrease in chemotactic sensitivity with the increase of the TAF concentration; the third term $\nabla \cdot (\rho n \nabla f)$ described the haptotactic motility, where ρ is a positive constant.

From [10–12], chemotaxis is in response to the TAF gradients and haptotaxis is in response to the fibronectin gradients. To derive the nonlinear PDEs governing the ECs motion, the total cell flux balance was considered in Eq. (1). Considering when the ECs migrate in the microenvironment towards the tumour, there are some uptake and binding of TAF molecular by the ECs. A simple TAF uptake function describing the TAF concentration was proposed in Eq. (2):

$$\frac{\partial c}{\partial t} = -\mu n c, \tag{2}$$

where μ is the consumption rate of TAF.

Endothelial cells are known to produce fibronectin with a rate β as they migrate and the degradation of fibronectin f depends upon the MDE density m at rate γ. The function is described in Eq. (3).

$$\frac{\partial f}{\partial t} = \beta n - \gamma \mathrm{m} f \tag{3}$$

The MDE is produced by each individual endothelial cell n at a rate of α. Once the MDF is produced, it diffuses locally with diffusion coefficient, and is spontaneously degraded at a rate ν. The function is defined as Eq. (4):

$$\frac{\partial m}{\partial t} = \alpha n + \varepsilon \nabla^2 m - \nu m \tag{4}$$

2.2 Initial Conditions and Parameter Settings

(1) TAF concentration distribution

Assume that a large initial concentration of TAF locates at the side of the tumour line source [13–16] and a low concentration locates at the parent vessel were considered in the original Anderson and Chaplains model [17]. The initial conditions of TAF concentration in a two-dimension domain $\omega(x, y)$were described as follows:

$$c(x, y, 0) = e^{-\frac{(1-x)^2}{\varepsilon_2}}, \quad (x, y) \in [0, 1] \times [0, 1], \tag{5}$$

where ε is a positive constant.

(2) Fibronectin concentration distribution

A large initial concentration of fibronectin is formed in and around the parent vessel in the original Anderson and Chaplain's model [13–16]. They assumed the largest concentration of fibronectin located at the side of the parent vessel and the lowest concentration located at the side of the tumour line source. The initial conditions of fibronectin concentration were described as follows:

$$f(x, y, 0) = ke^{-\frac{x^2}{\varepsilon_2}}, (x, y) \in [0, 1] \times [0, 1], \tag{6}$$

where $k < 1$ and ε is a positive constant.

The parameters in the aforementioned equations were set as: $D = 0.00035$, $\chi = 0.38$, $\rho = 0$; $(or \rho = 0.34 \, respectively)$, $\beta = 0.05$, $\gamma = 0.1 \, and \, \delta = 0.6$, $\alpha = 10^{-6}$, $\varepsilon = 0.01$, $\nu = 3$, $n_0 = 0.9$, $\varepsilon_1 = 0.45$, $\varepsilon_2 = 0.45$, $k_1 = 0.75$ [7, 8, 17].

3 The Proposed Improved Model in This Work

Based on the Anderson and Chaplain's model, we modified and improved their model in three aspects: i.e. involving ECs proliferation, ECM degradation and fibronectin concentration gradient.

3.1 Improved Model with the New Assumptions

In Anderson and Chaplain's model, there are not birth and death terms, which mean the total number of cell does not change. In fact, during the progress of angiogenesis, the density of the capillary network increases. This means that ECs density should increase in magnitude. So the proliferation of ECs happens during the approach. In addition, the fact that ECM is degraded by enzymes should be also taken into account.

We have created a proliferation term $\mu Mn\left(1 - \frac{n}{n_0}\right)$, which means that proliferation is proportional to the density of ECs, n, and is also limited by the density of itself autonomously, as described in $\left(1 - \frac{n}{n_0}\right)$. Here n_0 should be at the same scale of the non-dimensionlized initial ECs concentration n so we set $n_0 = 0.9$. We assume that the proliferation rate is also proportional to ECM because the proliferation of cells requires growth factors which rely on ECM. The ECM is a vast source of potent growth factors which can promote EC proliferation and subsequent invasion through a degraded ECM [18]. μ is the constant proliferation coefficient. By integrating the proliferation term into Eq. (1) we obtained Eq. (7).

$$\frac{\partial n}{\partial t} = D\nabla^2 n - \nabla \cdot (\chi(c)n\nabla c) - \nabla \cdot (\rho n \nabla f) + \mu M n\left(1 - \frac{n}{n_0}\right) \tag{7}$$

(1) TAF concentration distribution

Under hypoxia condition, the tumour secrets TAF to induce angiogenesis. The TAF is produced by the tumour and diffuses into the two dimensions surrounding tissue and ECM towards the vessels based on Ficks' diffusion law with diffusion constant D_c. At the same time, the decay of TAF occurs. Thus, the TAF concentration c satisfies Eq. (8). Once TAF is secreted and diffused, a steady-state concentration gradient between the tumour and vasculature can be set up. Therefore, the steady-state concentration gradient of TAF concentration c can be used as the initial condition for inducing angiogenesis:

$$\frac{\partial c}{\partial t} = \nabla^2 c - \theta c \tag{8}$$

(2) Degradation of ECM

ECM degradation as Mantzaris et al. [18] pointed out, it is necessary to account for the fact that the ECM should be degraded during the approach. In order to make way for the capillary network process, we assume that the ECM degradation rate follows the first-order dynamic reaction and is proportional to the MDE concentration. Thus, we modelled ECM degradation using Eqs. (9–11):

$$\frac{\partial f}{\partial t} = \beta n - \gamma m f \tag{9}$$

$$\frac{\partial m}{\partial t} = \alpha n + \varepsilon \nabla^2 m - \nu m \tag{10}$$

$$\frac{\partial M}{\partial t} = -\omega M m \tag{11}$$

M and m represent the concentration of ECM and MDE, respectively, ω is a positive degradation constant, and the minus represents the opposite direction of the ECM concentration change [19].

3.2 Parameter Settings

First, we should consider the length of the two-dimensional domain at an appropriate range. From the experiments of Gimbrone, an average distance from a tumour implant to the parent vessels in the cornea is between 1 and 2 mm [20, 21].

Therefore, a length scale of L = 2 mm was taken as the initial distance between the tumour and vessels from the experiments of Folkman et al. [22]. Folkman and Klagsbrun reported that angiogenesis was initiated when this distance was 2 mm. Therefore, we took a length scale of L = 2 mm for the domain to simulate the angiogenesis [22]. We assumed $\tau = \frac{L^2}{D_c} = \frac{2\times 2}{2.5} = 1.6$ day, so we took the time scaling as 1.6 day. The doubling time of the ECs was estimated at 18 h, so the proliferation coefficient of ECs can be estimated as $\mu = \frac{18}{24\times 1.6} = 0.47$. Considering the death of ECs, the proliferation coefficient of ECs should be less than 0.47, so we estimated it as $\mu = 0.3$.

After non-dimensionalization [17], the parameters in these equations are as follows: $M_0 = 0.9$, $D_c = 1, \theta = 0.003$ [19], $\mu = 0.3$ (estimated), $n_0 = 0.9$, Chaplain et al. [23] gave the MDE secrete rate coefficient as $\alpha = 10^{-6}$, and thus to make the simulation results of ECM visible and in an appropriate range, we estimate the ECM degradation coefficient to be $\omega = 2\times 10^6$.

3.3 Boundary Conditions

Appropriate initial and boundary conditions are demanded by the model. For Eq. (8), the boundary conditions for the vessel and tumour side satisfy the Diriohlet condition $c(0, y, t) = 0$, $c(1, y, t) = 1$ and on the left and right side satisfy the Neumann condition $\frac{\partial c}{\partial y} = 0$. For Eqs. (7), (9), (10) and (11), the boundary condition of the domain, subject to the non-flux conditions, satisfies the Neumann condition and can be presented as follows: $\frac{\partial n}{\partial x} = 0$, $\frac{\partial n}{\partial y} = 0$, $\frac{\partial m}{\partial x} = 0$, $\frac{\partial \mathrm{m}}{\partial \mathrm{y}} = 0$ $\frac{\partial M}{\partial x} = 0$, $\frac{\partial M}{\partial y} = 0$, and $\frac{\partial f}{\partial x} = 0$, $\frac{\partial f}{\partial y} = 0$.

3.4 Initial Conditions

For the solution of Eq. (7), the initial condition should be described. As we discussed above, the TAF is diffused from the edge of the tumour and reaches a steady state. Its initial condition of concentration can be determined by Eq. (8).

Recent studies [24–33] have implied that, for some cancers, the fibronectin density around the stroma is higher than that of the ECM. Accordingly, cancer marks its progression through increasing the ECM density, where the amount of fibrillar ECM proteins, including collagen and fibronectin, is enhanced [34]. In breast cancer, there is a nearly 10–20 fold increase in bulk stiffness in the tumour microenvironment [34].

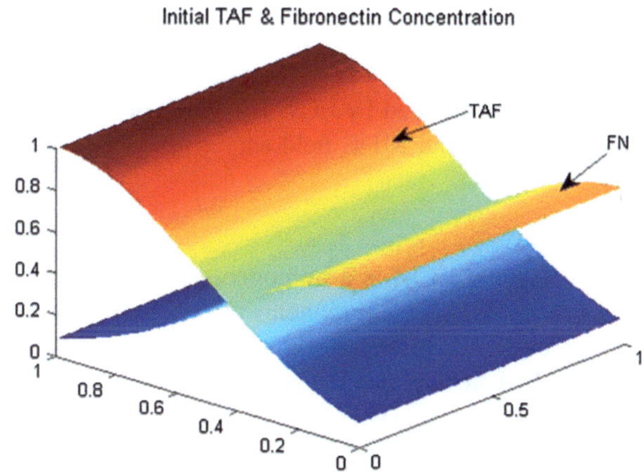

Fig. 1 The concentration of the TAF and fibronectin presented in Eqs. (5) and (6)

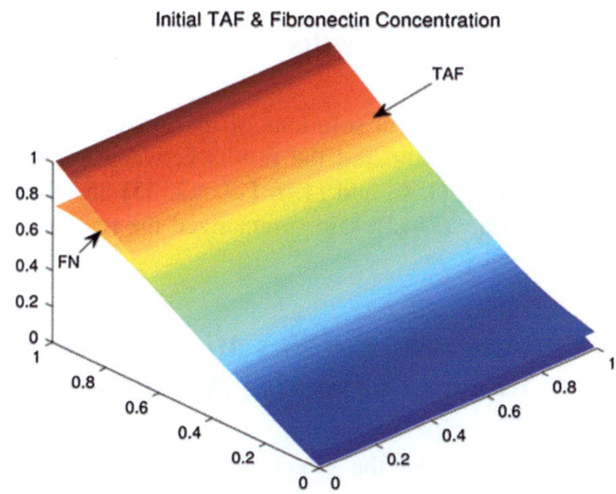

Fig. 2 Concentrations of the TAF and fibronectin from Eqs. (8) and (12) which are proposed in the current study

Based on the above, we assume the highest concentration of fibronectin is at the side of the tumour line source and the lowest concentration at the side of the parent vessel. Thus, we established the equation describing fibronectin concentration as follows:

$$f(x, y, 0) = ke^{-\frac{(1-x)^2}{\varepsilon_2}}, (x, y) \in [0, 1] \times [0, 1], \tag{12}$$

where $k < 1$, and ε is a positive constant.

From Fig. 1, we can observe that the gradient of TAF and the gradient of fibronectin are in the opposite direction. From Fig. 2, we can observe that the gradient of TAF and the gradient of fibronectin are in the same direction. We set four clusters of ECs at the side of a parent vessel. It can be represented as:

$$n(x, y, 0) = e^{-\frac{x^2}{\ell_3}}(1 - \sin^2(10\pi y)) \quad (13)$$

MDE is stimulated by ECs. The value of MDE equals to zero. Therefore, we can describe the initial MDE field as:

$$m(x, y, 0) = 0 \quad (14)$$

We assume ECM is a homogenous density field with an initially uniform concentration of ECM components. Daub and Merks [19] propose that ECM concentration is 0.9 after non-dimensionalization. This was taken and, therefore, the initial ECM field is presented as:

$$M(x, y, 0) = 0.9 \quad (15)$$

4 Simulation Results

All of the numerical solutions of the models were obtained with finite difference methods. We compared the results between the Anderson and Chaplain's model (refer to Eqs. (1–6) and (13)) [17, 23, 35] and the improved model in this study (refer to Eqs. (7–15)) and summarized the results as follows (Fig. 3).

4.1 *Comparison of ECs Migration Between Two Models Without Considering the Effect of Haptotaxis*

To separately test the effects of the chemotaxis and haptotaxis factors on the models, first we did not include the haptotaxis factor in the models. So the migration of ECs was mainly influenced with chemotaxis. Both the Anderson and Chaplain's model and our own improved model were simulated at the condition

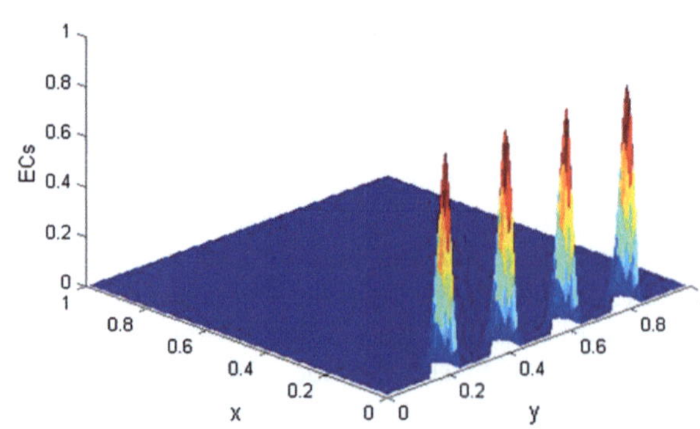

Fig. 3 The initial concentration of ECs from Eq. (13)

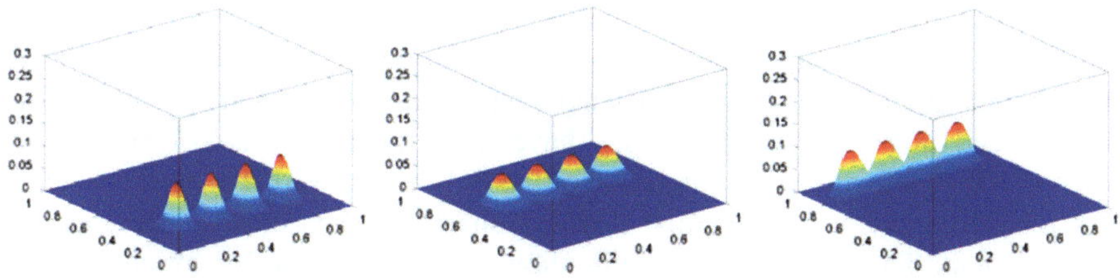

(a) $t = 1\tau$ (b) $t = 2\tau$ (c) $t = 3\tau$

Fig. 4 Spatio-temporal evolution of EC density from Anderson and Chaplain's model. The figure shows that the ECs migrate from the parent vessel ($x=0$) towards the tumour line source ($x=1$) with the influence of chemotaxis factor and without the influence of haptotaxis factor ($\rho=0$). **a** $t=1\tau$, **b** $t=2\tau$, **c** $t=3\tau$

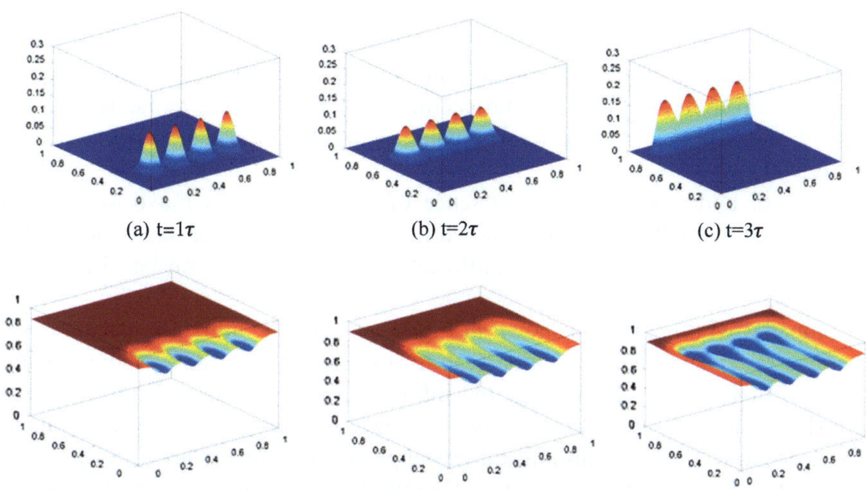

(a) t=1τ (b) t=2τ (c) t=3τ

(d) t=1τ (e) t=2τ (f) t=3τ

Fig. 5 Spatio-temporal evolution of EC density from the proposed improved model. **a–c** Show that the ECs migrate from the parent vessel ($x=0$) towards the tumour line source ($x=1$) with the influence of chemotaxis factor and without the influence of haptotaxis factor $\rho=0$. The initial four clusters of high cell density are drawn towards the line of the tumour. **d–f** Show that the ECM is hydrolyzed by the proteolytic enzyme secreted by ECs. The four clusters of ECs move forward to the line of the tumour and correspondingly leave four ditches behind them. **a** t = 1τ, **b** t = 2τ, **c** t = 3τ, **d** t = 1τ, **e** t = 2τ, **f** t = 3τ

without the influence of haptotaxis ($\rho=0$), and the results are shown in Figs. 4 and 5, respectively. From these Figs. 4 and 5, we can observe that ECs move towards the tumour line source with the same migration velocities for both two models. The moved distances covered by the ECs were the same at $t=1\tau$ (1.6 days), $t=2\tau$ (3.2 days), and $t=3\tau$ (4.8 days) for the two models. The difference between Figs. 4 and 5 is with regard to the ECs density. The ECs density is larger in Fig. 5 than that in Fig. 4 due to the term of proliferation in Eq. (7).

The same EC velocities seen in both models are due to the same chemotactic attraction. Here, chemotaxis determines the velocity of tip cell migration. We can observe that at $t = 2\tau$ (3.2 days), ECs have passed more than halfway through the domain, and at $t = 3\tau$ (4.8 days) they have migrated across almost 80% of the domain. No significant lateral migration of the ECs is observed. The four clusters of ECs retain the shape of the initial distribution, while the density decreases to a certain extent. This is because the migration is mainly controlled by chemotaxis, and at the same time diffuses outwards, leading to a small amount of lateral movement due to random motility.

4.2 *Comparison of ECs Migration Between Two Models with Considering the Effect of Haptotaxis*

Then we simulated the EC migration with the influence of both chemotaxis and haptotaxis. From the fact that the fibronectin gradient is opposite to the TAF shown in Fig. 1, the haptotaxis offsets the chemotaxis, and the influence on EC velocity is negative.

In Fig. 6, under the negative influence of haptotaxis, ECs move towards the tumour much more slowly than as shown in Fig. 4. ECs only covered one-third of the domain at $t = 3\tau$ (4.8 days). Significant overlaps of cells can be observed in Fig. 5c.

Compared with the results shown in Fig. 4, without the influence of haptotaxis, EC move faster and take three time scaling (t = 3τ) to cover 80% of the domain. However, in Fig. 6, with the negative influence of haptotaxis, ECs move much more slowly and take three time scaling (t = 3τ) to cover only 30% of the domain. This is because TAF and fibronectin are opposite in their gradient direction and counteract the movement of the ECs. Therefore, the negative influence of haptotaxis could slow down the velocity of ECs. On the other hand, if the fibronectin gradient is in the same direction as the TAF shown in Fig. 2, the influence of haptotaxis on velocity is enhanced.

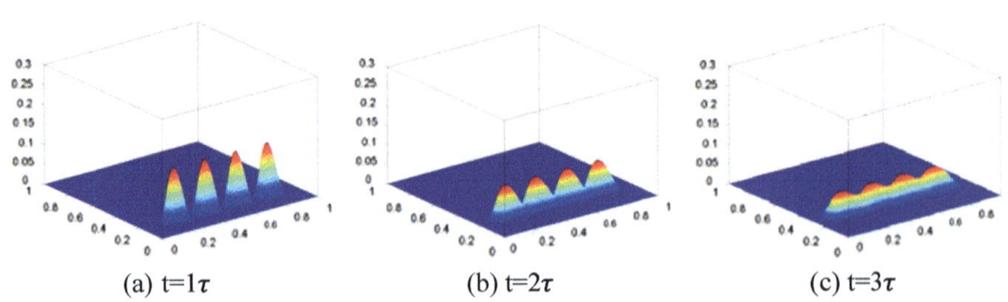

Fig. 6 Spatio-temporal evolution of ECs density in Anderson and Chaplain's model, when setting $\rho = 0.34$. **a** t = 1τ, **b** t = 2τ, **c** t = 3τ

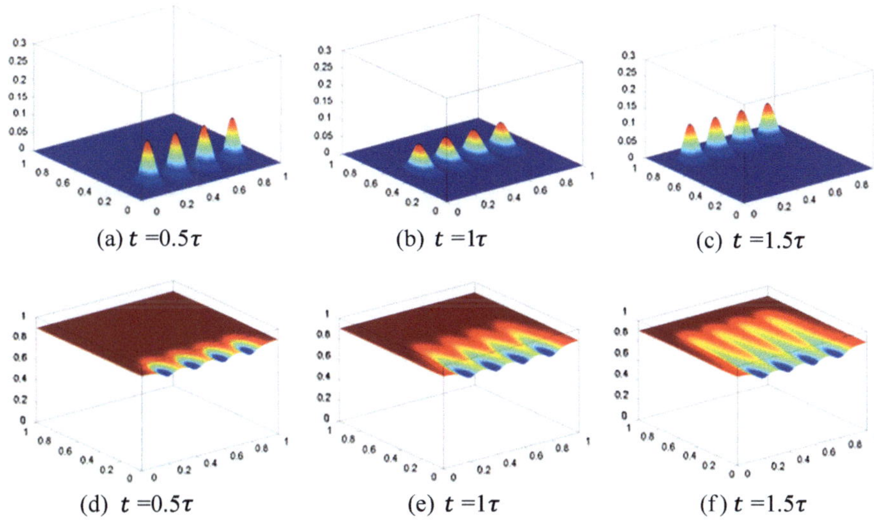

Fig. 7 Spatio-temporal evolution of ECs density from the improved model. **a–c** Show that the EC migration is influenced by chemotaxis and positive haptotaxis. **d–f** Show that the ECM is hydrolyzed by the proteolytic enzyme secreted by EC migration. ECs move forward to the tumour and correspondingly leave four ditches behind them. **a** $t = 0.5$, **b** $t = 1\tau$, **c** $t = 1.5\tau$, **d** $t = 0.5\tau$, **e** $t = 1\tau$, **f** $t = 1.5\tau$

Unlike with the results shown in Fig. 6, as shown in Fig. 7c, we considered the positive effect of haptotaxis on the velocity of ECs. We observed the four separate peaks of cells covered 80% of the domain at $t = 1.5\tau$ (2.4 days). The velocity of the ECs is the fastest because both the TAF and fibronectin gradients are in the same direction. The superposition of both TAF and fibronectin accelerates the EC migration and the ECM changes correspondingly.

4.3 *Effect of Proliferation on ECs Density*

Figure 8 shows the change in EC density with and without proliferation. In Fig. 8a, without proliferation, EC density is kept at conservation whereas with proliferation, it increases towards the tumour. In Fig. 8b, without proliferation, the trend of its average density for each cluster is declining in Eq. (1). This is due to the term of random motility which makes it diffuse, and its density per unit area is declining. With proliferation (see Eq. (7)), the trend of its average density for each cluster remains higher than that in Eq. (1).

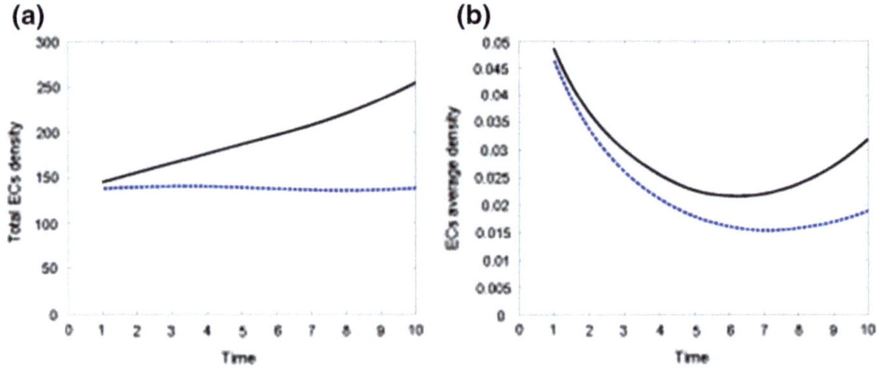

Fig. 8 The effect of proliferation on the total ECs density (**a**) and average ECs density (**b**). *Dash lines* show the results with proliferation

4.4 Results of ECM

The simulation results of ECM are shown in Figs. 5d–f and 7d–f, respectively. The ECM is proteolyzed by MDE secreted by ECs and degraded simultaneously when the ECs clusters move forward, forming a "ditch" in the ECM. This indicates that the density of the ECM decreases because of degradation. The front of the "ditches" is at the same distance from the parent vessel as that of ECs, which reveals that ECM approaches the tumour correspondingly with the ECs. It can be clearly seen that when the four clusters of ECs migrate forward, its enzymes, such as MMPs, degrade the ECM and leave the "ditch" behind them. Thus, we assume that the degradation can take place only at the front, and the four clusters of ECs are extremely actively, leaving the "ditches" behind it. It can also be clearly observed that in Fig. 5d–f, the "ditches" are deeper and wider than those in Fig. 7d–f. This indicates that the faster the ECs migrate, the shallower the "ditches" will be. The more slowly the ECs migrate, the deeper and wider are the "ditches".

5 Comparison of the Evolution of Capillary Network Between the Two Models

Agent-based modelling (ABM) is a computational modelling approach providing a systematic view of the simulation of action and interaction between autonomous individual entities [36]. Cellular automata (CA) is another framework for defining the interacting components in a system [37]. Since CA is not capable of representing a large amount of data transformation between entities due to the increasing complexity of the input symbols and the number of states, modelling tumour-induced angiogenesis requires the application of an agent-based approach

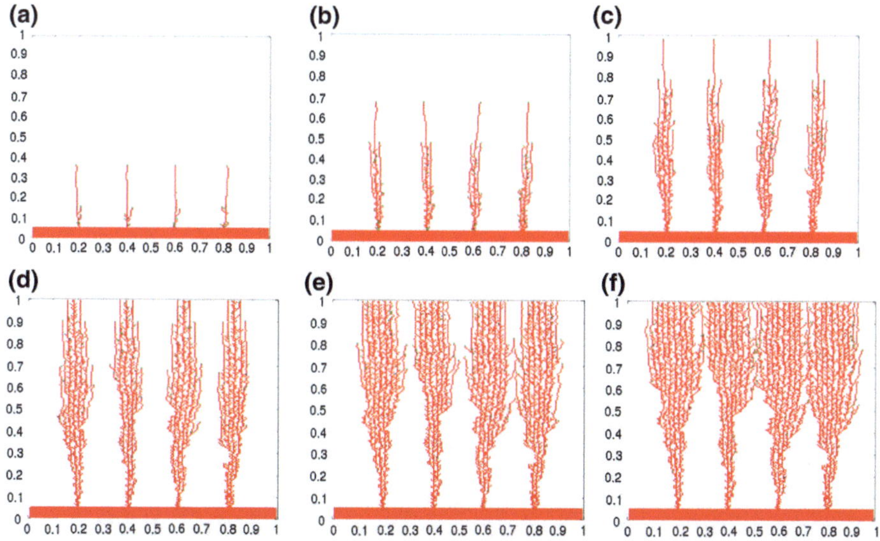

Fig. 9 Spatio-temporal evolution of capillary network with ABMs to discretize the continuum model where the ECM degradation is not involved. The figure shows ECs at the capillary sprout tips migrating from the parent vessel ($x = 0$) towards a tumour ($x = 1$)

which provides a more powerful state machine to allow for an intuitive exploration of the dynamics and the complexity of biological systems. We used ABM to discretize the mathematical model described above, so that we could obtain the feature of the capillary sprouts and individual behaviour. The discrete capillary sprouts are shown as follows.

In Fig. 9, we can observe that the four clusters of vessels develop from the parent vessel across the domain and finally reach the tumour. In front of the vessels are the sprout heads by the endothelial tip cells. In the simulation, the ECM degradation is not involved into the continuous model, so no background of ECM can be observed and it seems that the capillary network formed mechanically.

In Fig. 10, we can observe that the four clusters of vessels develop from the parent vessel across the domain and finally reach the tumour. In front of the vessels are the sprout heads by the endothelial tip cells. When the ECs approach, they secret MDE and proteolyze the ECM, making the density of ECM decline alone in the way the ECs passed. In this way, the endothelial tip cells remove barriers and make way for the vessels to develop. This can be observed from the background with different colours and contour lines. In the background, the local density of ECM adjacent to the vessels appears to change gradually through blue, green, and yellow to red towards the outside, which means the ECM density is the lowest at the vessels due to being proteolyzed. At the same time, the directions of the nascent sprouts are mainly guided by the TAF gradient and also influenced by the density of

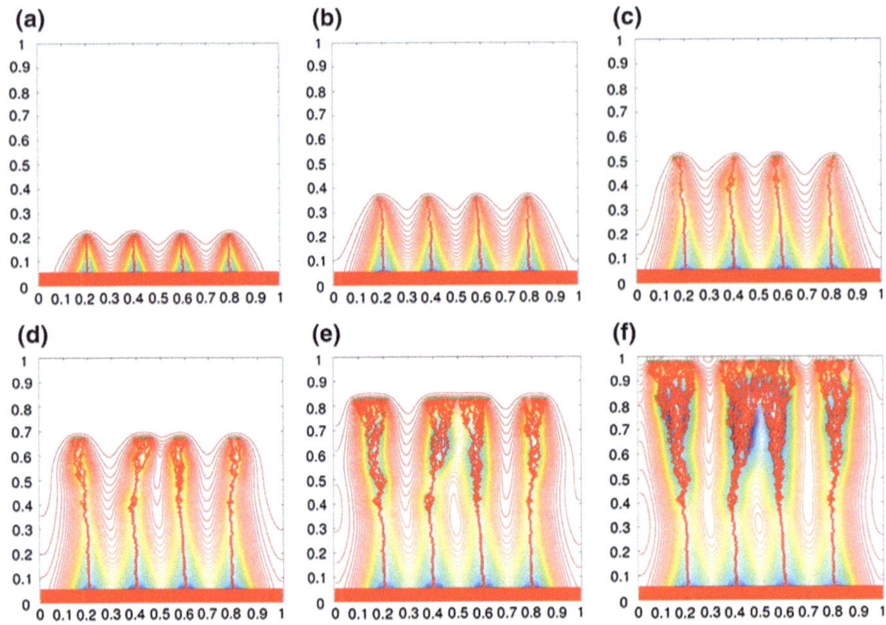

Fig. 10 Spatio-temporal evolution of capillary network with ABM to discretize the continuum model where the ECM degradation is involved. The figure shows the ECs at the capillary sprout tips migrating from the parent vessel ($x = 0$) towards a tumour ($x = 1$). The background formed with *contour lines* and *colours* shows the ECM density degraded by the MDE secreted from the ECs

local ECM, making the morphology of the vessels appear tortuous. We can also observe that the density of vessels is increasing, forming a "brush border" approaching the tumour, which demonstrates the density of endothelial tip cells increasing due to EC proliferation.

6 Discussion

The simulation results show an accelerated endothelial tip cell migration towards the tumour, which implies a possible mechanism that promotes angiogenesis in some subgroups of invasive tumours.

According to Shekhar, Pauley and Heppner, 'the invasive carcinoma is often associated with expansion of the tumour stroma and increased deposition of ECM' [38]. Such an increased deposition of ECM in tumours is known as desmoplasia and is similar to changes that are observed during organ fibrosis. According to Folkman, solid tumours vascularize more actively than the normal tissue [38]. The fibroblast is an important stromal cell type in cancer progression [24].

Therefore, we obtained the important points as follows.

1. Without haptotaxis, the clusters of EC density migrate directly across the domain to the tumour.
2. With negative haptotaxis in which the fibronectin gradient is opposite to the TAF gradient, the clusters of EC density migrate more slowly, with lateral movement.
3. With positive haptotaxis in which the fibronectin gradient is in the same direction as the TAF gradient, the clusters of EC density migrate most quickly. This means angiogenesis is promoted by the superposition of both haptotaxis and chemotaxis.

Tumours can be initiated and developed by the interactions between malignant cells and cancer-associated fibroblasts (CAFs) [25, 26]. Through modulating stromal ECM, CAFs produce directional paths of fibronectin in order for tumour invasion to occur [27]. Fibronectin is critical in many cell processes, such as cell adhesion, migration, growth and differentiation [39]. Fibronectin is expressed in many different cell types and its expression is activated during epithelial-mesechymal transition (EMT) [28]. This is a process which facilitates cell motility and invasion [29–31].

The increased expression of fibronectin has been observed by Lundberg [32] in a subgroup of colorectal cancer (CRC) classified as CIMP-negative tumours. They point out that fibroblasts induce CIMP-negative tumour cells to produce more fibronectin. They indicate that the density of fibroblasts is higher at the front of the tumour than the normal tissue. Similar results have been reported by Derya, Yilmaz and Aytekin in Small Cell Lung Cancer (SCLC), an aggressive form of lung cancer. The ECM surrounding SCLC cells contains a great amount of fibronectin [33]. Therefore, we think that there may be a density gradient of the fibronectin which is expressed as Eq. (12). This generates a positive haptotatic effect and augments the velocity of ECs shown in Figs. 2 and 7. The quick migration of ECs approaching the tumour means that angiogenesis occurs quickly when the invasive tumour produces a high concentration of fibronectin in the stroma and its adjacent microenvironment.

Previous studies have suggested that fibronectin could be generated by various types of cells, including tumour cells [40]. Fibronectin enhanced the secretion of MMPs and promoted the invasive migration of gallbladder cancer (GBC) cells. Stromal fibronectin is produced by tumour cells. Recent studies in human melanoma and ovarian cancer have revealed that MMPs could cleave fibronectin into shorter fragments, and facilitate the fibronectin integrin adhesive interactions. This process in turn reinforces the generation of MMPs, forming a positive feedback loop for the invasive migration of GBC cells [41].

Angiogenesis plays a major role in solid tumour growth, because it links the harmless avascular growth phase and the fatal vascular growth phase [7]. In our model, we assume that some subgroups of invasive tumour, such as CAFs or GBCs, can produce fibronectin and create a higher fibronectin concentration for the cancer cell migration, which results in a positive haptotatic effect and accelerates angiogenesis. Finally, the rapid growth of vascular tumours contributes to cancer cell progression known as metastasis for the establishment of a new colony in distant organs [42]. Therefore, we assume that there is a possible mechanism through which tumour-invasion and tumour-induced angiogenesis promote each other, making the tumour develop faster.

Firstly, we improved the aspect of the degradation of ECM because it should be degraded to make way for the development of the capillary network with the migration and proliferation of the approach of ECs to the tumour. Our simulation results provided above clearly show that ECM is degraded during angiogenesis. As a result, the degradation dynamics of ECM is catalyzed by the protease enzyme [18].

Second, we improved Anderson and Chaplain's work of tumour-induced angiogenesis in the aspect of fibronectin gradient direction. They assume that concentration of fibronectin is higher at the side of the parent vessel than at the side of the tumour, so they set Eq. (6). However, we think that there is another possibility in which the fibronectin concentration is higher at the tumour side than at the vessel side. This may happen in some invasive tumour situations, so we set Eq. (12). Our simulation results show that during angiogenesis the ECs move much faster as the TAF and fibronectin gradient are in the same direction. The synthetic effect of both chemotaxis and haptotaxis in the same direction accelerates the movement of the ECs.

According to Cao et al. [41], the overexpression of fibronectin will promote cancer progression. Their experimental results indicate that exogenous fibronectin significantly enhances and promotes proliferation and metastasis. Our simulation results show that the migration of ECs can be accelerated by the invasive tumour. This means that invasive tumours generate fibronectin, which enhances haptotaxis and promotes angiogenesis. According to the consensus theory, angiogenesis supplies oxygen and nutrients to promote tumour development. We propose that there is a possible mechanism of tumour-invasion and tumour-induced angiogenesis promoting each other, and making the tumour develop faster, and this at the very least becomes more harmful for some subgroups of tumours.

7 Conclusion

Tumour progression promotes overexpression of fibronectin. The overexpression of fibronectin generates a fibronectin concentration gradient which is in the same direction of TAF concentration gradient. The fibronectin gradient generates the positive haptotaxis which accelerates the ECs migration. The capillary vessels formed through accelerated angiogenesis supply oxygen and nutrients to the tumour and make it more invasive. Our conclusions agreed with this point and pointed out the migration of ECs can be accelerated by the invasive tumour. Thus we argue that there may be a possible mechanism for the cross promotion between tumour-invasion and tumour-induced angiogenesis, and we identify this as our future work. Moreover, the development of vascular systems also associates strongly with the behaviours of neural systems since both employ complex functions to accomplish network behaviours. We will also explore such connections between these aspects to motivate future research. Moreover, the proposed research also shows great potential in contributing to the understanding and explanation of how large recurrent neuronal assemblies develop and function in complex biological environments.

References

1. Adams, R.H., Alitalo, K.: Molecular regulation of angiogenesis and lymphangiogenesis. Nat. Rev. Mol. Cell Biol. **8**(6), 464–478 (2007)
2. Hanahan, D., Weinberg, R.A.: Hallmarks of cancer: the next generation. Cell **144**(5), 646–674 (2011)
3. Zhou, W., Wang, G., Guo, S.: Regulation of angiogenesis via Notch signaling in breast cancer and cancer stem cells. Biochim. Biophys. Acta **1836**(2), 304–320 (2013)
4. Blanco, R., Gerhardt, H.: VEGF and Notch in tip and stalk cell selection. Cold Spring Harbor Perspect. Med. **3**(1) (2013)
5. Folkman, J.: Tumor angiogenesis: therapeutic implications. New Engl. J. Med. **285**(21), 1182–1186 (1971)
6. Zacchigna, S., Almodovar, C.R., Carmeliet, P.: Similarities between angiogenesis and neural development: what small animal models can tell. Curr. Top. Dev. Biol. **80**(2007), 1–55 (2007)
7. McDougall, S.R., Anderson, A.R.A., Chaplain, M.A.J.: Mathematical modelling of dynamic adaptive tumour-induced angiogenesis: clinical implications and therapeutic targeting strategies. J. Theor. Biol. **241**(3), 564–589 (2006)
8. McDougall, S.R., Anderson, A.R.A., Chaplain, M.A.J., Sherratt, J.A.: Mathematical modelling of flow through vascular networks: implications for ibmour-induced angiogenesis and chemotherapy strategies. Bull. Math. Biol. **64**(4), 673–702 (2002)
9. Stephanou, A., McDougall, S.R., Anderson, A.R.A., Chaplain, M.A.J.: Mathematical modelling of flow in 2D and 3D vascular networks: applications to anti-angiogenic and chemotherapeutic drug strategies. Math. Comput. Model. **41**(10), 1137–1156 (2005)
10. Lacovara, J., Cramer, E.B., Quigley, J.P.: Fibronectin enhancement of directed migration of B-16 melanoma-cells. Cancer Res. **44**(4), 1657–1663 (1984)
11. Terranova, V.P., Diflorio, R., Lyall, R.M., Hic, S., Friesel, R., Maciag, T.: Human-endothelial cells are chemotactic to endothelial-cell growth-factor and heparin. J. Cell Biol. **101**(6), 2330–2334 (1985)
12. Mccarthy, J.B., Furcht, L.T.: Laminin and fibronectin promote the haptotactic migration of B-16 mouse melanoma-cells invitro. J. Cell Biol. **98**(4), 1474–1480 (1984)
13. Hynes, R.O.: Fibronectins. Sci. Am. **254**(6), 42–51 (1986)
14. Clark, R.A.F., Dvorak, H.F., Colvin, R.B.: Fibronectin in delayed-type hypersensitivity skin reactions—associations with vessel permeability and endothelial-cell activation. J. Immunol. **126**(2), 787–793 (1981)
15. Clark, R.A.F., Dellapelle, P., Manseau, E., Lanigan, J.M., Dvorak, H.F., Colvin, R.B.: Blood-vessel fibronectin increases in conjunction with endothelial cell-proliferation and capillary ingrowth during wound-healing. J. Investig. Dermatol. **79**(5), 269–276 (1982)
16. Paku, S., Paweletz, N.: 1st steps of tumor-related angiogenesis. Lab. Invest. **65**(3), 334–346 (1991)
17. Anderson, A.R.A., Chaplain, M.A.J.: Continuous and discrete mathematical models of tumor-induced angiogenesis. Bull. Math. Biol. **60**(5), 857–899 (1998)

18. Mantzaris, N.V., Webb, S., Othmer, H.G.: Mathematical modeling of tumor-induced angiogenesis. J. Math. Biol. **49**(2), 111–187 (2004)
19. Daub, J.T., Merks, R.M.H.: A cell-based model of extracellular-matrix-guided endothelial cell migration during angiogenesis. Bull. Math. Biol. **75**(8), 1377–1399 (2013)
20. Gimbrone Jr., M.A., Cotran, R.S., Leapman, S.B., Folkman, J.: Tumor growth and neovascularization: an experimental model using the rabbit cornea. J. Natl. Cancer Inst. **52**(2), 413–427 (1974)
21. Muthukkaruppan, V.R., Kubai, L., Auerbach, R.: Tumor-induced neovascularization in the mouse eye. J. Natl. Cancer Inst. **69**(3), 699–708 (1982)
22. Folkman, J., Klagsbrun, M.: Fundamental Aspects of Neoplasia. Springer (1975)
23. Chaplain, M.A.J., McDougall, S.R., Anderson, A.R.A.: Mathematical modeling of tumor-induced angiogenesis. Annu. Rev. Biomed. Eng. **8**, 233–257 (2006)
24. Elenbaas, B., Weinberg, R.A.: Heterotypic signaling between epithelial tumor cells and fibroblasts in carcinoma formation. Exp. Cell Res. **264**, 169–184 (2001)
25. Liu, M., Xu, J., Deng, H.: Tangled fibroblasts in tumor-stroma interactions. Int. J. Cardiol. **129**, 1795–1805 (2011)
26. Mueller, M.M., Fusenig, N.E.: Friends or foes—bipolar effects of the tumour stroma in cancer. Nat. Rev. Cancer **4**, 839–849 (2004)
27. Lee, H.O., Mullins, S.R., Franco-Barraza, J., Valianou, M., Cukierman, E., Cheng, J.D.: FAP-overexpressing fibroblasts produce an extracellular matrix that enhances invasive velocity and directionality of pancreatic cancer cells. BMC Cancer **11**, 245 (2011)
28. Cervantes-Arias, A., Pang, L.Y., Argyle, D.J.: Epithelial-mesenchymal transition as a fundamental mechanism underlying the cancer phenotype. Vet. Comp. Oncol. (2012)
29. Jia, D., Yan, M., Wang, X., Hao, X., Liang, L., Liu, L., Kong, H., He, X., Li, J., Yao, M.: Development of a highly metastatic model that reveals a crucial role of fibronectin in lung cancer cell migration and invasion. BMC Cancer **10**, 364 (2010)
30. Wei, P.L., Kuo, L.J., Huang, M.T., Ting, W.C., Ho, Y.S., Wang, W., An, J., Chang, Y.J.: Nicotine enhances colon cancer cell migration by induction of fibronectin. Ann. Surg. Oncol. **18**, 1782–1790 (2011)
31. Meng, X.N., Jin, Y., Yu, Y., Bai, J., Liu, G.Y., Zhu, J., Zhao, Y.Z., Wang, Z., Chen, F., Lee, K.Y.: Characterisation of fibronectin-mediated FAK signalling pathways in lung cancer cell migration and invasion. Br. J. Cancer **101**, 327–334 (2009)
32. Lundberg, I.: Fibroblasts and ECM in colorectal cancer. Umeå University (2012)
33. Derya, M., Yılmaz, I., Aytekin, M.: The role of extracellular matrix in lung diseases. Biol. Med. **6**(200) (2014)
34. Kumar, S., Das, A., Sen, S.: Extracellular matrix density promotes EMT by weakening cell-cell adhesions. Mol. BioSyst. **10**(4), 838–850 (2014)
35. Stephanou, A., McDougall, S.R., Anderson, A.R.A., Chaplain, M.A.J.: Mathematical modelling of the influence of blood rheological properties upon adaptative tumour-induced angiogenesis. Math. Comput. Model. **44**(1–2), 96–123 (2006)
36. Macal, C.M., North, M.J.: Tutorial on agent-based modelling and simulation. In: Book Tutorial on Agent-Based Modelling and Simulation, 2010 edn., pp. 151–162
37. Mallet, D.G., De Pillis, L.G.: A cellular automata model of tumor-immune system interactions. J. Theor. Biol. **239**(3), 334–350 (2006)
38. Kalluri, R., Zeisberg, M.: Fibroblasts in cancer. Nat. Rev. Cancer **6**(5), 392–401 (2006)
39. Pankov, R., Yamada, K.M.: Fibronectin at a glance. J. Cell Sci. **115**, 3861–3863 (2002)
40. Kamoshida, G., Matsuda, A., Miura, R., Takashima, Y., Katsura, A., Tsuji, T.: Potentiation of tumor cell invasion by co-culture with monocytes accompanying enhanced production of matrix metalloproteinase and fibronectin. Clin. Exp. Metastasis **30**(3), 289–297 (2013)
41. Cao, Y., Liu, X., Lu, W., Chen, Y., Wu, X., Li, M., Wang, X.-A., Zhang, F., Jiang, L., Zhang, Y., Hu, Y., Xiang, S., Shu, Y., Bao, R., Li, H., Wu, W., Weng, H., Yen, Y., Liu, Y.:

Fibronectin promotes cell proliferation and invasion through mTOR signaling pathway activation in gallbladder cancer. Cancer Lett. **360**(2), 141–150 (2015)
42. Chaplain, M.A.J., Lolas, G.: Mathematical modelling of cancer cell invasion of tissue: The role of the urokinase plasminogen activation system. Math. Models Methods Appl. Sci. **15** (11), 1685–1734 (2005)

Noise Reduction in ECG Signals Using Wavelet Transform and Dynamic Thresholding

Diptangshu Pandit, Li Zhang, Chengyu Liu, Nauman Aslam, Samiran Chattopadhyay and Chee Peng Lim

Abstract Biomedical signals produced by mobile sensors usually carry various noises. This poses great challenges for the subsequent signal processing and disease analysis. Thus, noise removal becomes an important step of signal processing. This research proposes a noise reduction algorithm which can be applied to noisy ECG (electrocardiogram) signals to obtain a higher signal-to-noise ratio (SNR) for further processing. The proposed algorithm utilises wavelet transform and dynamic thresholding to reduce specific types of noise embedded in raw ECG signals. To prove the efficiency of the proposed algorithm, we employ a half-hour-long real ECG signal and add different types of noise for the evaluation of the proposed algorithm. We also compare the results obtained using different families of wavelets and different decomposition levels. The experimental results show that the proposed algorithm is able to produce a higher SNR in the output signal than that in the raw test signals.

Keywords Noise reduction · ECG analysis · Wavelet transform

D. Pandit · L. Zhang (✉) · N. Aslam
Computational Intelligence Research Group, Faculty of Engineering and Environment, Department of Computing Science and Digital Technologies, University of Northumbria Newcastle, NE1 8ST, Newcastle, UK
e-mail: li.zhang@northumbria.ac.uk

C. Liu
Institute of Biomedical Engineering Shandong University Jinan, Jinan, China

S. Chattopadhyay
Department of Information Technology, Jadavpur University Kolkata, Kolkata, India

C.P. Lim
Institute for Intelligent Systems Research and Innovation, Deakin University, Waurn Ponds, VIC 3216, Geelong, Australia

A. Bhatti et al. (eds.), *Emerging Trends in Neuro Engineering and Neural Computation*, Series in BioEngineering,
DOI 10.1007/978-981-10-3957-7_10

1 Introduction

Electrocardiogram (ECG) represents different activities of the heart during the cardiac cycle [1]. Physicians use ECG signals to detect and diagnose multiple types of cardiac arrhythmias. In automated ECG processing, the raw signal has to go through multiple processing stages for detection of disease or abnormality. For instance, in the first stage, the signal is pre-processed for noise reduction. The next stage is feature extraction which is followed by the classification process. The pre-processing or noise reduction stage plays a very important role since noisy signals might affect the subsequent processes, and result in errors in the final outcome. Moreover, noise reduction or cancellation is not only useful for ECG signals, but also helpful in processing many other biomedical signals, such as Electroencephalogram (EEG), Electromyogram (EMG), Electrooculography (EOG), Galvanic skin response (GSR), Magnetoencephalogram (MEG) and Mechanomyogram (MMG). Figure 1 shows different types of noise in ECG signals, as taken from the MIT-BIH Arrhythmia database [2].

The noise removal process separates a valid part of the signal by cancelling unwanted artefacts which are not influenced by the electrical activities of the heart. The common noise categories include powerline interference, motion artefact, baseline wander, muscle contraction and electrode contact noise, as follows.

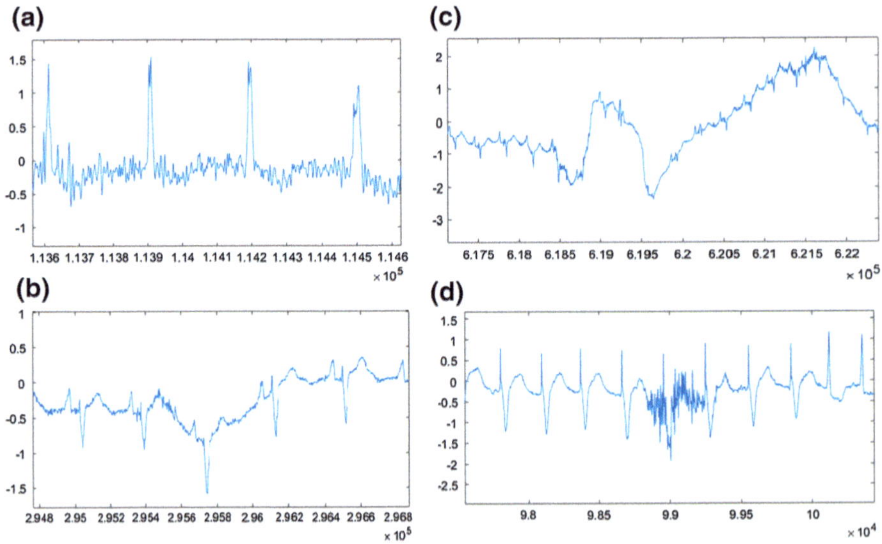

Fig. 1 Different types of noises in ECG signals, including **a** powerline interference, **b** motion artefact, **c** baseline wander and **d** EMG noise. Y axis represents the amplitude in mV and X axis denotes the sample number

- ***Powerline interference***

This type of noise consists of 50 Hz or 60 Hz (and multiples) pickups and harmonics. The amplitude of the noise varies up to 50% of the ECG amplitude peak [3]. The common causes include improper grounding, interference with other electrical equipment (e.g., motors, air conditioners), stray effect of the alternating current, etc. An example signal is shown in Fig. 1a.

- ***Motion Artefact***

The transient baseline changes in ECG signals due to the difference in the electrode-skin impedance with electrode motion are known as motion artefact [4]. This type of noise might cause very large spikes. It has a frequency typically less than 0.5 Hz except for the abrupt shifts (due to motions). A signal distorted by motion artefacts is illustrated in Fig. 1b.

- ***Baseline Wander***

Respiration or body movement causes this type of low frequency noise. Baseline wander causes problems for detecting R peaks of the signal. Because of this baseline wander, the T peak of the ECG signal could appear higher than the R peak [5]. Figure 1c shows an example of an ECG signal with baseline wander.

- ***Muscle Contraction***

Muscle contraction noise, also known as EMG noise, is caused by the electrical activities of body muscles [6]. This type of noise might have a frequency range from 20 to 10 kHz. The amplitude of this noise could go up to 10% of the QRS amplitude. A sample signal with this noise is illustrated in Fig. 1d.

- ***Electrode Contact Noise***

This category of noise appears due to the discontinuous connection between the ECG electrode and patient skin during signal acquisition.

As mentioned earlier, noise removal is a very important step and it highly impacts the subsequent ECG signal processing. It has drawn increasing research interests in the related field. Indeed, noise reduction using filtering has become a very popular research area in domain of signal processing. Different types of filters have been used for the ECG pre-processing such as bandpass filter, notch filter, finite impulse response (FIR) and infinite impulse response (IIR) filters, etc. A comprehensive study of employed notch filter is presented by Bai et al. [7]. In their work, the mean square error was used in performance comparison while demonstrating the adverse effect of higher order filter. Rani et al. [8] presented a comparison of multiple FIR and IIR filters and illustrated the superiority of the IIR filter over the FIR filter in terms of resource requirements. Other types of filters include adaptive and morphological filters [9, 10], median and bandpass filters [11, 12] and quantisation [13]. Recently, wavelet transform has gained popularity in automated ECG processing. As ECG is a non-stationary signal, wavelet methods are highly suitable for automatic ECG signal processing. Many wavelet noise removal techniques are proposed in the literature [2, 5, 14–19]. These include Fourier transform [20], Principal Component Analysis [21], Segmented-Beat Modulation [22], Independent Component Analysis [23] and Ensemble Empirical Mode Decomposition [24]. In this research, we propose a noise reduction technique using stationary wavelet transform (SWT) [25] and dynamic thresholding to deal with the removal of various types of ECG noises. We also demonstrate the effectiveness of the proposed algorithm by conducting an empirical evaluation with noisy ECG signals.

The rest of the article is structured as follows. In Sect. 2, we introduce the proposed noise reduction algorithm based on wavelet transform and dynamic thresholding. Section 3 presents the experimental study and outcomes. Finally, concluding remarks and directions for further work are presented in Sect. 4.

2 The Proposed Noise Reduction Algorithm

In this section, we propose a noise reduction method that incorporates five main steps as illustrated in Fig. 2. In the first step, the baseline is corrected using moving window average. Then, the signal is decomposed using discrete stationary wavelet transform. From the decomposed signal, thresholds are computed dynamically in the third step. Next, the computed thresholds are applied to the decomposed signal. Finally, the output signal is recomposed using inverse discrete stationary wavelet transform.

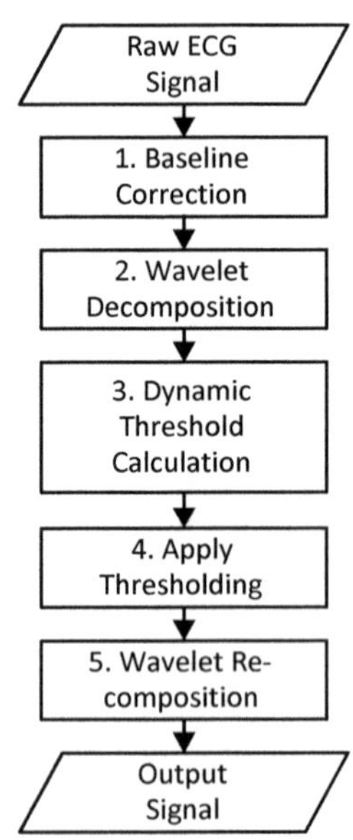

Fig. 2 Flowchart of the proposed noise reduction method

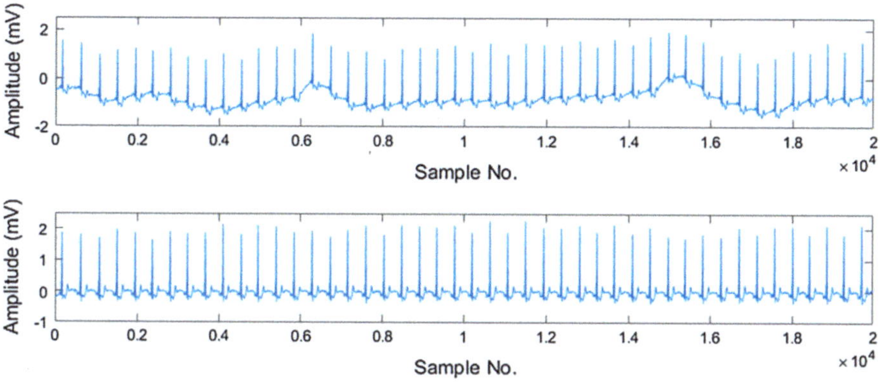

Fig. 3 An ECG signal with baseline problem (*top*) and a baseline corrected signal (*bottom*)

2.1 Baseline Correction

In this step, we correct any baseline-related deviations of the ECG signal. This is a necessary step as the subsequent dynamic threshold generation will be affected if the signal refers to a wrong baseline. We use a simple baseline removal technique, which subtracts the averaging window (around the current sample) of half a second from the current sample to remove baseline. The operation is defined as follows:

$$y_b(x) = \begin{cases} y(x) - \frac{1}{w}\sum_{i=1}^{w} y\left(x+i-\frac{w}{2}\right) & \textit{If } \frac{w}{2} \le x \le size(y) - \frac{w}{2} \\ y(x) & \textit{Otherwise} \end{cases} \quad (1)$$

$$w = round(fs \times 0.5) + round(fs \times 0.5)\%2 \quad (2)$$

where $y(x)$ denotes the raw ECG signal (with a baseline problem) and $yb(x)$ denotes the corrected signal. The window size w is an even number, which is close to the number of samples in half a second. An example using this technique is illustrated in Fig. 3, where the raw signal is shown on top of the figure and the corrected signal is plotted at the bottom.

2.2 Wavelet Decomposition

We use SWT method in this step. A detailed discussion on SWT can be found in [25]. The decomposition process depends on the type of wavelet family used and the decomposition level N. The output of the decomposed signal contains both the detailed coefficients for each level (up to N) and the detailed approximation coefficients of level N. An example using level 3 SWT decomposition is shown in

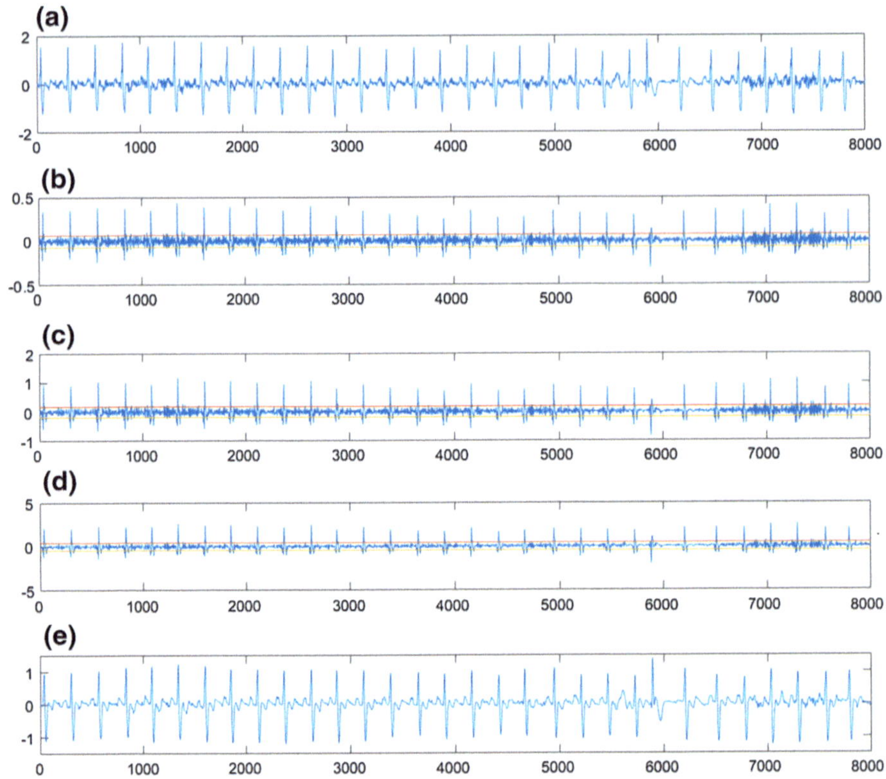

Fig. 4 **a** Noisy signal, **b–d** the decomposed signals (detailed coefficients and thresholds) and **e** the filtered signal

Fig. 4. The noisy (but with baseline corrected) signal (i.e. signal a) is illustrated in the top row of Fig. 4. The next three signals (b, c and d) in Fig. 4 represent resulting decomposed signals with the last row representing the filtered signal (i.e. signal e).

2.3 Threshold Calculation

In this stage, we calculate amplitude approximation thresholds for each decomposed signal. The threshold is calculated using a certain percentage value *tp*. This indicates that only about *tp*% of the sample values exist, and they are above the output cut-off threshold *ta*. The following algorithm is used for the threshold calculation.

In Fig. 4, an example is provided where the threshold values are calculated using $tp = 10\%$. The resulting *ta* is marked as positive and negative straight (orange and yellow) lines for each decomposed signal (b, c and d).

```
Algorithm 1: Cut-off threshold calculation
Input:
tp      //Threshold percentage
y(x)    //The decomposed ECG signal
Output:
ta      //Cut-off threshold for the decomposed ECG signal
Begin
{
  //Find the average of the mean of the absolute signal and the peak of the signal as the starting threshold
  t = [(Σ_{i=1}^{L} abs(y(i)))/L + max(y)]/2   //L is the number of elements in the signal
  dr = 0.001;                    //Decrement amount
  s = true;
  While s = true do
        t = t − dr ;
        n = Σ_{i=1}^{L} pos(abs(y(i)));  // pos(x) returns x if x > 0; otherwise it returns 0.
        If n/L > tp
                s = false;
        End
  End
Output ta
}
End
```

2.4 Threshold Application

Next, the calculated threshold is applied to all the decomposed signals. The following formula is used to apply the threshold to the corresponding decomposed signal.

$$y_a(x) = \mathrm{sign}(y(x)) \times \mathrm{srn}(|y(x)| - T), \tag{3}$$

where sign(x) returns the sign of x; srn(x) returns x if the value of $x > 0$; otherwise it returns 0; y_a represents the output signal after applying the threshold on the input signal, y. In Fig. 4b, c, d, the threshold values are marked as two straight lines along the X axis (for each signal) with both having the same positive and negative amplitudes. All the values within the threshold regions will be reduced to 0.

2.5 Wavelet Re-Composition

The filtered decomposed signals are used in this stage to reconstruct the final filtered ECG signal. This step involves inverse discrete stationary wavelet transform. Detailed inner workings of this process can be found in [25].

3 Experimental Study and Results

In this section, we conduct a series of experiments using different wavelet families as well as different levels of decomposition. All of our experiments are conducted using MATLAB. In this section, we discuss the experimental data, the experiment method and analyse the results obtained.

3.1 Experimental Data

We used 'signal 100' from the MIT-BIH Arrhythmia [2] database as the raw test signal. Different types of noise were added to this ECG signal, which result in varied signal-to-noise ratio (SNR) for each of the generated test signals. The added noise included both 60 Hz sinusoidal signal and Gaussian white noise. A set of 20 noisy signals was generated by integrating time varying and amplitude varying noise signals with the raw ECG signal (i.e. signal 100). Each of the generated noisy signals has increasing noise power, resulting in SNR starting from −0.0182 dB up to −4.2091 dB. The base waveform contains a half-hour-long real ECG recording sampled at 360 Hz. Each of the generated noisy signals had the same length containing different degrees of power and variation of noise. Table 1 summarises SNR values for all noisy test signals.

3.2 Evaluation Results

We tested the proposed algorithms using the 'Daubechies' family of wavelets (DB2, DB3 and DB4) with the decomposition levels of 3, 4 and 5. Two tests were conducted. In the first test, we assumed $tp = 10\%$. The first test results are shown in Table 2. The same results are depicted pictorially by plotting them in Figs. 5 and 6. In Fig. 5, we compared the SNR output using the same family of wavelet ('DB1') and varying decomposition levels for each test signal. It can be observed that increasing the decomposition level greatly improves the final SNR output. In Fig. 6, we compared the outputs using the same level of decomposition but different families of wavelet for each signal. It can be seen that the outputs from different

Table 1 A summary of SNR values in dB for all the test signals

Signal No.	1	2	3	4	5	6	7	8	9	10	11
SNR	−0.01	−0.06	−0.14	−0.24	−0.38	−0.53	−0.71	−0.91	−1.12	−1.35	−1.58
Signal No.	10	11	12	13	14	15	16	17	18	19	20
SNR	−1.35	−1.58	−1.83	−2.08	−2.34	−2.60	−2.86	−3.12	−3.39	−3.65	−3.91

Table 2 Comparison of the final SNR (in dB) of the output signals using the proposed system with $tp = 10\%$

Wavelet family >		DB1			DB2			DB3			DB4			DB5		
Signal No.	Initial SNR	Decomposition level			Decomposition level			Decomposition level			Decomposition level			Decomposition level		
		3	4	5	3	4	5	3	4	5	3	4	5	3	4	5
1	−0.01	0.39	0.86	2.04	0.46	1.39	2.60	0.54	1.96	3.28	0.76	2.37	3.96	0.79	2.49	4.16
2	−0.06	0.44	0.93	2.12	0.50	1.44	2.66	0.56	1.99	3.34	0.77	2.40	4.00	0.81	2.52	4.21
3	−0.14	0.50	1.01	2.23	0.53	1.48	2.72	0.59	2.04	3.40	0.79	2.42	4.04	0.83	2.56	4.27
4	−0.24	0.57	1.11	2.35	0.57	1.53	2.80	0.61	2.07	3.44	0.79	2.44	4.07	0.84	2.59	4.33
5	−0.38	0.64	1.21	2.48	0.61	1.59	2.88	0.63	2.10	3.50	0.79	2.46	4.10	0.85	2.62	4.37
6	−0.53	0.70	1.32	2.61	0.65	1.66	2.98	0.65	2.14	3.57	0.79	2.46	4.12	0.85	2.63	4.41
7	−0.71	0.77	1.42	2.74	0.69	1.72	3.07	0.67	2.18	3.64	0.79	2.48	4.15	0.85	2.65	4.45
8	−0.91	0.82	1.53	2.87	0.72	1.78	3.16	0.69	2.22	3.71	0.79	2.49	4.19	0.84	2.66	4.48
9	−1.12	0.87	1.62	2.99	0.75	1.83	3.24	0.71	2.26	3.78	0.79	2.51	4.23	0.83	2.67	4.52
10	−1.35	0.91	1.72	3.11	0.76	1.88	3.31	0.72	2.28	3.83	0.78	2.52	4.28	0.83	2.69	4.56
11	−1.58	0.94	1.81	3.23	0.77	1.92	3.39	0.72	2.31	3.90	0.77	2.54	4.32	0.81	2.70	4.60
12	−1.83	0.97	1.89	3.34	0.78	1.96	3.45	0.72	2.33	3.95	0.76	2.55	4.36	0.80	2.70	4.63
13	−2.08	0.98	1.96	3.44	0.77	1.99	3.51	0.71	2.34	3.98	0.74	2.54	4.38	0.77	2.70	4.66
14	−2.34	0.99	2.03	3.53	0.75	2.02	3.57	0.69	2.34	4.02	0.71	2.54	4.41	0.74	2.69	4.68
15	−2.60	0.99	2.09	3.61	0.73	2.04	3.62	0.67	2.34	4.05	0.68	2.53	4.42	0.71	2.68	4.69
16	−2.86	0.98	2.14	3.69	0.70	2.06	3.67	0.64	2.33	4.07	0.64	2.51	4.43	0.67	2.66	4.69
17	−3.12	0.97	2.19	3.76	0.67	2.07	3.71	0.60	2.32	4.09	0.59	2.49	4.42	0.62	2.63	4.68
18	−3.39	0.95	2.23	3.83	0.63	2.08	3.74	0.55	2.30	4.10	0.54	2.46	4.42	0.56	2.59	4.67
19	−3.65	0.92	2.26	3.89	0.58	2.07	3.77	0.50	2.28	4.11	0.49	2.43	4.41	0.50	2.56	4.65
20	−3.91	0.88	2.28	3.93	0.53	2.07	3.80	0.45	2.26	4.11	0.42	2.39	4.39	0.44	2.51	4.62

families of wavelet (for the same level of decomposition) vary for the signals with low noise. However, similar results can be observed when the signal contains a high amount of noise.

In the second test, we focused on the noisiest signal (test signal no. 20) and the least noisy signal (test signal no. 1) for experimentation. We applied wavelets decomposition at multiple levels (levels 2 to 7), each using DB1 to DB7 families of wavelet on both signals. A detailed comparison of the test results is shown in Table 3. Figures 7 and 8 show the comparison graphically for test signal 1 and test signal 20, respectively. For the input signal with a low noise level, the performance of our algorithm increases corresponding to a higher family of wavelet and a higher level of decomposition. This is also the case for processing signals with more noise. However, the improvement of SNR is comparatively less as compared with those in the previous situation (see Fig. 8).

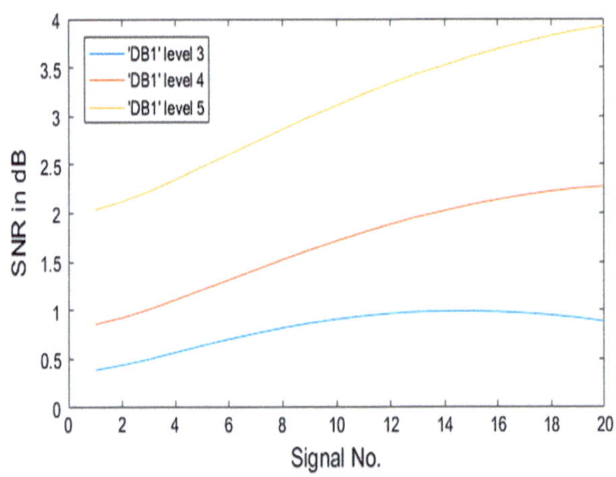

Fig. 5 Comparison of the SNR of the output signals using multiple levels of 'DB1' wavelet transform

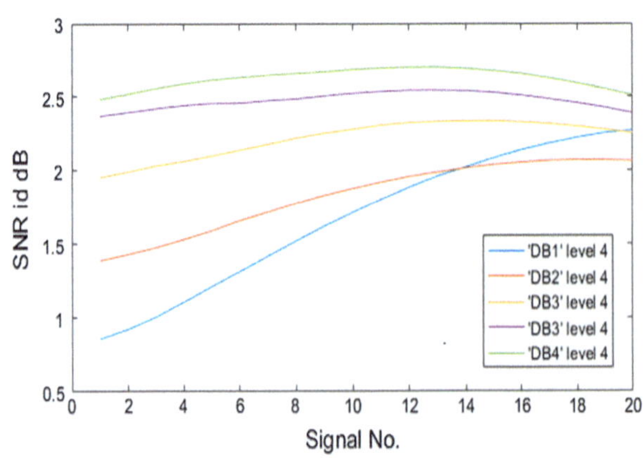

Fig. 6 Comparison of the SNR of the output signals using multiple families and level 4 wavelet decomposition

Table 3 Comparison of the final SNR (in dB) of the output of the proposed algorithm using two noisy signals

Wavelet family	Decomposition levels (Signal 1)						Decomposition levels (Signal 20)					
	2	3	4	5	6	7	2	3	4	5	6	7
DB1	0.151	0.391	0.861	2.042	2.631	2.926	−0.444	0.883	2.275	3.927	4.983	5.614
DB2	0.092	0.464	1.392	2.603	3.253	3.674	−0.848	0.527	2.069	3.797	4.916	5.691
DB3	0.080	0.536	1.957	3.281	4.042	4.582	−0.980	0.447	2.256	4.113	5.315	6.231
DB4	0.077	0.759	2.372	3.959	4.833	5.527	−1.043	0.424	2.390	4.392	5.659	6.716
DB5	0.075	0.790	2.485	4.159	5.060	5.829	−1.081	0.440	2.513	4.624	5.954	7.142
DB5	0.076	0.991	2.908	4.750	5.790	6.733	−1.105	0.460	2.641	4.819	6.220	7.525
DB6	0.074	1.085	3.168	5.106	6.254	7.334	−1.124	0.484	2.764	5.000	6.468	7.890

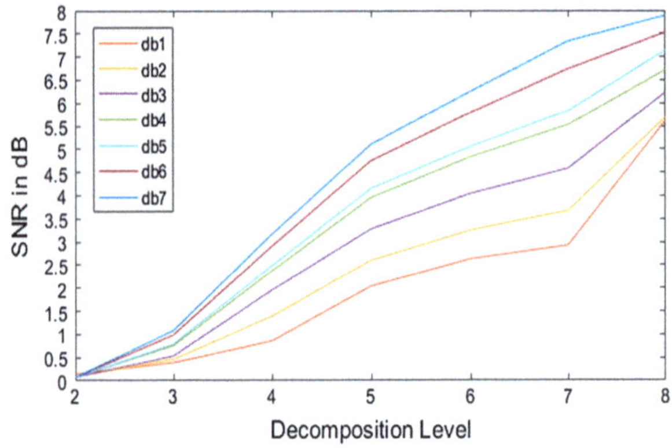

Fig. 7 Comparison of the SNR of the output signal using multiple families and levels of wavelet transform using the test signal 1

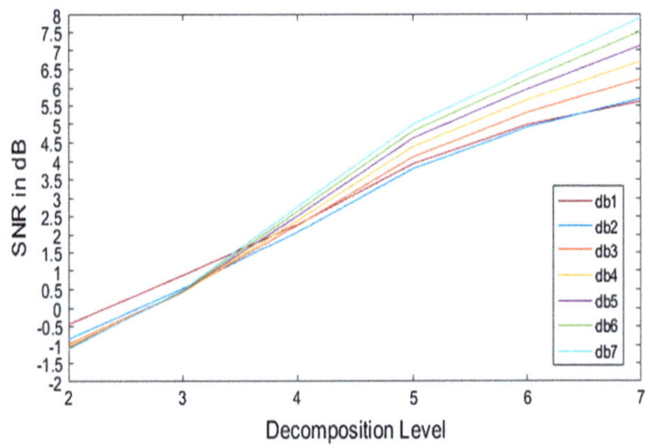

Fig. 8 Comparison of the SNR of the output signal using multiple families and levels of wavelet transform using test signal 20

4 Conclusions

In this study, we have proposed an effective noise reduction algorithm for ECG signals using baseline removal, discrete SWT and dynamic thresholding. The empirical results indicate that the efficiency of the proposed algorithm depends on the wavelet family used and the applied decomposition level. We have evaluated the proposed algorithm using both a half-hour-long real ECG signal from the MIT-BIH Arrhythmia database and the generated noisy signals. The proposed algorithm can also be applied to any other one-dimensional signals (e.g., EEG signals) for the noise removal.

In future work, we aim to combine the proposed noise reduction algorithm with peak detection methods pertaining to different wavelets and classification algorithms to identify abnormal ECG signals for heart disease detection. Furthermore, the proposed algorithm can be utilised for other biomedical signals, such as EEG and EMG, with appropriate modifications according to the characteristics of different types of signals.

Acknowledgements This research is supported by European Union (EU) sponsored (Erasmus Mundus) cLINK (Centre of excellence for Learning, Innovation, Networking and Knowledge) project (EU Grant No. 2645).

References

1. Hall, J.E., Guyton, A.C.: Textbook of medical physiology. J. Chem. Inf. Model. **53**, 160 (2011)
2. Moody, G.B., Mark, R.G.: The impact of the MIT-BIH arrhythmia database. IEEE Eng. Med. Biol. Mag. **20**, 45–50 (2001). doi:10.1109/51.932724
3. Thalkar, S.: Various techniques for removal of power line interference from ECG signal. Int. J. Sci. Eng. Res. **4**, 12–23. http://www.ijser.org (2013). Accessed 15 April 2016
4. Friesen, G.M., Jannett, T.C., Jadallah, M.A., Yates, S.L., Quint, S.R., Nagle, H.T.: A comparison of the noise sensitivity of nine QRS detection algorithms. IEEE Trans. Biomed. Eng. **37**, 85–98 (1990). doi:10.1109/10.43620
5. Kaur, M., Seema, B.S.: Comparison of different approaches for removal of baseline wander from ECG signal. In: Proceedings of the International Conference & Workshop on Emerging Trends in Technology—ICWET '11, p. 1290 (2011). doi:10.1145/1980022.1980307
6. Raphisak, P., Schuckers, S.C., Curry, A.D.J.: An algorithm for EMG noise detection in large ECG data. Comput. Cardiol. **2004**(1), 369–372 (2004). doi:10.1109/CIC.2004.1442949
7. Bai, Y.-W., Chu, W.-Y., Chen, C.-Y., Lee, Y.-T., Tsai, Y.-C., Tsai, C.-H.: Adjustable 60 Hz noise reduction by a notch filter for ECG signals. In: 21st IEEE Instrumentation and Measurement Technology Conference, pp. 1706–1711 (2004). doi:10.1109/IMTC.2004. 1351410
8. Rani, S., Kaur, A., Ubhi, J.S.: Comparative study of FIR and IIR filters for the removal of Baseline noises from ECG signal. Int. J. Comput. Sci. Inf. Technol. **2**, 1105–1108 (2011)
9. An-dong, W., Lan, L., Qin, W.: An adaptive morphologic filter applied to ECG de-noising and extraction of R peak at real-time. In: AASRI Conference on Computational Intelligence and Bioinformatics, pp. 474–479 (2012). doi:10.1016/j.aasri.2012.06.074
10. Tadejko, P. Rakowski, W.: Mathematical morphology based ECG feature extraction for the purpose of heartbeat classification. In: 6th International Conference on Computer Information Systems and Industrial Management Applications, CISIM 2007, pp. 322–327 (2007). doi:10. 1109/CISIM.2007.47
11. Pan, J., Tompkins, W.J.: A Real-Yime QRS detection algorithm. IEEE Trans. Bio-Med. Eng. Biomed. Eng. BME-32 230–236 (1985). doi:10.1109/TBME.1985.325532
12. Tamil, E.M., Kamarudin, N.H., Salleh, R., Idris, M.Y.I., Noor, M., Tamil, A.M.: Heartbeat Electrocardiogram (ECG) signal feature extraction using Discrete Wavelet Transforms (DWT). In: CSPA, 2008, pp. 1112–1117 (2008)
13. Chouhan, V.S., Mehta, S.S.: Detection of QRS complexes in 12-lead ECG using adaptive quantized threshold. Int. J. Comput. Sci. Netw. Secur. **8**, 155–163 (2008)
14. Kadambe, S., Murray, R., Paye, G.: Boudreaux-Bartels Wavelet transform-based QRS complex detector. IEEE Trans. Biomed. Eng. **46**, 838–848 (1999). doi:10.1109/10.771194
15. Kabir, M.A., Shahnaz, C.: Denoising of ECG signals based on noise reduction algorithms in EMD and wavelet domains. Biomed. Signal Process. Control **7**, 481–489 (2012). doi:10. 1016/j.bspc.2011.11.003
16. Garg, G., Gupta, S., Singh, V., Gupta, J.R.P., Mittal, A.P.: Identification of optimal wavelet-based algorithm for removal of power line interferences in ECG signals. In: India International Conference on Power Electronics, IICPE 2010 (2011). doi:10.1109/IICPE.2011. 5728090

17. Khan, M., Aslam, F., Zaidi, T., Khan, S.A.: Wavelet based ECG denoising using signal-noise residue method. In: 5th International Conference on Bioinformatics and Biomedical Engineering, iCBBE 2011, pp. 1–4. IEEE (2011). doi:10.1109/icbbe.2011.5780263
18. Li, Z., Ni, J., Gu, X.: A denoising framework for ECG signal preprocessing. In: 6th International Conference on Internet Computing for Science and Engineering, ICICSE 2012, pp. 176–179. IEEE (2012). doi:10.1109/ICICSE.2012.59
19. Tiwari, R., Dubey, R.: Analysis of different denoising techniques of ECG signals. Int. J. Emerg. Technol. Adv. Eng. **4**, 2–6 (2014)
20. Sadhukhan, D., Mitra, M.: ECG noise reduction using Fourier coefficient suppression. In: Proceedings of the 2014 International Conference on Control, Instrumentation, Energy and Communication (CIEC), pp. 142–146. IEEE (2014). doi:10.1109/CIEC.2014.6959066
21. Bortolan, G., Christov, I., Simova, I., Dotsinsky, I.: Noise processing in exercise ECG stress test for the analysis and the clinical characterization of QRS and T wave alternans. Biomed. Signal Process. Control **18**, 378–385 (2015). doi:10.1016/j.bspc.2015.02.003
22. Agostinelli, A., Giuliani, C., Burattini, L.: Extracting a clean ECG from a noisy recording: a new method based on segmented-beat modulation. Comput. Cardiol. **2014**(41), 49–52 (2014)
23. Kuzilek, J., Kremen, V., Soucek, F., Lhotska, L.: Independent component analysis and decision trees for ECG holter recording de-noising. PLoS ONE **9**, e98450 (2014). doi:10.1371/journal.pone.0098450
24. Liu, G., Luan, Y.: An adaptive integrated algorithm for noninvasive fetal ECG separation and noise reduction based on ICA-EEMD-WS. Med. Biol. Eng. Comput. **53**, 1113–1127 (2015). doi:10.1007/s11517-015-1389-1
25. Nason, G., Silverman, B.: The stationary wavelet transform and some statistical applications. Wavelets Stat. 281–299 (1995). doi:10.1007/978-1-4612-2544-7_17

Development of a Co-evolutionary Radial Basis Function Neural Classifier by a *k*-Random Opponents Topology

Bee Yan Hiew, Shing Chiang Tan and Way Soong Lim

Abstract The interest of the research in this paper is to introduce a novel competitive co-evolutionary (ComCoE) radial basis function artificial neural network (RBFANN) for data classification. The motivation is to derive a compact and accurate RBFANN by implementing an interactive "game-based" fitness evaluation within a ComCoE framework. In the CoE process, all individual RBFANNs interact with each other in an intra-specific competition. The fitness of each RBFANN is evaluated by measuring its interaction/encounter with *k* number of other randomly picked RBFANNs in the same population through a quantitative yet *subjective* manner under a *k*-random opponents topology. To calculate the fitness value, both the hidden nodes number and classification accuracy of each RBFANN are taken into consideration. To obtain a potential near optimal solution, the proposed model performs a global search through ComCoE approach and then performs a local search that is initiated by a scaled conjugate backpropagation algorithm to fine-tune the solution. Results from a benchmark study show high effectiveness of the co-evolved model with a *k*-random opponents topology in constructing an accurate yet compact network structure.

Keywords Competitive co-evolutionary algorithm · Radial basis function artificial neural network · Classification

B.Y. Hiew (✉) · S.C. Tan
Faculty of Information Science and Technology,
Jalan Ayer Keroh Lama, Bukit Beruang, 75450 Melaka, Malaysia
e-mail: vivianhby@yahoo.com

S.C. Tan
e-mail: sctan@mmu.edu.my

W.S. Lim
Faculty of Engineering and Technology, Multimedia University,
Jalan Ayer Keroh Lama, Bukit Beruang, 75450 Melaka, Malaysia
e-mail: wslim@mmu.edu.my

A. Bhatti et al. (eds.), *Emerging Trends in Neuro Engineering and Neural Computation*, Series in BioEngineering,
DOI 10.1007/978-981-10-3957-7_11

1 Introduction

Artificial neural networks (ANNs) are generally accepted by the computing community as a branch of machine-learning technique that can perform data classification with promising results [1]. ANNs could be integrated with other machine-learning techniques to enhance the recognition performance. For example, to train an ANN with good results, two popular techniques of evolutionary computation (EC), i.e. co-evolutionary algorithm (CoEA) and evolutionary algorithm (EA), have been applied. These two techniques are similar in the sense that their mechanisms are based on the ideas of natural evolution whereby the mechanisms encompass a selection phase as well as genetic operations, e.g. crossover and mutation [2]. Nevertheless, conceptually, a CoEA and an EA are not similar. In CoEAs, interaction and adaptation to changes occur among individuals themselves [3]. In contrast, in EAs, no such interaction exists among individuals in the evolutionary process. In addition, in EAs, individuals are evaluated independently with an objective fitness measure [4]. Conversely, in CoEAs, the fitness of an individual is obtained in a subjective manner [4], for which the evaluation involves a function of measuring the interaction between the individual and others in the population [3].

Cooperative co-evolution and competitive co-evolution are two fundamental types of co-evolution. In cooperative co-evolution, its core idea engages "divide and conquer" [5]. The initial step is the division of a problem to (smaller) sub-problems (where their potential solutions are defined and assigned to different populations). Then, these populations are evolved concurrently and continuously from one generation to another generation to refine the solutions. Finally, selected individuals (that represent the best solutions from each population) are merged to form a complete solution [5]. In competitive co-evolution, individuals in a population generally interact by either one of two modes: (1) intra-specific competition; (2) inter-specific competition [6]. For the former, individuals from the same population (same species) compete among each other. For the latter, competition occurs among the members from different populations (different types of species). Notably, it is possible to design a single-population CoE model [4, 5] where individuals from the same population evolve and "interact" via intra-specific competition individuals [4]. A single-population CoE model has been applied to game playing and artificial life simulations [7]. For instance, under a tournament topology in a game, a fitness value is rewarded to an individual after competing/playing with an opponent. Two commonly used tournament topologies in games are single elimination tournament and round robin [4].

In the recent literature, CoE model was applied in many research areas. For example, in machine learning, to improve the population diversity, a multipopulation CoE approach was proposed [8]. In [9], a cooperative CoE genetic fuzzy finite state machine was introduced to identify and improve the classification results of epilepsy (i.e. a neurological disorder that could cause cognitive damage and depression). Chandra et al. [10] introduced a novel architecture to encode two-dimensional time-series data into Elman Recurrent ANNs. Elman Recurrent ANNs comprised a single input neuron and were trained by a combination of cooperative co-evolution and backpropagation through-time algorithms to perform time-series prediction [10]. A novel competitive CoE team-based particle swarm optimiser algorithm, which has overcome a saturation problem when adapting ANN weights, was introduced to train teams of multi-agent that are essentially the neuro-controlled soccer players [11]. Artificial Neuron–Glia Networks (ANGNs) requires manual tuning of parameters configuration. By means of cooperative CoE genetic algorithm (CCGA), the proposed ANGN-CCGA in [12] has improved ANGNs by automatically learning all parameters. In [13], a CoE particle swarm optimizer mimicking parasitic behaviours of the host–parasites interactions in the natural ecosystem was proposed to perform function optimization. Apart from tackling the flexible job shop scheduling problem, a cooperative CoE algorithm was proposed to find a solution for task relative order and machine assignment [14]. Recurrent ANNs that are trained with cooperative co-evolution were utilised to predict rapid intensification in tropical cyclones in the South Pacific region [15]. A competitive cooperative CoE technique was introduced to train recurrent ANNs to perform chaotic time-series prediction [16]. The authors proposed two versions of competitive methods (i.e. two- and three-island competitive methods) to embark on competitive co-evolution [16]. Three hybrid algorithms which integrate the heuristic algorithm and co-evolution based on distance-adaptive policy were proposed in [17]. In [18], a radial basis function, ANN classifier was built through a subspace learning process involving optimization of the ANN's structure and feature subspace vectors with a cooperative CoE algorithm.

In this paper, a competitive co-evolutionary (ComCoE) RBFANN is introduced. From the literature, ComCoE ANN is relatively less explored as compared to cooperative CoE ANN(s). The construction of a ComCoE ANN is based on a common setup [19–25], i.e. (1) where a data sample population and a classifier population were involved in the evolutionary process (e.g. [19, 20, 23–25]; (2) normally only one objective was utilised, i.e. classification accuracy (for instance [19–25]).

In this paper, a ComCoE RBFANN model (single-population-based) is introduced which deploys a k-random opponents topology during fitness evaluation process for improving data classification performance. The proposed model is named as ComCoE(kRO) RBFANN.

The structure of the remaining of the paper is as follows: In Sect. 2, the model of ComCoE(kRO) RBFANN is explained. Section 3 presents results from experiments. In Sect. 4, conclusions are given.

2 The Proposed ComCoE(*k*RO) RBFANN Model

In the proposed model, a population of RBFANNs (developed from the Netlab toolbox [26]) is initialised before the competitive co-evolution embarks. During the process of co-evolution, the competitions/games among these RBFANNs are carried out/played under a tournament topology regulated by *k*-random opponents. Section 2.1 describes the proposed ComCoE process. The explanation of the fitness evaluation process under the *k*-random opponents topology is given in Sect. 2.2.

2.1 The Proposed CoE Process

In the proposed model, competitive co-evolution was applied to adapt RBFANNs by searching as well as refining the networks' width, σ_j, centre, $\mathbf{c}_j$, weight, $\boldsymbol{\omega}_{uj}$ as well as bias, ω_{u0}. To discover fitter chromosomes, a global-local search algorithm is executed where ComCoE is utilised to carry out a global search whereas a scaled conjugate backpropagation algorithm is employed to carry out a local search. The ComCoE(*k*RO) RBFANN model is constituted according to the following procedure:

1. **Population initialisation, chromosome representation and initial fitness evaluation**. One population of L RBFANNs (chromosomes) is initialised on a random basis. The number of hidden nodes, h ($h \in [1, 10]$) set in every RBFANN is different. Each value of hidden nodes number, h is a pseudorandom integer value (within a certain range) drawn from the discrete uniform distribution. Every RBFANN's initial fitness, f, is computed. Section 2.2 presents the fitness evaluation/computation of a RBFANN in details. Then, the RBFANNs are ranked from the best to the worst according to their fitness scores. Each RBFANN is encoded as a chromosome, z_i where z_i is the ith chromosome corresponding to the ith RBFANN. An RBFANN's width, σ_j, centre, $\mathbf{c}_j$, weight, $\boldsymbol{\omega}_{uj}$ together with bias, ω_{u0}, are combined to create a chromosome,z_i

$$z_i = \left[\mathbf{c}_j \sigma_j \boldsymbol{\omega}_{uj} \omega_{u0}\right] \tag{1}$$

2. **Mutation**. All chromosomes are performed with the Gaussian mutation (the main search operator). To control the mutation operation, a user defined probability value, i.e. a mutation rate MR, is utilised. Gaussian mutation is executed once MR is higher than the probability value and the following offspring, z_i' is generated.

$$\eta_i' = \eta_i \exp\left(\tau' N(0,1) + \tau N(0,1)\right) \tag{2}$$

$$z_i' = z_i + \eta_i' N(0,1), \tag{3}$$

where τ' and τ are equal to $\left(\sqrt{2n}\right)^{-1}$ and $\left(\sqrt{2\sqrt{n}}\right)^{-1}$ respectively in this work; η_i signifies the variance vector, $N(0,1)$ refers to a normally distributed one dimensional random number with mean and variance equal to zero and one respectively.

3. **Fitness evaluation**. Every z_i' (in the chromosome form) is decoded back into the structure of an RBFANN and the network will go through a fitness evaluation as explained in Sect. 2.2. Once the fitness evaluation is completed, each RBFANN is converted to a chromosome.
4. **The construction of a new population**. The chromosomes are sorted in a sequence based on descending fitness scores. A new population is formed by these sorted chromosomes (offspring). The process returns to step 2 from step 4. The procedure from step 2 to step 4 is carried on for a fixed number of generations (i.e. *G*).
5. Before embarking on phase-2 search, the best candidate chromosome obtained from the last generation of the global search from phase-1 (steps 1 to 4) is picked. The chromosome is decoded to an RBFANN followed by the local training through the scaled conjugate (SCG) backpropagation algorithm.
6. Using a testing dataset, evaluate the trained RBFANN from step 5.

2.2 Fitness Evaluation

A key difference between CoE and a typical EA is in fitness evaluation. In the EAs, an individual performance is measured using an objective fitness. On the contrary, one's fitness in the CoE framework is measured in a subjective manner through interaction between the chromosome and the others [27]. In this study, an interactive approach to fitness evaluation by adopting a *k*-random opponents tournament topology is proposed. In the tournament of *k*-random opponents, each individual plays with *k* other different (which are randomly picked) individuals [28]. In this approach, only one complete cycle of *k*-random opponents tournament is implemented during the fitness evaluation phase in every generation The explanation is as follows.

During the fitness evaluation phase in every generation, every individual maintains the following information:

- a counter of k that indicates the number of games an individual chromosome has played. In this work, k is reset as 5 before starting the k-random opponents tournament in a generation so that each individual could play five games in the k-random opponents tournament in a generation.
- the opponents' record (whom it has played with). This record is reset as null before beginning the k-random opponents tournament in a generation.
- fitness score. This score is reset as 0 before the k-random opponents tournament starts.

When an individual RBFANN I is to be evaluated, an opponent RBFANN (with $k > 0$) is picked randomly from the population to play with individual RBFANN I under a condition that no individual plays with individual RBFANN I more than one time. This is to ensure that each individual plays with different opponent in each game. In each game, the following indicators are used to compare the performance of two RBFANNs:

- The number of hidden nodes, h
- The training classification accuracy

To carry out an accurate classification task is the main goal of a classifier [29]. The concern of devising a low complexity classifier (e.g. a neural classifier with small number of hidden nodes) is only taken into account after a classifier's main goal is achieved (can classify correctly). Hence, the award of point(s) to the number of hidden nodes and classification accuracy are different with a priority to emphasis on classification accuracy. An RBFANN is awarded 1 point whenever its hidden node number is lower than its opponent. An RBFANN is awarded 2 points if its classification accuracy is higher than its opponent.

At the end of each game played under the k-random opponents tournament, the k value for both individual RBFANN I and the opponent are decremented. If the k counter of an individual RBFANN reaches 0, then that individual is "removed" from that population and is no longer considered as a candidate of opponent in the subsequent games under the k-random opponents tournament. A new opponent for individual RBFANN I is chosen, and this process continues until the k value of individual RBFANN I has reached 0. Thereafter, a new player RBFANN J is chosen, and the fitness evaluation goes on using the same way practiced by individual RBFANN I. At times, an individual RBFANN K may have no opponent (with $k > 0$) because none is available in the population to play games with the individual RBFANN K. When this occurs, this causes the individual RBFANN K could not complete its remaining k games. For an example, k value of individual RBFANN $K = 3$ (which means individual RBFANN K has remaining 3 games to play), the number of remaining opponents (with $k > 0$) available for playing games with the individual RBFANN K is null. Hence, to complete remaining three games of individual RBFANN K, no leftover opponent is available. Under such circumstances, the individual RBFANN K will play with any opponent disregards its k value randomly from among the "removed" individuals. The selected opponent should not play against the individual RBFANN K more than once. When all

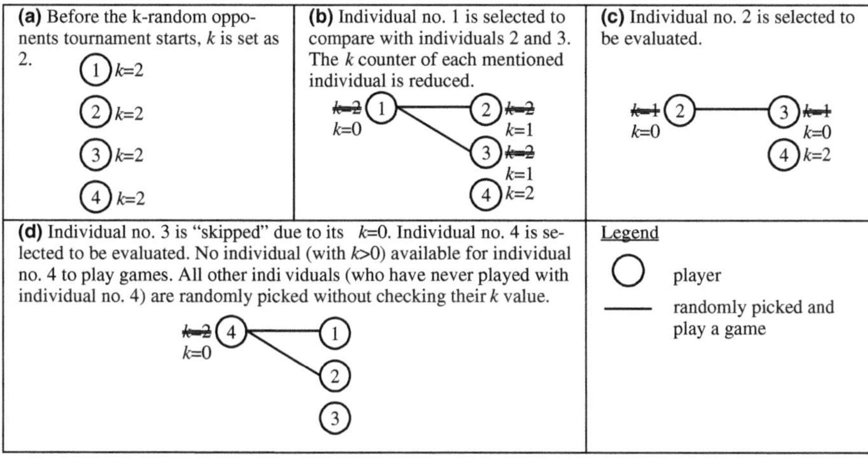

Fig. 1 An example of how a complete cycle of *k*-random opponents tournament works during the fitness evaluation phase in a generation

individuals have played *k* games, the fitness evaluation process which implements a cycle of *k*-random opponents tournament, is completed. [Note: the fitness score attained by every individual is accumulated from all games that it has played in a complete cycle of *k*-random opponents tournament.]

Figure 1 depicts an example of how a complete cycle of *k*-random opponents tournament works during the fitness evaluation phase in a generation. In this example, four players ($L = 4$) are initially set with $k = 2$. This means each member will compete with two other opponents without encountering the same opponents. In Fig. 1b–d, the *k* value of an individual is reduced by one after each game played from one fitness evaluation session in a generation; the updated *k* value is written at the bottom of initial *k* value (which is double strikethrough after each game).

3 Experiments

The performance between ComCoE(*k*RO) RBFANN and the baseline RBFANN and also with other CoE ANNs is evaluated and compared. In the experiments, the benchmark datasets applied (shown in Table 1 and were normalised to [0, 1] in our experiments) were taken from UCI Machine Learning Repository [30]. The average attribute value was calculated and then assigned to missing attribute values. A 10-fold cross-validation was utilised.

A baseline RBFANN was developed from the Netlab toolbox [26]. A 10-fold cross-validation strategy was imposed. Besides, to train/test the baseline RBFANN, the same partition of datasets used in the training and testing process of the proposed model were utilised. For each baseline RBFANN, its hidden node number

Table 1 The benchmark datasets

No.	Datasets	Classes	Attributes	Number of instances	Missing attribute values (%)
1	Breast	2	9	699	0.3
2	Heart	2	13	270	0
3	Iris	3	4	150	0
4	Pima	2	8	768	0
5	Sonar	2	60	208	0
6	Wine	3	13	178	0

Table 2 Comparison results between ComCoE(*k*RO) RBFANN and its baseline RBFANN

Dataset	Proposed model		Baseline RBFANN	
	Acc. (%)	Average #*h*	*Acc.* (%)	Average #*h*
Breast	96.3	2.0	93.10	5.9
Heart	84.1	2.7	82.96	7.5
Iris	96.7	2.9	90.00	6.8
Pima	77.1	1.5	75.40	5.3
Sonar	70.8	2.0	53.38	5.1
Wine	93.3	3.5	53.82	7.2

(between 1 and 10) was set at random. That number is also a pseudorandom integer value picked from a discrete uniform distribution. The RBFANN was trained via the SCG backpropagation algorithm for 100 iterations.

The performance of ComCoE(*k*RO) RBFANN and its baseline RBFANN is compared in Table 2. In general, although the number of average #*h* of the proposed model is lower, its classification accuracy (*Acc.*) results are significantly better than its baseline. The results show that while performing with high classification accuracy, the proposed model can produce a less complex network structure than its baseline.

The proposed method is compared with other CoE ANNs, i.e. MPSON [31], CO^2RBFN [32] and CO-RBFNN [33]. MPSON is one of the generic EAs; The model of CO^2RBFN engages evolutionary cooperative–competitive process; and CO-RBFNN is entirely involved in cooperative CoE mode. MPSON and CO^2RBFN used a 10-fold cross-validation in their experiments [3, 4]. In CO-RBFNN, the datasets were divided into training (50%), validation (25%) and testing (25%) subsets [8]. The results are shown in Table 3. The lowest #*h* and the highest *Acc.* of a method in each classification task are bolded. From Table 3, it shows that the proposed model provides the least complex network architecture while its classification accuracies are at least comparable to MPSON, CO^2RBFN and CO-RBFNN.

Table 3 Comparison between ComCoE(*k*RO) RBFANN and other (co-)evolutionary-based RBFANN classifiers

Dataset	Proposed model[a]		MPSON [31][a]		CO^2RBFN [32][a]		CO-RBFNN [33][b]	
	Acc. (%)	Average #*h*	*Acc.* (%)	Average #*h*	*Acc.* (%)	Average #*h*	*Acc.* (%)	Average #*h*
Breast	96.3	**2.0**	**97.66**	5.8	97.08	5.0	96.94	11.6
Heart	84.1	**2.7**	**85.20**	5.3	–	–	82.99	12.7
Iris	**96.7**	**2.9**	89.11	10.0	96.27	6.0	–	–
Pima	77.1	**1.5**	**77.34**	7.9	75.95	4.0	76.98	23.1
Sonar	70.8	**2.0**	–	–	75.09	8.0	**75.99**	22.2
Wine	93.3	**3.5**	75.67	8.3	96.74	7.0	**96.89**	6.57

[a]Single 10-fold cross-validation
[b]50% training, 25% validation, 25% testing
Note: "–" denotes unavailable results; The lowest #*h* and the highest *Acc.* of a method in each classification task are bolded

4 Conclusions

A ComCoE ANN model (with single-population) which adopts tournament-based fitness evaluation in its design is introduced. The proposed model, called as ComCoE(*k*RO) RBFANN, has adopted *k*-random opponents tournament topology in creating a *subjective* fitness evaluation. By utilising benchmark datasets from the UCI repository, ComCoE(*k*RO) RBFANN was evaluated. The comparison results between ComCoE(*k*RO) RBFANN and other CoE ANNs as well as its baseline ANN affirmed that the ComCoE(*k*RO) RBFANN can provide high classification accuracy with a compact network architecture. The future works include: (1) extension of the existing proposed model by taking into account several populations of different or same neural classifiers, which apply either the same or different EAs; (2) evaluation of the applicability of ComCoE(*k*RO) RBFANN to the real life problems.

References

1. Zhang, G.P.: Neural networks for classification: a survey. IEEE Trans. Syst. Man Cybern. Part C Appl. Rev. **30**(4), 451–462 (2000). doi:10.1109/5326.897072
2. Jaskowski, W.: Algorithms for Test-Based Problems. Institute of Computing Science, Poznan University of Technology, Poznan, Poland (2011)
3. Ilhem, B., Julien, L., Patrick, S.: A survey on optimization metaheuristics. Inf. Sci. **237**(2013), 82–117 (2013). doi:10.1016/j.ins.2013.02.041
4. Popovici, E., Bucci, A., Wiegand, R.P., De Jong, E.D.: Coevolutionary principles. In: Grzegorz Rozenberg, T.B., Kok, J.N. (eds.) Handbook of Natural Computing, pp. 987–1033. Springer, Berlin (2010)
5. Wiegand, R.P.: An analysis of cooperative coevolutionary algorithms. Doctoral Dissertation, George Mason University, Fairfax, Virginia (2004)

6. Jorgensen, S.E., Fath, B.: Encyclopedia of Ecology, Five-Volume Set: Online version. Elsevier Science, Amsterdam (2008)
7. Jaśkowski, W., Krawiec, K., Wieloch, B.: Fitnessless coevolution. In: Paper presented at the Proceedings of the 10th Annual Conference on Genetic and Evolutionary Computation, Atlanta, GA, USA (2008)
8. Shi, J., Gong, M., Ma, W., Jiao, L.: A multipopulation coevolutionary strategy for multiobjective immune algorithm. Sci. World J. **1**, 539128 (2014)
9. de la Cal, E., Villar, J.R., Vergara, P.M., Sedano, J., Herrero, A.A.: Preliminary cooperative genetic fuzzy proposal for epilepsy identification using wearable devices. In: 10th International Conference on Soft Computing Models in Industrial and Environmental Applications, Burgos, Spain, 2015, pp. 49–63. Springer International Publishing (2015)
10. Chandra, R., Deo, R., Omlin, C.W.: An architecture for encoding two-dimensional cyclone track prediction problem in coevolutionary recurrent neural networks. In: International Joint Conference on Neural Networks, Vancouver, Canada (2016)
11. Scheepers, C., Engelbrecht, A.P.: Analysis of stagnation behaviour of competitive coevolutionary trained neuro-controllers. In: IEEE Symposium on Swarm Intelligence, Orlando, FL, USA (2014)
12. Pablo, M., Oscar, I., Enrique, F., Cedrón, F., Alejandro, P., Ana, B.P.: Artificial neuron-glia networks learning approach based on cooperative coevolution. Int. J. Neural Syst. **25**(4), 15500121–155001219 (2015)
13. Qin, Q., Cheng, S., Zhang, Q., Li, L., Shi, Y.: Biomimicry of parasitic behavior in a coevolutionary particle swarm optimization algorithm for global optimization. Appl. Soft Comput. **32**(7), 224–240 (2015)
14. Palacios, J.J., González-Rodríguez, I., Vela, C.R., Puente, J.: Coevolutionary makespan optimisation through different ranking methods for the fuzzy flexible job shop. Fuzzy Sets Syst. **278**(C), 81–97 (2015)
15. Chandra, R.: Coevolutionary recurrent neural networks for prediction of rapid intensification in wind intensity of tropical cyclones in the south pacific region. In: Proceedings of the 22nd International Conference on Neural Information Processing, Istanbul, Turkey, 2015. Neural Information Processing, pp. 43–52. Springer International Publishing, Switzerland. doi:10.1007/978-3-319-26555-1_6
16. Chandra, R.: Competition and collaboration in cooperative coevolution of Elman recurrent neural networks for time-series prediction. IEEE Trans. Neural Netw. Learn. Syst. **26**(12), 1 (2015)
17. Ding, Z., Xu, Z., Zeng, X.: Hybrid routing and spectrum assignment algorithms based on distance-adaptation combined coevolution and heuristics in elastic optical networks. Opt. Eng. **53**(4), 0461051–0461059 (2014)
18. Tian, J., Li, M., Chen, F., Feng, N.: Learning subspace-based RBFNN using coevolutionary algorithm for complex classification tasks. IEEE Trans. Neural Netw. Learn. Syst. **27**(1), 47–61 (2016)
19. Paredis, J.: Steps towards coevolutionary classification neural networks. In: Brooks, R., Maes, P. (eds.) Proceedings Artificial Life IV, pp. 102–108. MIT Press/Bradford Books (1994)
20. Paredis, J.: Coevolutionary life-time learning. In: Proceedings of the 4th International Conference on Parallel Problem Solving from Nature, pp. 72–80. Springer (1996)
21. Juillé, H., Pollack, J.B.: Co-evolving intertwined spirals. In: Proceedings of the Fifth Annual Conference on Evolutionary Programming, 1996. Evolutionary Programming V: Proceedings of the Fifth Annual Conference on Evolutionary Programming, pp. 461–468. MIT Press, Cambridge, MA
22. Juillé, H., Pollack, J.B.: Dynamics of co-evolutionary learning. In: Proceedings of the Fourth International Conference on Simulation of Adaptive Behavior, pp. 526–534. MIT Press, Cambridge, MA (1996)
23. Douglas, A.A., Helio, J.C.B., Nelson, F.F.E.: Coevolution of data samples and classifiers integrated with grammatically-based genetic programming for data classification. In: Paper

Presented at the Proceedings of the 10th Annual Conference on Genetic and Evolutionary Computation, Atlanta, GA, USA (2008)
24. Stoean, C., Stoean, R., Preuss, M., Dumitrescu, D.: Coevolution for classification. Technical Report CI-239/08. Collaborative Research Center on Computational Intelligence, University of Dortmund (2008)
25. McLntyre, A.R., Heywood, M.I.: Toward co-evolutionary training of a multi-class classifier. In: Proceedings of the 2005 IEEE Congress on Evolutionary Computation, 2–5 Sept 2005, pp. 2130–2137. IEEE Press, Edinburgh, UK (2005)
26. Nabney, I.T.: Netlab toolbox. Available via Aston University (2011). http://www1.aston.ac.uk/eas/research/groups/ncrg/resources/netlab/. Accessed 3 Mar 2011
27. Watson, R., Pollack, J.: Coevolutionary dynamics in a minimal substrate. In: Spector, L., Goodman, E., Wu, A. et al. (eds.) Proceedings of the Genetic and Evolutionary Computation Conference, pp. 702–709. Morgan Kaufmann (2001)
28. Panait, L., Luke, S.A.: Comparison of two competitive fitness functions. In: Proceedings of the Genetic and Evolutionary Computation Conference, pp. 503–511. Morgan Kaufmann Publishers Inc. (2002)
29. Mitchell, T.M.: Machine Learning. McGraw-Hill (1997)
30. Bache, K., Lichman, M.: UCI machine learning repository. Available via University of California, Irvine, School of Information and Computer Sciences (2013). http://archive.ics.uci.edu/ml
31. Sultan Noman, Q., Siti Mariyam, S., Siti Zaiton Mohd, H., Maslina, D., Eiman, A.-S.: Memetic multiobjective particle swarm optimization-based radial basis function network for classification problems. Inf. Sci. **239**, 165–190 (2013). doi:10.1016/j.ins.2013.03.021
32. Perez-Godoy, M., Rivera, A., Berlanga, F., Del Jesus, M.: CO^2RBFN: an evolutionary cooperative-competitive RBFN design algorithm for classification problems. Soft. Comput. **14**(9), 953–971 (2010)
33. Li, M.Q., Tian, J., Chen, F.Z.: Improving multiclass pattern recognition with a co-evolutionary RBFNN. Pattern Recognit. Lett. **29**(4), 392–406 (2008). doi:10.1016/j.patrec.2007.10.019

Mining Outliers from Medical Datasets Using Neighbourhood Rough Set and Data Classification with Neural Network

Pey Yun Goh, Shing Chiang Tan and Wooi Ping Cheah

Abstract In this paper, a neighbourhood rough set is modified and applied as a data pre-processing method to select samples from a data set before training with a radial basis function neural network (RBFN). Data samples that are not selected for training is considered as outliers. Four medical datasets from a famous repository were used and results were compared in terms of number of training samples and accuracy between the proposed model and RBFN. The results are encouraging where classification accuracy of the proposed model is improved after outlier removal. Results are compared with other classification models as well using a medical dataset. The proposed model is competitive to give high classification accuracy.

Keywords Neural network · Rough set · Outlier · Medical data

1 Background of the Study

Since many decades ago, a lot of machine-learning and data-mining technologies have been introduced to support the decision of diagnostic and prognostic tasks. However, there are many open questions pending to answer by delivering more effective solutions, for instance, about the prediction capability of machine learning and its power in knowledge discovery for decision support in the medical domain [1–3]. One of the challenging tasks in mining medical data is to find outliers [2]. Outliers may carry information that is deviated from the norm of the data that could lead to a doubt whether they are generated by a different mechanism [4]. Outliers

P.Y. Goh (✉) · S.C. Tan · W.P. Cheah
Multimedia University, Jln. Ayer Keroh Lama, 75450 Melaka, Malaysia
e-mail: pygoh@mmu.edu.my

S.C. Tan
e-mail: sctan@mmu.edu.my

W.P. Cheah
e-mail: wpcheah@mmu.edu.my

A. Bhatti et al. (eds.), *Emerging Trends in Neuro Engineering and Neural Computation*, Series in BioEngineering,
DOI 10.1007/978-981-10-3957-7_12

could affect the prediction capability of a data-mining model. As such, pre-processing medical data using an outlier detection method may be helpful to reduce the negative impacts of outliers on the model's prediction capability. Rough set theory (RST) is efficient for finding hidden patterns in data [5, 6]. It has been actively applied in various application domains, to name a few, such as medical, finance and image processing [5]. These hidden patterns could be outliers. The application of RST for outlier detection can be seen in [6–9]. A limitation of RST is it is designed to deal with categorical data only. A generalized model, i.e. neighbourhood rough set (NRS), was proposed by Hu et al. [10] who extended RST to process numerical data. In [10], NRS was proposed to identify a subset of features effective for data classification. In this paper, NRS is modified to determine outliers in a data set. It is assumed that by pre-processing input samples with the proposed outlier detection method, the classification performance of a predictive model such as a neural network could be better and more reliable. Neural network is well known and popular as one of the classification tools in medical domain with significant advantages [11–14]. In this research, we have chosen radial basis function network (RBFN) as the classification model due to its favourable characteristics: fast in learning [15], simple topological structure [16] and a universal approximator for information processing [14, 15]. However, to obtain the optimum parameter values of RBFN is tedious [11]. To ease the problem at this point, we use the exact design RBFN in MATLAB, with the function named as *newrbe*, where number of hidden nodes that build up is similar to the number of input samples. The proposed NRS will pre-process the data samples from a training set before constructing an RBFN with *newrbe*.

The remaining sections of this paper are organized as follows. Related works are presented in Sect. 2. An overview of the NRS, RBFN and the description of proposed methods are presented in Sect. 3. Experiment and results that follow are explained in Sect. 4. The empirical study includes a performance comparison between the proposed RBFN and its original model as well as with other classification methods using several benchmark data sets from the UCI machine-learning repository [17]. The paper ends in Sect. 5 with a few concluding remarks and notes for future work.

2 Related Works

RST is commonly applied to pre-process datasets before a neural network is used. Such hybridization of RST and neural network compensates the weaknesses of each other [18, 19]. Here are the reasons: RST is good in dealing with inconsistent data but it is sensitive to noise which makes it generalize poorly; neural network generalizes better but it may have a complex structure when a training dataset has high number of dimensions.

One of the popular application areas of RST is in medical or clinical field. Some related works in soft computing focus on reducing the number of features of high dimensional medical datasets. For example, Anitha [18] applied rough neurons (a pair of neurons with upper and lower approximation) to reduce features while a neural network trained by a back propagation algorithm to predict embryonal tumour; Eissa [19] applied RST with granular to reduce features and a neural network trained by a back propagation algorithm to classify Hepatitis C; and Ratnarparkhi and Ghongade [20] applied fuzzy rough to extract and reduce the number of decision rules while multilayer perceptron to classify the Electrocardiogram (ECG) which helps the diagnose of heart disease. Wang et al. [21] developed self-organizing maps from a breast cancer dataset, then reduced features through genetic algorithm and finally induced rules using rough sets. Durairaj and Nandhakumar [22] is another related work but instead of feature selection, RST is applied to fill up missing data in in vitro fertilization data. Feature reduction is common in the integration work of rough set and neural network but input reduction is not. Our work is focused on reducing the number of input data instead of the number of features. RST is applied to identify whether or not the input data is an outlier, i.e. it contains meaningless information. After pre-processing with RST, a dataset with no outliers is sent to an RBFN for learning and classification.

3 The Methods

3.1 Overview of NRS

NRS is a rough set model proposed by Hu et al. [10] to deal with feature selection. It is applicable to both categorical and numerical attributes. The description of NRS is below.

Assume an information system, $ISyst = \langle Uni, Fields \rangle$, where *Uni* is the universe with a nonempty finite set of samples $\{x_1, x_2, x_3, \ldots, x_n\}$, and Fields represents the characteristics of the samples, denoted as $\{f_1, f_2, f_3, \ldots, f_m\}$. There are two types of attributes in the system, i.e. information *InforF* and decision attributes *DecF* where $Fields = InforF \cup DecF$. The granularity level is controlled using a neighbourhood parameter, δ. The neighbourhood granule for attribute Z of sample x_i, $\delta_Z(x_i)$, is determined through the distant function, D. A general distant metric, i.e. Minkowsky distance is used

$$D_k(x_1, x_2) = \left(\sum_{i=1}^{m} |d(x_1, f_i) - d(x_2, f_i)|^k \right)^{1/k}, \quad (1)$$

where $x_1, x_2 \in Uni$; Fields=$\{f_1, f_2, \ldots f_m\}$; m = dimensional space; $d(x, f_i)$ implies the value of sample x in the i-th attribute f_i; k = type of distant function (with $k = 1$, it is Manhattan; $k = 2$, it is Euclidean; $k = \infty$, it is Chebychev). The neighbourhood information granule $\delta_Z(x_i)$ is defined as

$$\delta_Z(x_i) = \{x_i | x_j \in Uni,\ D^Z(x_i, x_j) \leq \delta\} \tag{2}$$

For $\forall x_1, x_2, x_3 \in U$, the following conditions must be fulfilled:

(a) $D(x_1, x_2) \geq 0, D(x_1, x_2) = 0$ if and only if $x_1 = x_2$;
(b) $D(x_1, x_2) = D(x_2, x_1)$;
(c) $D(x_1, x_3) \leq D(x_1, x_2) + D(x_2, x_3)$.

The definitions below explain how this neighbourhood information granule can deal with categorical and numerical attributes.

Definition 1 Let Z_1 = numerical attributes, where $Z_1 \subseteq Fields$ and Z_2 = categorical attributes, where $Z_2 \subseteq Fields$. The neighbourhood granule of $\delta_{Z_1}(x)$, $\delta_{Z_2}(x)$ and $\delta_{Z_1 \cup Z_2}(x)$ are

(a) $\delta_{Z_1}(x) = \{x_i | D_{Z_1}(x, x_j) \leq \delta, x_i \in Uni\}$ that deals with numerical attributes;
(b) $\delta_{Z_2}(x) = \{x_i | D_{Z_2}(x, x_i) = 0, x_i \in Uni\}$ that deals with categorical attributes;
(c) $\delta_{Z_1 \cup Z_2}(x) = \{x_i | D_{Z_1}(x, x_i) \leq \delta, \wedge D_{Z_1}(x, x_i) = 0, x_i \in Uni\}$, where ^ denotes an "And" operator. It deals with a mixture of numerical and categorical attributes.

Then, the neighbourhood relation, *Neighrel* on the *Uni* can be written as a relation matrix

$$M(Neighrel) = (r(x_i, x_j))_{n \times n} \tag{3}$$

where $x_i, x_j \in Uni$, $r(x_i, x_j) = 1$ when $D(x_i, x_j) \leq \delta$ or else $r(x_i, x_j) = 0$.

Neighrel fulfils $r(x_i, x_j) = 1$ and $r(x_i, x_j) = r(x_j, x_i)$, i.e. reflexivity and symmetry respectively and this relation can be considered as similarity relations. This relation shows that samples with a closer distant that fulfil the conditions are categorized together and these samples are within the same neighbourhood. Thus, the neighbourhood granules are close to each other. The definition of neighbourhood approximation space is

Definition 2 Let $\langle Uni, Neighrel \rangle$ equal to an approximation space of the neighbourhood, with *Uni* = universe (a set of samples); *Neighrel* = neighbourhood relation. The lower and upper approximations of X in $\langle Uni, Neighrel \rangle$, with $X \subseteq Uni$, are defined as

$$\begin{aligned} \underline{NeighrelX} &= \{x_i | \delta(x_i) \subseteq X, x_i \in Uni\}, \\ \overline{NeighrelX} &= \{x_i | \delta(x_i) \cap X \neq \emptyset, x_i \in Uni\} \end{aligned} \tag{4}$$

Hence, $\underline{NeighrelX} \subseteq \overline{NeighrelX}$. With these, the roughness degree of X can be shown through the boundary region, *Boundary*

$$Boundary = \overline{NeighrelX} - \underline{NeighrelX} \quad (5)$$

For implementation, an inclusion degree is needed. This degree generalizes the implementation of upper and lower approximation. It enables the model to become more robust against noisy data. Let information attribute denoted by *InforF*, decision attribute denoted by *DecF*, with $InforF \subseteq Uni$ and $DecF \subseteq Uni$. Although *InforF* could contain multiple information attributes, but here we assume that it has only one attribute. The definition of inclusion degree for *InforF* in *DecF* with respect to *InforF* is

$$IncluDegree(InforF, DecF) = \frac{\text{Card}(InforF \cap DecF)}{\text{Card}(InforF)}, \text{ where } InforF \neq \varnothing \quad (6)$$

For example, $InforF = \{x_1, x_4, x_5, x_7, x_8\}$, $DecF = \{x_1, x_4, x_6\}$, the cardinality of $InforF \cap DecF$ is 2, cardinality of *InforF* is 5. Therefore, $Inclusion(InforF, DecF) = 0.4$. Hence, the definitions of lower and upper approximation are as below

Definition 3 Let $X \subseteq Uni$ in (*Uni*, *Field*, *Neighrel*), the variables' correctness, i.e. lower and upper approximations of *X* are

$$\begin{aligned} \underline{Neighrel^{id}}X &= \{x_i | Inclusion\ (\delta(x_i), X) \geq id, x_i \in Uni\}, \\ \overline{Neighrel^{id}}X &= \{x_i | Inclusion\ (\delta(x_i), X) \geq 1 - id, x_i \in Uni\}, \end{aligned} \quad (7)$$

where $1 \geq id \geq 0.5$, *id* is the inclusion degree. Assume that *id* is set as minimum 0.5, based on the example above, $IncluDegree(InforF, DecF) = 0.4$ does not fulfil the condition of both lower and upper approximations. Therefore, *InforF* is not selected as the information feature to predict the decision.

3.2 Overview of RBFN

The original RBFN, which was firstly proposed by Moody and Darken [23], contains two layers: the hidden and the output layers. The hidden layer is formed by the hidden units that aim to compute the response R_i for each data input *x* (see Fig. 1). In [23], Gaussian response function was used. The hidden units are defined by nonlinear functions, and all responses are combined linearly in the output layer. This linear combination is parameterized by the weights, *w* between the hidden and output layers. *K*-mean clustering algorithm is used to optimize the centres and the radii functions in the hidden layer while supervised training is applied to compute the weights in the output layer.

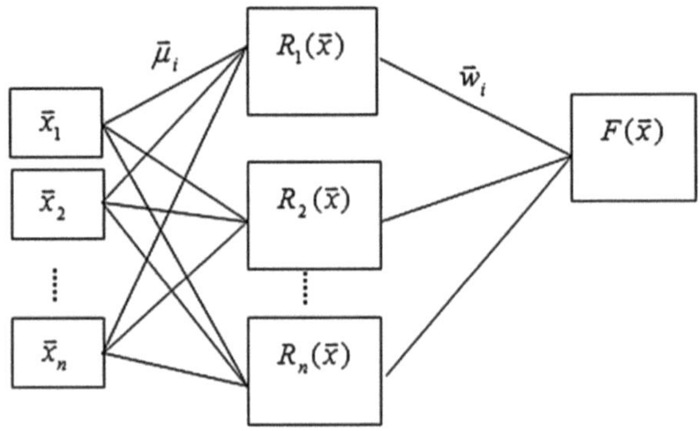

Fig. 1 The architecture of RBFN

Table 1 A heterogeneous data set

Data, *X*	Numerical attribute, Z_1	Categorical attribute, Z_2	Decision attribute, *DecF*
x_1	0.71	1	No
x_2	0.95	1	Yes
x_3	0.74	2	Yes
x_4	0.41	2	No
x_5	0.92	1	Yes
x_6	0.30	3	No

3.3 The Proposed Method

NRS [10] was introduced to find a subset of attributes. In this work, it is modified to select samples. The proposed idea will be explained through an example of a heterogeneous data set defined in Table 1.

Assume that the neighbourhood distant parameter, $\delta = 0.1$, the equivalence classes of each sample input x can now be identified. For example, x_2 under Z_1 has relationship R, i.e. $Rx_{21} = \{x_2, x_5\}$ by considering the distant function between x_2 with x_2, x_2 with x_5 are within ± 0.1. However, it has no relationship NR with $\{x_1, x_3, x_4, x_6\}$, i.e. $NRx_{21} = \{x_1, x_3, x_4, x_6\}$; under Z_2, $Rx_{22} = \{x_1, x_2, x_5\}$ but $NRx_{22} = \{x_3, x_4, x_6\}$. Two different classes are available for decision, i.e. No and Yes. A relationship exists for $DecF_{no}$ is $\{x_1, x_4, x_6\}$ but not for $\{x_2, x_3, x_5\}$ while for $DecF_{yes}$, a relationship exists for $\{x_2, x_3, x_5\}$ but not for $\{x_1, x_4, x_6\}$. Correctness matching ($CorM$) for each input x_i per attribute Z_j relevant to a class can be defined as:

$$CorM_{ij} = \mathrm{Card}(Rx_{ij} \cap DecF_{class}R) + \mathrm{Card}(NRx_{ij} \cap DecF_{class}NR) \quad (8)$$

The higher the $CorM$, the better the input data to approximate the decision feature. The probability of $CorM$ can be obtained using Eq. 9

$$\mathrm{P}(CorM_{ij}) = \frac{CorM_{ij}}{n}, \quad (9)$$

where n is the total number of input. P(*CorM*) of each attribute Z_j is then averaged using Eq. 10

$$E(CorM_j) = \frac{\sum_{i=1}^{n} \mathrm{P}(CorM_{ij})}{n} \tag{10}$$

Consider Z_1, by applying Eq. 9, the probability for each sample under Z_1 is: $x_1 = 0.5$, $x_2 = 0.83$, $x_3 = 0.5$, $x_4 = 0.67$, $x_5 = 0.83$, and $x_6 = 0.5$. As the rank value is applied on attribute, we accumulate the probability of correctness matching of each sample and average the value per attribute. The rank value is based on $E(CorM_j)$ to reflect the significant weightage of each attribute. Higher rank value is obtained when $E(CorM_j)$ is high. Based on Table 1, we can obtain $E(CorM_1) = 0.64$, $E(CorM_2) = 0.56$, then Z_1 is ranked as 2, and Z_2 is ranked as 1. Finally, the degree of correctness matching for each sample (*CorMD*) can now be obtained

$$CorMD_i = \frac{\sum_{j=1}^{M} (rank_j * CorM_{ij})}{\sum_{j=1}^{M} rank_j} \tag{11}$$

Then, potential outliers can be defined using the RST through lower and upper approximation

$$\begin{aligned} \underline{Neighrel^{id}}Out &= \{x_i | CorMD\,(\delta(x_i),\ Out) \le id, x_i \in Uni\}, \\ \overline{Neighrel^{id}}Out &= \{x_i | CorMD\,(\delta(x_i),\ Out) \le 1 - id, x_i \in Uni\} \end{aligned} \tag{12}$$

Based on Eq. 11, *CorMD* of each sample is: 0.44, 0.78, 0.50, 0.61, 0.78 and 0.56 respectively. If the parameter of inclusion degree is set as $id = 0.5$, then x_1 is an outlier based on Eq. 12. This information is used to remove all input samples below *id*. The selected data samples are then sent to RBFN for classification. The proposed model is called as neighbourhood rough set with modified Correctness to RBFN (NRSC-RBFN).

4 Experiments

Four medical datasets are used to benchmark the results of NRSC-RBFN. These datasets from UCI Machine Learning Repository [17] are: Blood Transfuse (BT), Haberman (Haber), Heart, and Indian Liver (Liver). All datasets have binary classes with number of samples 748, 306, 267 and 583 and number of attributes 5, 3, 44 and 8 respectively. Experiments were carried out using a 10-fold cross validation to calculate the performance of the proposed model. The neighbourhood parameter is set as 0.1 and inclusion degree is 0.5. The RBFN component of NRSC-RBFN is built using a *newrbe* in Matlab with a default setting of spread as 1. This network could be built automatically where only input samples and target values are needed. The experiment is implemented on a computer platform with these specifications: operating system Windows 7, Intel Core (TM) CPU i5-2410M and 4.0 GB RAM. The performance is compared between the RBFN (that is built using *newrbe*) and NRSC-RBFN in terms of the number of input samples and accuracy rate.

Table 2 shows the results before and after the implementation of NRSC. NRSC-RBFN performs with higher accuracy rates and has a lower number of input samples as compared to original RBFN. The percentage of accuracy improvement of NRSC-RBFN in BT is 4.54, Liver 9.87, Haber 9.08 and Heart 17.21. NRSC is an effective method to identify outliers at the data pre-processing stage. The percentage of outliers that is found from each data set is 26.01, 10.14, 40.01 and 31.92 respectively. These data samples are removed from each training set. RBFN that is built by *newrbe* will have the number of hidden nodes similar to the number of input samples. Thus, after NRSC reduced the number of input samples, the resulting hidden nodes in RBFN are also reduced. The samples that have been removed from a training set are considered as outliers.

The performance of NRSC-RBFN is also compared with other classification models in [24] using a Liver data set. It is claimed that outliers exist in the Liver data [25]. As such, when applying ordinary classification model, the accuracy performance is not good (see Table 3). The compared models include Naïve Bayes (NB), Decision Tree (DT), Multilayer Perceptron (MLP) and *k*-Neural Network (*k*-NN). Details of these models can be referred from [24]. Authors in [24] applied 10-fold classification. As such, we ran the experiments using the similar method as well.

Table 2 The results of RBFN and NRSC-RBFN

		BT	Liver	Haber	Heart
RBFN	Accuracy	70.33	85.33	56.91	61.04
	# training sample	673.20	524.70	275.40	240.30
NRSC-RBFN	Accuracy	74.87	95.20	65.99	78.25
	# training sample	497.50	471.50	165.20	163.60

Table 3 Classification performance of various classifiers (Acc. = average test accuracy) from a 10-fold cross validation strategy. (The results of NB, DT, MLP, *k*-NN are taken from [20])

Methods	Acc.
NB	53.90
DT	69.40
MLP	67.90
k-NN	65.30
RBFN (*newrbe*)	85.33
NRSC-RBFN (*newrbe*)	95.20

The results are compared with NRSC-RBFN in terms of prediction accuracy to see whether the proposed model can produce compatible results. The present of outliers causing the ordinary classification methods, i.e. NB, DT, MLP and k-NN has on average around 64% in prediction accuracy. Although RBFN has higher accuracy, i.e. 85.33 but after outlier removal, better performance is achieved with the average accuracy equal to 95.20. It is motivated that NRSC-RBFN outperform all the other classification methods Such comparison is not to show the superiority of NRSC-RBFN but just to provide another alternative way to improve performance when apply classification, i.e. outlier removal.

5 Conclusion

In this study, a NRSC-RBFN is proposed to mine the outliers from a data set. NRSC-RBFN could perform with encouraging results where the proposed method could detect outliers from a training set before proceeding to training with RBFN. By referring to the experimental results, we notice that these outliers if are not removed from the training set can affect the prediction ability of RBFN. NRSC-RBFN is also compared with other classification methods where its performance is found competitive in the benchmark study. In the future, additional experiments will be carried out using other benchmark and real datasets. It is as well to consider other RBFN that contains the nature of incremental learning instead of batch training neural network to see the impact of outlier. Other interesting research in the future is to integrate attribute selection with sample selection for identifying outliers from high volume datasets in multiclass.

References

1. Bellazzi, R., Zupan, B.: Predictive data mining in clinical medicine: current issues and guidelines. Int. J. Med. Inform. **77**, 81–97 (2008)
2. Devi, R.D.H., Devi, M.I.: Outlier detection algorithm combined with decision tree classifier for early diagnosis of breast cancer. Int. J. Adv. Eng. Technol. **7**(2), 93–98 (2016)
3. Soni, J., Ansari, U., Sharma, D.: Predictive data mining for medical diagnosis: an overview of heart disease prediction. Int. J. Comput. Appl. **17**(8), 43–48 (2011)
4. Hawkins, S., He, H., Williams, G., Baxter, R.: Data warehousing and knowledge discovery. In: Kambayashi, Y., Winiwarter, W., Arikawa, M. (eds.) DaWaK 2002 Aix-en-Provence. 4th International Conference, France, September, 2002 Proceedings, pp. 170–180. Springer, Heidelberg (2002)
5. Pawlak, Z.: Rough sets and data analysis. In: Fuzzy Systems Symposium on Soft Computing in Intelligent Systems and Information Processing. Proceedings of the 1996 Asian, pp. 1–6 (1996)
6. Liu, D., Li, T., Ruan, D., Zhang, J.: Incremental learning optimization on knowledge discovery in dynamic business intelligent systems. J. Glob. Optim. **51**(2), 325–344 (2011)

7. Maciá-Pérez, F., Berna-Martinez, J.V., Oliva, A.F., Ortega, M.A.A.: Algorithm for the detection of outliers based on the theory of rough sets. Decis. Support Syst. **75**, 63–75 (2015)
8. Jiang, F., Sui, Y., Cao, C.: Some issues about outlier detection in rough set theory. Expert Syst. Appl. **36**(3), 4680–4687 (2009)
9. Albanese, A., Pal, S., Petrosino, A.: Rough sets, kernet set, and spatiotemporal outlier detection. IEEE Trans. Knowl. Data Eng. **26**(1), 194–207 (2014)
10. Hu, Q., Yu, D., Liu, J., Wu, C.: Neighborhood rough set based heterogeneous feature subset selection. Inf. Sci. (NY) **178**(18), 3577–3594 (2008)
11. Alexandridis, A., Chondrodima, E.: A medical diagnostic tool based on radial basis function classifiers and evolutionary simulated annealing. J. Biomed. Inform. **49**, 61–72 (2014)
12. Dreiseitl, S., Ohno-Machado, L.: Logistic regression and artificial neural network classification models: a methodology review. J. Biomed. Inform. **35**(5–6), 352–359 (2002)
13. Wasan, S.K., Bhatnagar, V., Kaur, H.: The impact of data mining techniques on medical diagnostics. Data Sci. J. **5**, 119–126 (2006)
14. Downs, J., Harrison, R.F., Kennedy, R.L., Cross, S.S.: Application of the fuzzy ARTMAP neural network model to medical pattern classification tasks. Artif. Intell. Med. **8**(4), 403–428 (1996)
15. Meng Joo, E., Shiqian, W., Juwei, L.: Face recognition using radial basis function (RBF) neural networks. In: Proceedings of the 38th IEEE Conference in Decision and Control, vol. 3, pp. 2162–2167 (1999)
16. Kang, L., Jian-Xun, P., Er-Wei, B.: Two-stage mixed discrete-continuous identification of Radial Basis Function (RBF) Neural models for nonlinear systems. IEEE Trans. Circuits Syst. I Regul. Pap. **56**(30), 630–643 (2009)
17. Bache, K., Lichman, M.: UCI machine learning repository. University of California, School of Information and Computer Science (2013). http://archive.ics.uci.edu/ml
18. Anitha, K.: Rough neural network. GSTF J. Math. Stat. Oper. Res. (JMSOR) **3**(2), 59–62 (2016)
19. Eissa, M., Elmogy, M., Hashem, M.: Rough–Granular neural network model for making treatment decisions of Hepatitis C. In: 9th International Conference on Informatics and Systems (INFOS), pp. 19–26 (2014)
20. Ratnaparkhi, A., Ghongade, R.: A frame work for analysis and optimization of multiclass ECG classifier based on rough set theory, In: International Conference on Advances in Computing, Communications and Informatics (ICACCI 2014), pp. 2740–2744 (2014)
21. Wang, Z., Zhang, X., Yang, W.: Rule induction of breast cancer medical diagnose based on combination of rough sets, artificial neural network and genetic algorithm. In: 2016 Chinese Control and Decision Conference (CCDC, pp. 5707–5711 (2016)
22. Durairaj, M., Nandhakumar, R.: An integrated methodology of artificial neural network and rough set theory for analyzing IVF data. In: 2014 International Conference on Intelligent Computing Applications (ICICA), pp. 126–129 (2014)
23. Moody, J., Darken, C.J.: Fast learning in networks of locally-tuned processing units. Neural Comput. **1**(2), 281–294 (1989)
24. Jin, H., Kim, S., Kim, J.: Decision factors on effective liver patient data prediction. Int. J. Bio-Sci. Bio-Technol. **6**(4), 167–178 (2014)
25. Bahramirad, S., Mustapha, A., Eshraghi, M.: Classification of liver disease diagnosis: a comparative study. In: Second International Conference on Informatics and Applications (ICIA), pp. 42–46 (2013)

A Modified Functional Link Neural Network for Data Classification

Toktam Babaei, Chee Peng Lim, Hamid Abdi and Saeid Nahavandi

Abstract The functional link neural network (FLNN) increases the input dimension by functionally expanding the input features. In this paper, modifications to the FLNN are proposed for undertaking data classification tasks. The main objective is to optimize the FLNN by formulating a parsimonious network with less complexity and lower computational burden as compared with the original FLNN. The methodology consists of selecting a number of important expanded features to build the FLNN structure. It is based on the rationale that not all the expanded features are equally important in distinguishing different target classes. As such, we modify the FLNN in a way that less—relevant and redundant expanded input features are identified and discarded. In addition, instead of using the back-propagation learning algorithm, adjustment of the network weights is formulated as an optimisation task. Specifically, the genetic algorithm is used for both feature selection as well as weight tuning in the FLNN. An experimental study using benchmark problems is conducted to evaluate the efficacy of the modified FLNN. The empirical results indicate that even though the structure of the modified FLNN is simpler, it is able to achieve comparable classification results as those from the original FLNN with fully expanded input features.

1 Introduction

Artificial neural networks, or simply neural networks, have emerged as a promising model for tackling different problems, e.g., prediction [1–3], pattern recognition and classification [4–7], and optimization [8]. They have been successfully used in many applications, ranging from medical and industrial to financial sector [9–11].

In a multilayer perceptron (MLP) neural network with single or multiple hidden layers, the input features undergo a nonlinear transformation before being propagated

T. Babaei (✉) · C.P. Lim · H. Abdi · S. Nahavandi
Institute for Intelligent Systems Research and Innovation (IISRI), Deakin University, Waurn Ponds, VIC 3216, Australia
e-mail: tbabaeij@deakin.edu.au

A. Bhatti et al. (eds.), *Emerging Trends in Neuro Engineering and Neural Computation*, Series in BioEngineering,
DOI 10.1007/978-981-10-3957-7_13

to the next layer. Without loss of generality, assume an MLP with one hidden layer and one output layer. A mapping function from the hidden layer to the output node can be formulated as a quadratic optimization task. Pao et al. [12] proposed a non-linear transformation method known as the functional links that generates expanded features from the original inputs (see Fig. 1). They conducted several experiments, and showed that the proposed functional link neural network (FLNN) could serve as an accurate approximator for mapping the input and output data samples [12, 13]. Ingelnik et al. [14] presented mathematical proof for the weight learning process in the first transition as well as mathematical information to implement such neural network.

As proposed by Klassen and Pao [12], the FLNN can be used for data classification and prediction tasks with a faster convergence speed and a lighter computational load as compared with the MLP network. This is because that the FLNN has a flat structure, without any hidden layers, in contrast to the stacked structure of the MLP network.

Although the FLNN model has only one layer of trainable weights, it is able to undertake nonlinear classification and regression problems. This is owning to the functional expansion units embedded in the FLNN. These functional expansion units(or nodes) effectively enhance the input features by expanding them into a higher dimensional space, allowing the boundary (either linear or nonlinear) to be approximated by hyperplanes in the expanded feature space [12]. Several FLNN variants have been developed using different functional units. As an example, in a recent work [15], different FLNN models were devised and their performances in financial forecasting have been compared.

In [16], an FLNN-based model was used to predict machinery noise in the mining industry. In [17], a FLNN-based classifier was benchmarked against different common classifiers, including kNN (k-nearest neighbor), C4.5 decision tree, and MLP.

FLNN-based models have also been shown as useful tools for solving nonlinear equations. In [18], a variant of the FLNN which uses Hermite polynomial expansion was proposed. It was then used to solve a challenging problem in nonlinear dynamics known as the Duffing–van der Pol oscillator equation. In [19], a Chebyshev polynomial expansion FLNN was proposed and used for solving Emden–Fowler type of equations. Comparing the results with those from the conventional neural network model, the FLNN required less CPU time to produce similar performances.

Research towards improving FLNN models can be broadly categorized into two: (i) improving the FLNN learning algorithms; (ii) reducing the FLNN model complexity. A number of studies in the first area are as follows. In [20] the gradient-based back-propagation learning algorithm was replaced with a modified artificial bee colony algorithm. The proposed FLNN was able to overcome the limitations of gradient decent and achieve better classification rates, as compared with the original FLNN. In [21], Harmony search (HS) was integrated with the gradient decent algorithm to improve the learning capability of the original FLNN. In [22], the original FLNN was trained with another meta-heuristic algorithm, i.e., the firefly algorithm. The resulting FLNN was used for time series forecasting. The predictive accuracy and processing time were better than those from the original FLNN. In [23], a FLNN with a hybrid particle swarm optimisation (PSO)—back-propagation learning algorithm for data classification was proposed. An improved version of this model, was developed in succession [17]. The same group of researchers attempted to decrease the complexity and computational load of the FLNN using a genetic algorithm (GA) to select a subset of input features from the original feature space [24]. In [25], a distributed learning strategy to improve the learning property of a random vector FLNN was presented. The proposed model performed with promising results in problems requiring big data processing.

While enhancing the input dimension enables the FLNN model to handle nonlinear classification and regression problems, it increases the resulting network complexity. A number of studies [24, 26–29], have been conducted to tackle this problem. The main idea is to select some important input features and to reduce the input dimension. Most of the investigations focus on reducing the original features by incorporating different feature selection methods into the FLNN model. As an example, the gain ratio method has been used to select the best subset of features from the original feature set in [26], and differential evolution method to train the FLNN weights. In [17] a model called Hybrid-FLNN (HFLNN), has been proposed. An optimal set of features is selected from the original feature set using the GA. The HFLNN model is trained using the back-propagation learning algorithm.

A different approach to overcome the problem associated with the high-dimensional input space in the original FLNN was presented in [28]. The rational is to select and optimize the expanded input features, rather than the original features themselves. An evolutionary algorithm was used for this purpose, and the network is known as the evolutionary FLNN (or EFLNN). Training of the EFLNN is accomplished using the back-propagation learning algorithm. In this study, we further improve the EFLNN [28], such that the need for using the back propagation algorithm is removed. Specifically, the GA is used for selection of the expanded input features as well as for training the network. In this case, a new combined search problem covering both expanded input features and network weights is considered. The proposed model, known as reduced FLNN (rFLNN), requires fewer functional links as compared with those in the original FLNN. A number of benchmark classification problems are used to evaluate rFLNN, and the results are compared with those published in the literature.

This paper is organized as follows. In Sect. 2, the background knowledge of this research is given. In Sect. 3, the proposed rFLNN model is described in detail. A series of experiments is presented in Sect. 4, with the results discussed and analyzed. Concluding remarks are presented in Sect. 5.

2 Background

In this section, the background for understanding the proposed FLNN-based models is described.

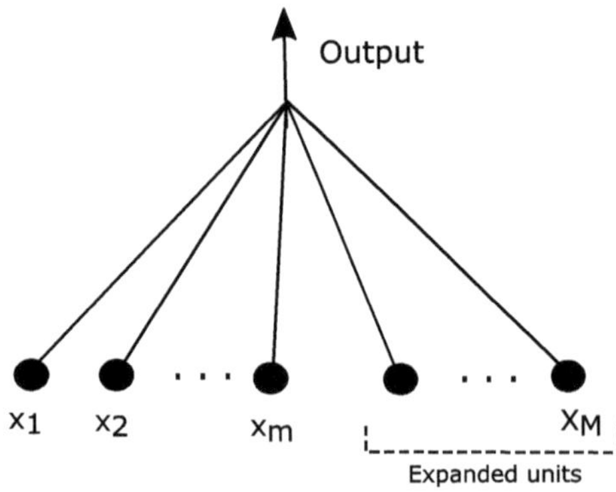

Fig. 1 Topology of a functional link neural network (FLNN)

2.1 The Functional Link Neural Network

The general topological structure of the FLNN is shown in Fig. 1. Consider that the FLNN is used to solve a classification task. The training set consists of N data samples. Each input–output sample pair can be represented by (x, y) where $x = [x_1, x_2, \ldots, x_m]$ represents an m dimensional input feature vector and y provides the true class label of that particular input.

The FLNN extends the original m-dimensional input space to an $M = (n + 1)m$ dimensional space by expanding each input feature to $(n + 1)$ secondary features using a set of basis functions $\phi = [\phi_0, \phi_1, \phi_2, \ldots, \phi_n]$ as follows:

$$\phi(x_i) = [\phi_0(x_i), \phi_1(x_i), \phi_2(x_i), \ldots, \phi_n(x_i)] \tag{1}$$

where n is the expansion degree. The set of expansion functions has the following characteristics:

- ϕ_0 is a linear function
- ϕ_i, $2 \leq i \leq n$ are linearly independent functions
- $sup_n(\sum_{i=2}^{n}(\| \phi_i \|)^2 \prec \infty$

As an example, using the trigonometric basis function, the first feature, x_1 is expended to degree n, as follows:

$$\phi(x_1) = x_1, \sin \pi x_1, \cos \pi x_1, \sin 2\pi x_1, \cos 2\pi x_1, \ldots, \sin n\pi x_1, \cos n\pi x_1 \tag{2}$$

As such, the original feature space grows up to an $(2n + 1)m$-dimensional space.

A number of FLNN models have been proposed using different basis functions [30]. They include the Chebyshev FLNN [23], Legendre FLNN, and Laguerre FLNN. Another important choice is polynomial functions upto n degrees. This type of FLNN which is a type of random vector FLNN has a good mathematical proof

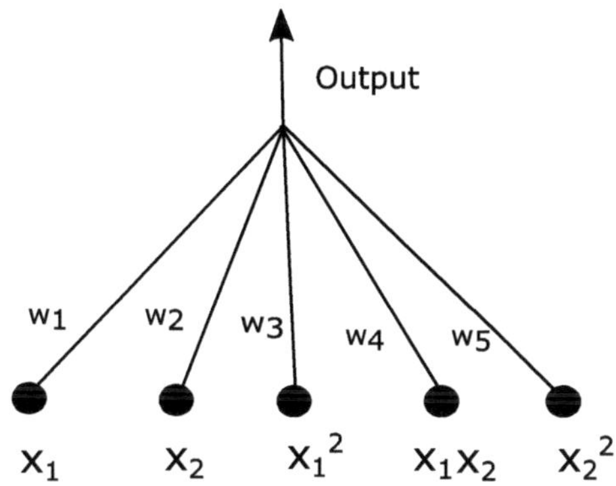

Fig. 2 Topology of the FLNN represented by Eq. 4

[14] and is easier to implement. In a polynomial FLNN model, the feature space is expanded to

$$\frac{(m+n)!}{m!n!} \tag{3}$$

where n is the highest degree of the polynomial function. As an example, in the case of a two-dimensional input feature, the expanded feature space to a degree of 2 is

$$[x_1, x_1^2, x_1x_2, x_2^2, x_2]. \tag{4}$$

The corresponding FLNN model is shown in Fig. 2.

Let the expanded input feature vector be represented by X, and the network weight vector by W, which is found by some learning algorithm, e.g., back-propagation. The FLNN generates output $\hat{y}$ by applying an activation function ρ, to the weighted sum of the expanded input feature vector, z as follows:

$$\hat{y} = \rho(z)$$

where the weighted sum of the expanded input feature, z, is calculated by following matrix multiplication:

$$z = XW$$

Then, z is propagated to an activation function to generate the output. Different types of output functions can be used, as shown in Table 1. In this study, we use the tanh(.) function (Eq. 6) as the activation function of the output nodes. As such, the output is given by

$$\hat{y} = \tanh(z) \tag{5}$$

$$\tanh(z) = \frac{1 - e^{-2z}}{1 + e^{-2z}} \tag{6}$$

Despite its simplicity, the FLNN generates a large number of expanded units, many of them usually do not contribute toward discriminating different target classes.

Table 1 Activation functions for generating the output

Activation function ($\rho(.)$)
$a + bz$
ae^{-bz}
$a + b/z$
$a + bln(z)$
az^b
$1/(a + be^{-z})$
$ae^{b/z}$
$\tanh(z) = \frac{e^{2z}-1}{e^{2z}+1}$

In the case of the polynomial FLNN, which is used in this study, the expanded features grow according to Eq. 3. For most real-world problems, this a large number even if it is expanded up to a degree of 2. It affects the FLNN performance, as well as increase its computational burden. This is the issue that we aim to tackle by using the GA.

2.2 *Training Algorithm*

The back-propagation learning algorithm is used for the FLNN model in [31]. Training the FLNN involves finding the weight matrix $[W]$ that minimizes a cost function, $C(.)$, over the training data set, D_N, i.e.,

$$[W] = arg\,\min_{[W]} C(D_N, [W])$$

The mean squared error (MSE) is considered as the cost function

$$MSE = \frac{1}{N}\sum_{N} e^2$$

where the error $e = (y - \hat{y})$ of each data sample is the difference between the target output and the predicted output of the FLNN for that particular sample. The sum covers all training samples.

The weights are initialised randomly. Then the back-propagation algorithm [32] is used to minimize the cost function using the gradient decent method as follows:

$$w_j^{new} = w_j^{old} - \alpha \frac{\partial C}{\partial w_j}$$

Considering Eqs. 1–6, knowing that $\frac{\partial \rho(s)}{\partial s} = 1 - s^2$, and applying the chain rule, the back-propagation algorithm alters every weight according to

$$w_j^{new} = w_j^{old} + \alpha \sum_{i=1}^{N} X_{ji}(1 - \hat{y}_i^2)e_i$$

where α is the learning rate, between 0 and 1. The process of updating the weights continues until convergence is achieved.

After completing the learning process, the trained FLNN model can be used for evaluation.

2.3 *Genetic Algorithm*

The GA [33, 34] is one of the most prominent evolutionary algorithms introduced to date. The GA first forms an initial population of candidate solutions to solve a given problem. It evolves the population for achieving better solutions using operators inspired by natural genetic operations such as crossover and mutation in accordance with the Darwinian principle of survival of the fittest. The GA is robust and well-suited for tackling optimisation problems in high-dimensional, discontinuous search spaces, where conventional search methods like gradient descent algorithms may fail to find the global optimal solution. The conventional GA operating on fixed length strings, has been successfully used to optimize neural network structures [35]. The GA has also been used to evolve weights, links, and structures, as well as features and rules of different machine learning models [36, 37].

The operation of the GA is as follows. It maintains a pool of chromosomes, i.e., $P = \{ch_1, ch_2, \ldots, ch_p\}$ where p is the population size. Each chromosome comprises q number of genes and encodes a solution. The genes can take either binary or real values as follows:

$$ch_i = (ch_{i1}, ch_{i2}, \ldots, ch_{iq})$$

The GA can be initialised randomly or by a customized initial population. Each chromosome is evaluated using a fitness (objective or cost) function $f(.)$. Fitter chromosomes are allowed to bear offspring into the next generation. Crossover and mutation are two reproduction operators that produce a child (offspring) from two selected parents. Crossover creates a new offspring by swapping certain genes between two parents, while mutation perturbs certain randomly selected genes in the produced offspring.

3 Model Description

In this section, the proposed rFLNN model is described. Since rFLNN can be considered as a modification to the EFLNN in [28], this model is described as well. In general, the FLNN structural complexity is proportional to the number of expanded input

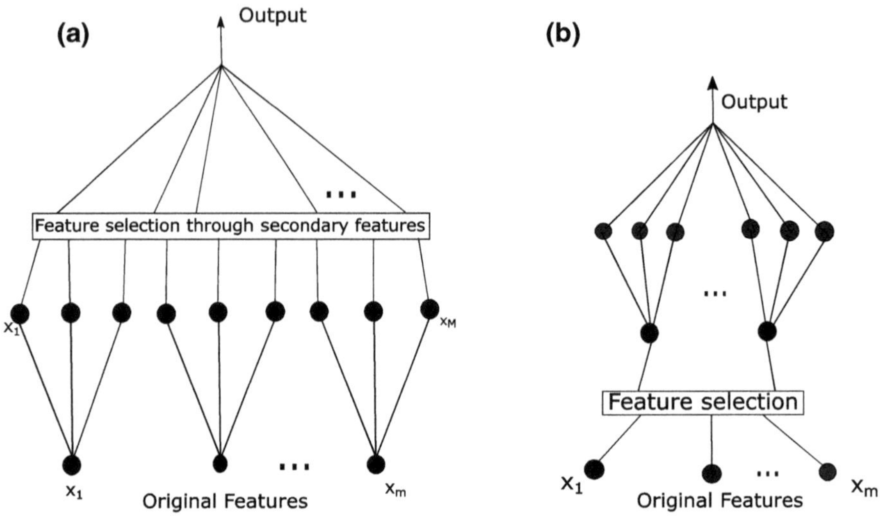

Fig. 3 Optimizing the FLNN structure by selecting through **a** the expanded features; **b** the original features

features. As mentioned earlier, several studies including [17, 24, 26–28], attempted to decrease the network complexity using different feature selection strategies to reduce the original input features before expansion. However, this approach eliminates certain original features (Fig. 3b). This is a shortcoming because while an original feature by itself may not be important in constructing a decision boundary, its expanded, higher order terms may be useful in building the decision boundary in the expanded feature space.

On the other hand, since the FLNN expands all features in the same way (e.g., Eq. 4), it is possible that a number of expanded features are redundant, therefore increasing its computational cost and comprising its performance. Moreover, it is expected that different expanded features have different importance. Embedding a suitable feature selection method after the expansion process, instead of focusing on feature selection in the original feature space (Fig. 3a), can tackle the highlighted shortcomings. As a result, the feature selection algorithm can search through both the expanded and original features. This allows the expanded features to be used for constructing the decision boundary, even when their corresponding original features are eliminated. The EFLNN uses with polynomial expansion functions, in which the GA is used to select the important features in the expanded feature space, and the back-propagation learning algorithm is employed to optimize the EFLNN weights.

The GA chromosomes are encoded such that each gene takes the value of either 0 or 1, indicating whether the respective expanded feature is active or inactive. Therefore, the length of the chromosomes is equal to the number of expanded features. As an example, in case of the expanded features shown in Eq. 4, each chromosome is represented by a 5-dimensional binary valued tuple, like $ch_i = \langle 10100 \rangle$. This chromosome indicates that only x_1, and x_1x_2 are selected to build the EFLNN model. The selected expanded features then undergo the back-propagation learning algorithm. Finally, the EFLNN model is constructed using the weights adjusted by the back-propagation algorithm.

The initial GA population is composed of individuals that represent the FLNN with individual original features (excluding the bias term). This helps the GA to search more efficiently in identifying the irrelevant features [28].

For the fitness function, a combination of classification accuracy and network complexity is formulated as in Eq. 7.

$$f(ch_i) = Aq - \gamma \frac{\mu}{M}, \tag{7}$$

where Aq is classification accuracy of the EFLNN model encoded by ch_i, and μ is the number of selected expanded features. Here, γ is a trade-off factor set to 0.01 in order to ensure that classification accuracy is prioritized in the process of achieving the minimum number of expanded features.

3.1 rFLNN

In EFLNN, the weights are computed using the back-propagation training algorithm. In essence, it is possible to combine both weight tuning and expanded feature selection processes as one optimisation task. As stated earlier, one advantage of the FLNN as compared with other neural networks, especially the MLP network, is that it has only one layer of tunable weights. By exploiting this advantage, the rFLNN model is proposed. The GA is employed to undertake both weight tuning and expanded feature selection processes in one operation. To evolve the rFLNN model, it is required to introduce an appropriate chromosome structure to encode the weight values, in addition to indicating the status (on or off) of the expanded features. As an example, a chromosome with the following structure:

$$ch_i = \langle 1, 0, 1, 0, 0, 0.25, 0, 0.7, 0, 0 \rangle ,$$

represents that the first and third features are selected, where their weights are 0.25 and 0.7, respectively.

Accordingly, customized crossover and mutation operators are defined to handle the reproduction operation in rFLNN. The crossover points are selected randomly from the weight part of the chromosome, and their corresponding weight values are exchanged according to the normal crossover operation. The corresponding status genes, pertaining to the expanded features are also carried forward to the offspring. For the mutation operator, the mutation points are selected from the genes that encode the status of the expanded features. If the status is flipped from 0 to 1, a random value from the feasible range is generated as the corresponding weight. If it is flipped from 1 to 0, the corresponding weight is set to zero.

Parameters setting Selecting the degree of the polynomial functions is a key question. Some studies suggest finding the appropriate degree through evolution, or defining the fitness function such that it penalizes polynomial functions with higher degrees [28]. But, as shown in [28, 38], polynomial functions with a degree higher than 2 lead to the risk of over-fitting in the FLNN. As a result, we keep the polynomial function up to a degree of 2 in this study. The fitness function is the same as shown in Eq. 7, i.e., it penalizes classification error and the number of network weights. The population size is set to a few times of the number of expanded features given by Eq. 3. Meanwhile, the crossover rate is set to 0.8, after several trials over the [0 1] range.

4 Performance Analysis

To evaluate the rFLNN model, three benchmark classification problems have been used. The key characteristics of the data sets are shown in Table 2. All the data sets are from the University of California at Irvine (UCI) machine learning repository [39]. For comparative purposes, some of them are extracted from [40]. In all experiments, the data sets are divided into two sets, one for training and another for test. As the GA and back-propagation methods are stochastic in nature, multiple experiments are required to evaluate the respective models, so that statistical measurements can be computed. As such, each experiment has been executed 30 times to compute the average results. The results are compared with those reported in the literature.

4.1 Datasets

Table 2 shows a summary of the data sets used in this experimental study. A description of the data sets is as follows:

Cancer1: The Wisconsin breast cancer data set has 699 samples of breast cancer tumors. Each sample has nine real-valued features representing information from microscopic examination. The samples belong to two classes: benign (65%) and

Table 2 Summary of the data sets used for the experimental study

Dataset	Samples	Features	Classes	Samples per class
Cancer1	699	9	2	458/241
Diabetes1	768	8	2	481/252
Ionosphere	351	34	2	225/126

malignant tumors. This data set has missing values, all in feature 6. In this study, we use the Cancer1 data set from [40], in which the missing values have been replaced by the mean value.

Ionosphere: This data set contains 351 samples, each with 34 continuous features, collected from a radar system in Goose Bay, Labrador. The target classes are good or bad ionospheric signals depending on whether the samples show some kind of structure with respect to the earth ionosphere.

Diabetes1: This data set contains 768 samples pertaining to the diagnosis results of female diabetes patients (positive or negative). There are 268 (34.9%) samples in the positive class and 500 (65.1%) samples in the negative class. Each sample is composed of 8 real-valued clinical information.

4.2 Results

To evaluate different models, their generalization ability and structural complexity, i.e., test accuracy and number of expanded features selected, are compared. Since all are binary-classification problems, all FLNN-based models have only one output unit. Each expanded feature is linked to the output unit though one weight. As such, the structural complexity is represented by the number of weights. Tables 3, 4, 5, 6, 7 and 8 present all the results. A description is as follows.

Table 3 Test error results for the Cancer1 data set

Model	Test error rate (%)	
rFLNN	Average	1.65
	Best	1.15 (0.57)
EFLNN [28]	Average	1.15
EPNET [28]	Best	1.38
EFNL [28]	Best	1.15
FNNCA [28]	Best	1.45
HMLP [28]	Best	1.15

Table 4 Structural complexity for the Cancer1 data set

Model	Average number of weights
rFLNN	12
EFLNN [28]	11
EPNET [28]	41
EFNL [28]	13
FNNCA [28]	38
HMLP [28]	100

Table 5 Test error results for the Diabetes1 data set

Model	Test error rate	
rFLNN	Average	24.61
	Best	22
EFLNN [28]	Average	21.92
Kohonen [28]	Average	22.4
BP [28]	Average	22.4
RBF [28]	Average	24.3
LVQ [28]	Average	27.2

Table 6 Structural complexity for the Ionosphere data set

Model	Average number of weights
rFLNN	21.5
EFLNN [28]	12

Table 7 Test error results for the Ionosphere data set

Model	Test error rate (%)	
rFLNN	Average	4.68
	Best	5.3
EFLNN [28]	Average	5.73
HOP [28]	Best	5.3
C4.5 [28]	Best	6.0
IB3 [28]	Best	3.3

Table 8 Structural complexity for the Ionosphere data set

Model	Average number of weights
rFLNN	29
EFLNN [28]	29.7
HOP [28]	34

Cancer1 In this data set, the first 350 samples are considered for training and the remaining 175 samples for test. All values are normalized within [0 1]. The average result and 95% confidence interval, as well as the best result of rFLNN over 30 runs are given in Table 3. Table 4 shows the average number of selected expanded features of rFLNN. For comparison, the results from different models in [28] are included in the respective tables.

Diabetes1 The data set from [40] is used. It has 768 samples. The first 576 samples are employed for training, and the remaining 174 samples for test. Tables 5 and 6 show the results of rFLNN and those in the literature. They include EFLNN [28]

and EPNET (an evolutionary neural network) [41], Kohonen self-organizing map [42], back-propagation, and RBF (radial basis function) neural networks, and LVQ (learning vector quantization) [43]. It should be noted that the EFLNN and EPNET results are obtained using the same way as adopted in this experiment, while other results are obtained using the 12-fold cross-validation method, as reported in [28].

Ionosphere This data set has 34 input features, leading to 629 expanded features using polynomial functions with a degree of 2. A total of 276 samples are randomly selected for training, and the remaining 75 samples for test. The results of rFLNN are compared with different models reported in [28], i.e., EFLNN, C4.5, IB3, and a higher order perceptron (HOP) model with a constructive method for feature selection [28, 44].

4.3 Discussion

From the results, it is obvious that all models have used fewer number of features, as compared with the numbers of original or expanded features, for classification. In other words, it is useful to apply a mechanism to identify less important features (either original or expanded) to improve the FLNN performance and computational load. A common way to achieve this aim is using constructive algorithms. However, some neural network-based constructive models such as FNNCA [45] or HOP [44] are prone to the local minima problem. This problem can be mitigated using evolutionary algorithms such as the proposed rFLNN model and EFLNN [28]. In rFLNN, the training procedure is accomplished during the evolution process, therefore eliminating the need for gradient-based training algorithms. From the results it is found that rFLNN is useful in most classification cases. While the rFLNN performance is comparable with those from other models, it is inferior as compared with the EFLNN [28].

This is expected as the search landscape of rFLNN is more complex than that of the EFLNN, since both selection of expanded features and weight tuning processes are combined as one optimisation task. This highlights the need for a more efficient search algorithm to improve the rFLNN performance.

5 Conclusions

In this paper, a modified FLNN model for undertaking data classification tasks have been described. The rationale is, while the expansion process is vital in the FLNN model, many redundant expanded features are produced. In the proposed rFLNN model, fewer than half of the expanded features are selected to build the resulting structure. The proposed model utilizes the GA for tackling both weight tuning and selection of expanded features processes as a combined search problem. As a result, it is not necessary to use the back-propagation algorithm for weight tuning in rFLNN. The experiments show that rFLNN performs reasonably well in terms of classification accuracy with a less complex network structure. In this study, the polynomial basis functions are used as the functional expansion links in constructing rFLNN. However, other basis functions can be used to build the proposed rFLNN model. In future work, we will carry out more experiments using more complicated problems. An in-depth theoretical analysis on the scalability of rFLNN will be conducted. We will also investigate mother evolutionary algorithms for constructing rFLNN. A comprehensive performance comparison between rFLNN and other FLNN variants, especially EFLNN [28] and related models, for classification tasks will be conducted.

References

1. Dash, P., Satpathy, H., Liew, A., Rahman, S.: A real-time short-term load forecasting system using functional link network. IEEE Trans. Power Syst. **12**(2), 675–680 (1997)
2. Tsai, C.F., Wu, J.W.: Using neural network ensembles for bankruptcy prediction and credit scoring. Expert Syst. Appl. **34**(4), 2639–2649 (2008)
3. Shahzadeh, A., Khosravi, A., Nahavandi, S.: Improving load forecast accuracy by clustering consumers using smart meter data. In: International Joint Conference on Neural Networks (IJCNN), Killarney, pp. 1–7, July 2015
4. Miller, A., Blott, B., et al.: Review of neural network applications in medical imaging and signal processing. Med. Biol. Eng. Comput. **30**(5), 449–464 (1992)
5. Bishop, C.M.: Neural Networks for Pattern Recognition. Oxford University Press (1995)
6. Ripley, B.D.: Pattern Recognition and Neural Networks. Cambridge University Press (2007)
7. Lippmann, R.P.: Pattern classification using neural networks. IEEE Commun. Mag. **27**(11), 47–50 (1989)
8. Cochocki, A., Unbehauen, R.: Neural Networks for Optimization and Signal Processing. Wiley (1993)
9. Khan, J., Wei, J.S., Ringner, M., Saal, L.H., Ladanyi, M., Westermann, F., Berthold, F., Schwab, M., Antonescu, C.R., Peterson, C., et al.: Classification and diagnostic prediction of cancers using gene expression profiling and artificial neural networks. Nat. Med. **7**(6), 673–679 (2001)
10. Meireles, M.R., Almeida, P.E., Simões, M.G.: A comprehensive review for industrial applicability of artificial neural networks. IEEE Trans. Ind. Electron. **50**(3), 585–601 (2003)
11. Kaastra, I., Boyd, M.: Designing a neural network for forecasting financial and economic time series. Neurocomputing **10**(3), 215–236 (1996)
12. Pao, Y.H., Takefuji, Y.: Functional-link net computing: theory, system architecture, and functionalities. IEEE Comput. J. **25**(5), 76–79 (1992)
13. Pao, Y.-H., Park, G.-H., Sobajic, D.J.: Learning and generalization characteristics of the random vector functional-link net. Neurocomputing **6**(2), 163–180 (1994)
14. Igelnik, B., Pao, Y.-H.: Additional perspectives on feedforward neural-nets and the functional-link. In: Proceedings of the 1993 International Joint Conference on Neural Networks (IJCNN93), Nagoya, pp. 2284–2287 (1993)
15. Bebarta, D.K., Rout, A.K., Biswal, B., Dash, P.K.: Forecasting and classification of Indian stocks using different polynomial functional link artificial neural networks. In: Proceedings of the Annual IEEE India Conference (INDICON), Kochi, pp. 178–182 (2012)
16. Nanda, S.K., Tripathy, D.P.: Application of functional link artificial neural network for prediction of machinery noise in opencast mines. Adv. Fuzzy Syst. **2011**, 4 (2011)

17. Dehuri, S., Roy, R., Cho, S.-B., Ghosh, A.: An improved swarm optimized functional link artificial neural network (ISO-FLANN) for classification. J. Syst. Softw. **85**(6), 1333–1345 (2012)
18. Mall, S., Chakraverty, S.: Hermite functional link neural network for solving the Van der Pol Duffing oscillator equation. Neural Comput. **28**(8), 1574–1598 (2016)
19. Mall, S., Chakraverty, S.: Numerical solution of nonlinear singular initial value problems of Emden-Fowler type using Chebyshev neural network method. Neurocomputing **149**(Part B), 975–982 (2015)
20. Hassim, Y., Ghazali, R.: Optimizing functional link neural network learning using modified bee colony on multi-class classifications. In: Advances in Computer Science and its Applications, vol. 279, Sec. 23, pp. 153–159. Springer, Berlin (2014)
21. Naik, B., Nayak, J., Behera, H., Abraham, A.: A harmony search based gradient descent learning-FLANN (HS-GDL-FLANN) for classification. In: Computational Intelligence in Data Mining, vol. 2, pp. 525–539, Springer (2015)
22. Bebarta, D., Venkatesh, G.: A low complexity FLANN architecture for forecasting stock time series data training with meta-heuristic firefly. In: Proceedings of the International Conference on Computational Intelligence in Data Mining, Odisha, pp. 377–385 (2016)
23. Dehuri, S., Cho, S.-B.: A comprehensive survey on functional link neural networks and an adaptive PSO-BP learning for CFLNN. Neural Comput. Appl. **19**(2), 187–205 (2010)
24. Dehuri, S., Cho, S.-B.: A hybrid genetic based functional link artificial neural network with a statistical comparison of classifiers over multiple datasets. Neural Comput. Appl. **19**(2), 317–328 (2010)
25. Scardapane, S., Wang, D., Panella, M., Uncini, A.: Distributed learning for random vector functional-link networks. Inf. Sci. **301**, 271–284 (2015)
26. Dash, C.S.K., Dehuri, S., Cho, S.-B., Wang, G.-N.: Towards crafting a smooth and accurate functional link artificial neural networks based on differential evolution and feature selection for noisy database. Int. J. Comput. Intell. Syst. **8**(3), 539–552 (2015)
27. Dehuri, S., Mishra, B.B., Cho, S.-B.: Genetic feature selection for optimal functional link artificial neural network in classification. In: Proceedings of the 9th International Conference on Intelligent Data Engineering and Automated Learning (IDEAL), Daejeon, pp. 156–163 (2008)
28. Sierra, A., Macias, J., Corbacho, F.: Evolution of functional link networks. IEEE Trans. Evol. Comput. **5**(1), 54–65 (2001)
29. Dehuri, S., Cho, S.-B.: Evolutionarily optimized features in functional link neural network for classification. Expert Syst. Appl. **37**(6), 4379–4391 (2010)
30. Mili, F., Hamdi, M.: A comparative study of expansion functions for evolutionary hybrid functional link artificial neural networks for data mining and classification. In: International Conference on Computer Applications Technology (ICCAT), Sousse, pp. 1–8 (2013)
31. Liu, L., Manry, M., Amar, F., Dawson, M., Fung, A.: Image classification in remote sensing using functional link neural networks. In: Proceedings of the IEEE Symposium on Image Analysis and Interpretation, Southwest, pp. 54–58 (1994)
32. Jang, J.-S.R., Sun, C.-T., Mizutani, E.: A Computational Approach to Learning and Machine Intelligence. Prentice Hall, New Jersey (1997)
33. Goldberg, D.E.: Genetic Algorithms in Search, Optimization and Machine Learning, 1st edn. Addison-Wesley Longman, Boston, MA (1989)
34. Goldberg, D.E.: The Design of Innovation: Lessons from and for Competent Genetic Algorithms, 1st edn. Springer Science & Business Media, Dordrecht (2002)
35. Chalmers, D.J.: The evolution of learning: an experiment in genetic connectionism. In: Proceedings of the Connectionist Models Summer School, San Mateo, CA (1990)
36. Leung, F.H.F., Lam, H.K., Ling, S.H., Tam, P.K.S.: Tuning of the structure and parameters of a neural network using an improved genetic algorithm. IEEE Trans. Neural Netw. **14**, 79–88 (2003)
37. Kaya, M., Alhajj, R.: Genetic algorithm based framework for mining fuzzy association rules. Fuzzy Sets Syst. **152**, 587–601 (2005)

38. Chen, C.L.P., LeClair, S.R., Pao, Y.-H.: An incremental adaptive implementation of functional-link processing for function approximation, time-series prediction, and system identification. Neurocomputing **18**, 11–31 (1998)
39. Lichman, M.: UCI Machine Learning Repository (2013). http://archive.ics.uci.edu/ml/
40. Prechelt, L.: Proben1 a set of neural network benchmark problems and benchmarking rules. Technical Report 21/94, Fakultt fr Informatik, Univ. Karlsruhe, Karlsruhe, Germany, Sept 1994
41. Yao, X., Liu, Y.: A new evolutionary system for evolving artificial neural networks. IEEE Trans. Neural Netw. **8**(3), 694–713 (1997)
42. Michie, D., Spiegelhalter, D.J., Taylor, C.C.: Machine Learning, Neural and Statistical Classification. Ellis Horwood (1994)
43. Ster, B., Dobnikar, A.: Neural networks in medical diagnosis: comparison with other methods. In: Proceedings of the International Conference EANN, pp. 427–430 (1996)
44. Thimm, G.: Optimization of high order perceptrons. Ph.D. dissertation, 1633, cole Polytechnique Fdrale de Lausanne, Lausanne, Switzerland, June 1997
45. Setiono, R., Hui, L.C.K.: Use of a quasinewton method in a feedforward neural-network construction algorithm. IEEE Trans. Neural Netw. **6**, 273–277 (1995)

Experimental Study of Elman Network in Temporal Classification

Shih Yin Ooi, Shing Chiang Tan and Wooi Ping Cheah

Abstract Elman network is an extension of multilayer perceptron (MLP), where it introduces single hidden layer architecture, as well as an additional state table to store the time units for hidden neurons. This additional state table allows it to do the sequential prediction which is not possible in MLP. To examine its general performance as a temporal classifier, a Weka version of Elman network is exploited on 11 public temporal datasets released by UCI Machine Repository.

Keywords Elman network · Temporal classification · Machine learning

1 Preliminary of Elman Network

Elman network is also known as simple recurrent network (SRN), and was first published by Jeff Elman in his founding paper on 1990 [1]. The main interest is to extend the usage of multilayer perceptron (MLP) in language processing. The usage and robustness of MLP is limited in this context, because the sequence has never been taken into consideration when classifying an object. Thereafter, Elman proposed a simplified version of MLP with an additional state table to store the time units for hidden neurons (depicted in Fig. 1), so that each time before they made a decision, the sequence will be taken into consideration.

S.Y. Ooi (✉) · S.C. Tan · W.P. Cheah
Faculty of Information Science and Technology, Multimedia University,
Jalan Ayer Keroh Lama, 75450 Melaka, Malaysia
e-mail: syooi@mmu.edu.my

S.C. Tan
e-mail: sctan@mmu.edu.my

W.P. Cheah
e-mail: wpcheah@mmu.edu.my

A. Bhatti et al. (eds.), *Emerging Trends in Neuro Engineering and Neural Computation*, Series in BioEngineering,
DOI 10.1007/978-981-10-3957-7_14

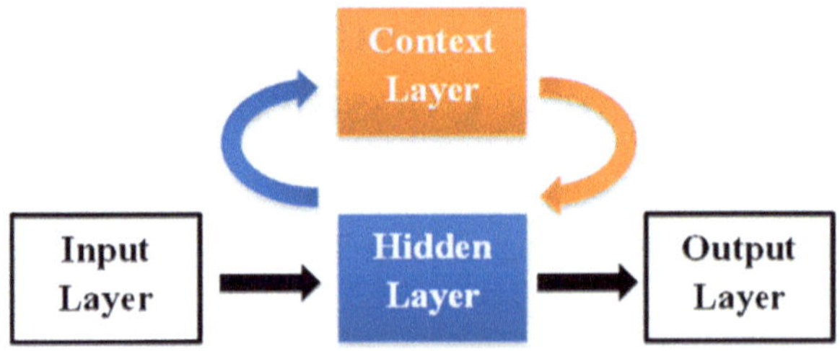

Fig. 1 Elman network architecture

As shown in Fig. 1, an Elman network is formed with only one recursive hidden layer. The output values as well as its sequence value from the hidden neurons will be stored in the context layer. With a recursive connection from context layer to hidden layer with a fixed weight value of 1, the values stored in the context layer will be additionally fed to the hidden neurons in the next time unit [1]. By doing so, the network is able to maintain a sequential state table, and allowing them to do sequential or temporal prediction.

As defined in the founding paper [1], the structure of Elman network algorithm can be summarized as below:

i. Let $w(0)$ denote a weight vector, and is set to a random values within $(1,1)$; η denotes the learning rate; k representing the repetitions counter, and set the initial point to $k = 0$. Start the context node with the value of 0.5.
ii. When the initialization of epoch = k, the current value of weight vector will be stored as $w_{current} = w(k)$.
iii. Let the network represented as $n = 1, 2, 3, \ldots, N$:
    a. Generate a training sample, t^n from the dataset x^n; the partial derivatives, $\frac{\partial E^n}{\partial w_i}$ can be obtained through error backpropagation.
    b. The new weight (w_i) now is equivalent to $w_i(k+1) = w_i(k) - \eta \frac{\partial E^n}{\partial w_i}$, where E^n represent the square error of this network.
    c. Hidden node's value is stored in context layer.
    d. Increment epoch, $k = k + 1$.
iv. The process is recursively performed until the termination value of epoch is reached.

Elman network was gaining wide attention especially in the field of cognitive science, because it is the first learner to handle the time-varying structure in the speech stream. However, the usage of Elman network is not just limited to speech recognition. It is also used to model the word perception [2], intrusion detection system [3], chaotic time series prediction [4], hydraulic servo system [5], wind speed forecasting [6], medical [7], temperature drift of a tuning fork micro-electromechanical system (MEMS) gyroscope [8], and many more.

2 Literature Review of Elman Network

As mentioned earlier, Elman network has been applied to handle different prediction and detection tasks. Liou et al. [2] slightly amended the encoding methodology of Elman network, so that it can be used to iteratively re-encode and generate a set of useful semantic (temporal) representations for words. By iteratively merging the sequence of words, the meaning can be extracted. The proposed method was tested on Shakespeare's writing corpus. The promising results showed the positive ability of this method in discovering the semantic meaning for each tested word through an inductive learning Elman network. Wang et al. [7] improved the Elman network by adjusting its self-feedback parameter with a given constant, in the application of X-ray radiation measurement. The improved Elman network managed to improve the precision in detecting radiation.

Elman network is also adopted by several researchers to form a hybrid model of neural network. The main purpose of using Elman network is due to its ability to bridge the past events to the current prediction and detection. Tong et al. [3] adopted Elman network in their proposed hybrid neural networks (with an RBF network) to detect both anomaly network traffics as well as misuse network traffics. Elman was adopted to learn the past traffic logs, so that the false alarm of intrusion detection system can be reduced. Tested on DARPA 1999 dataset, this hybrid model has shown a set of promising results, whereby the detection rates of 93%, 95.3%, 87.5%, and 100% were reported for anomaly detection, misuse detection, denial-of-service (DOS), and probing, respectively. Ardalani-Farsa et al. [4] also embedded Elman network with a NARX network in the chaotic time series prediction. Tested on two chaotic systems, Mackey–Glass and Lorenz equations and one real-life Sunspot time series, this proposed hybrid model managed to give a more accurate prediction as compared to other in the literature. Liu et al. [5] amalgamated Elman network with support vector regression (SVR) in predicting the performance degradation and also to forecast the health state of hydraulic servo system. Tested on a set of synthetic hydraulic servo data, the Elman network managed to provide a more accurate output of hydraulic servo system. Liu et al. [6] provided a hybrid model by using secondary decomposition algorithm (SDA) and Elman network to predict the wind speed. Elman network was adopted to enable the multistep wind speed prediction. Tested on five real forecasting cases, this hybrid model managed to outperform Elman network when it was using alone. Chong et al. [8] combined the usage of genetic algorithm (GA) and Elman network to generate a higher precision for temperature drift of a tuning fork micro-electromechanical system (MEMS) gyroscope.

3 Elman Network as a Temporal Classifier

While many are adopting it to solve some specific issues, it is even more interesting to examine its robustness as a general temporal classifier. Hereafter, a Weka version of Elman network is applied to classify 11 temporal databases, retrieved from UCI Machine Learning Repository (http://archive.ics.uci.edu/ml/datasets). Table 1 provides brief information of the datasets. All evaluations are conducted on a 10-fold cross validation.

3.1 D1: Activity Recognition from Single Chest-Mounted Accelerometer

This dataset was contributed by [9]. The uncalibrated data was collected from a wearable accelerometer mounted on the chest, with 52 Hz sampling frequency. Users were required to perform seven different activities during the data collection period, including: (i) working at computer; (ii) standing up, walking, and going up and down the stairs; (iii) standing; (iv) walking; (v) going up and down the stairs; (vi) walking and talking with someone; and (vii) talking while standing. In this

Table 1 Summary of temporal datasets

	Temporal dataset	# Instances	# Attributes	# Classes
D1	Activity recognition from single chest-mounted accelerometer [9]	402,843 (for 3 users)	5	7
D2	ADL recognition with Wrist-Worn accelerometer dataset	79,438	4	14
D3	Bach choral harmony dataset [10]	5665	17	72
D4	Daily and sports activities dataset [11]	142,500	46	19
D5	EEG eye state	10,075	15	2
D6	EMG physical action dataset	128,886	9	20
D7	Gesture phase segmentation	19,780	19 (raw)* 33 (processed)*	5
D8	Grammatical facial expressions dataset	27,936	302	18
D9	Ozone level detection	3890	74	2
D10	User identification from walking activity [9]	149,332	5	22
D11	Wall following robot navigation	16,368	3 (2 sensors)* 5 (4 sensors)* 25 (24 sensors)*	4

dataset, we have selected 3 out of 15 users for the experimental purposes. The results are tabled in Table 2.

3.2 D2: ADL Recognition with Wrist-Worn Accelerometer Dataset

This dataset consists of 14 simple ADL recording from a total of 16 volunteers, with a single triaxial accelerometer attached to the right wrist of the volunteer. In this dataset, we have selected five sets from each activity for the experimental purposes. The results are tabled in Table 3.

3.3 D3: Bach Choral Harmony Dataset

This dataset contains some music information, and all of the chord labels were manually annotated by a human expert. Full database is used in the experiment, and the results are tabled in Table 4.

Table 2 Experimental results for three users in the database of *D1*: Activity Recognition from Single Chest-Mounted Accelerometer

	User 1	User 2	User 3
Classification accuracy	94.6708%	90.7421%	94.019%
Kappa statistic	0.9183	0.8793	0.9166
Mean absolute error	0.0418	0.0652	0.0449
Root mean squared error	0.1122	0.1516	0.1094
Relative absolute error	22.219%	33.407%	24.8687%
Root relative squared error	36.5742%	48.5489%	36.3901%

Table 3 Experimental results for the database of *D2*: ADL Recognition with Wrist-Worn Accelerometer Dataset

Classification accuracy	94.6708%
Kappa statistic	0.9183
Mean absolute error	0.0418
Root mean squared error	0.1122
Relative absolute error	22.219%
Root relative squared error	36.5742%

Table 4 Experimental results for the database of *D3*: Bach Choral Harmony Dataset

Classification accuracy	22.6518%
Kappa statistic	0.156
Mean absolute error	0.0167
Root mean squared error	0.0909
Relative absolute error	89.5092%
Root relative squared error	94.0537%

3.4 D4: Daily and Sports Activities Dataset

This dataset contains 19 activities performed by eight users, collected with a sensor unit calibrated at 25 Hz sampling frequency. Total signal duration allocated for each activity is 5 min. All activities performed by a user are used in the experiment, and the results are tabulated in Table 5.

3.5 D5: EEG Eye State

This dataset contains a set of EEG eye state dataset, and all of them were from one continuous EEG measurement with the Emotiv EEG Neuroheadset. Full dataset is used in the experiment, and the results are tabled in Table 6.

3.6 D6: EMG Physical Action Dataset

This dataset contains a set of 10 normal and 10 aggressive human physical actions, collected using the Delsys EMG wireless apparatus on four subjects. A subject is used in the experiment, and the results are tabled in Table 7.

Table 5 Experimental results for the database of *D4*: Daily and Sports Activities Dataset

Classification accuracy	72.6422%
Kappa statistic	0.6885
Mean absolute error	0.0623
Root mean squared error	0.1598
Relative absolute error	62.5142%
Root relative squared error	71.5552%

Table 6 Experimental results for the database of *D5*: EEG Eye State

Classification accuracy	54.6862%
Kappa statistic	0.0397
Mean absolute error	0.485
Root mean squared error	0.5015
Relative absolute error	98.0225%
Root relative squared error	100.8337%

Table 7 Experimental results for the database of *D6*: EMG Physical Action Dataset

Classification accuracy	25.2897%
Kappa statistic	0.1697
Mean absolute error	0.1601
Root mean squared error	0.2818
Relative absolute error	88.9729%
Root relative squared error	93.9478%

3.7 D7: Gesture Phase Segmentation

This dataset contains the features extracted from seven videos with people gesticulating. Each video is represented by a raw file and a processed file. Full dataset is used in the experiment, and the results are tabled in Table 8.

3.8 D8: Grammatical Facial Expressions Dataset

This dataset contains a set of grammatical facial expressions from Brazilian Sign Language (Libras), which were recorded using Microsoft Kinect sensor. Two subjects are used in the experiment, and the results are tabled in Table 9.

3.9 D9: Ozone Level Detection

This dataset contains two ground ozone level datasets, which are 8 h peak set and 1 h peak set. Both of them were collected at the Houston, Galveston, and Brazoria area. Full dataset is used in the experiment, and the results are tabulated in Table 10.

Table 8 Experimental results for the database of *D7*: Gesture Phase Segmentation

	Raw	Processed
Classification accuracy	58.1759%	29.2236%
Kappa statistic	0.4334	−0
Mean absolute error	0.1865	0.2548
Root mean squared error	0.3046	0.3585
Relative absolute error	73.0368%	99.7159%
Root relative squared error	85.2373%	100.3051%

Table 9 Experimental results for the database of *D8*: Grammatical Facial Expressions Dataset

	User A	User B
Classification accuracy	44.9082%	35.7392%
Kappa statistic	0.3982	0.3007
Mean absolute error	0.0778	0.0886
Root mean squared error	0.1928	0.2068
Relative absolute error	75.2769%	85.1636%
Root relative squared error	84.7704%	90.6717%

3.10 D10: User Identification from Walking Activity

This dataset contains walking patterns 22 participants collected over a predefined path, with an Android smartphone positioned in the chest pocket. Full dataset is used in the experiment, and the results are tabled in Table 11.

3.11 D11: Wall Following Robot Navigation

This dataset contains the moving data of SCITOS G5 robot, collected using three different number of sensors. Full dataset is used in the experiment, and the results are tabled in Table 12.

3.12 Summary of Results

Classification accuracies plotted by Elman network in the aforementioned 11 temporal datasets are illustrated in Fig. 2. Elman network has shown its capability to make a classification by considering the sequential patterns, however, its performances are greatly deteriorated in *D3*, *D4*, *D6*, *D8*, and *D10* where the number of classes is very large. The performance is not stable too when the number of attributes is large, i.e.: *D4*, *D7-Processed*, *D8*, and *D9*.

Table 10 Experimental results for the database of *D9*: Ozone Level Detection

	One-hour	Eight-hour
Classification accuracy	95.572%	93.6859%
Kappa statistic	0	0
Mean absolute error	0.0756	0.1151
Root mean squared error	0.2017	0.2409
Relative absolute error	88.6327%	97.048%
Root relative squared error	98.0227%	99.0564%

Table 11 Experimental results for the database of *D10*: User Identification from Walking Activity

Classification accuracy	23.4638%
Kappa statistic	0.1529
Mean absolute error	0.0764
Root mean squared error	0.1943
Relative absolute error	91.3236%
Root relative squared error	95.0214%

Table 12 Experimental results for the database of *D11*: Wall Following Robot Navigation

	2 sensor reading	4 sensor reading	24 sensor reading
Classification accuracy	89.8644%	83.1195%	77.8776%
Kappa statistic	0.8428	0.7294	0.6535
Mean absolute error	0.0602	0.1059	0.167
Root mean squared error	0.1686	0.2299	0.2956
Relative absolute error	18.1806%	31.9813%	50.4094%
Root relative squared error	41.4243%	56.4847%	72.6422%

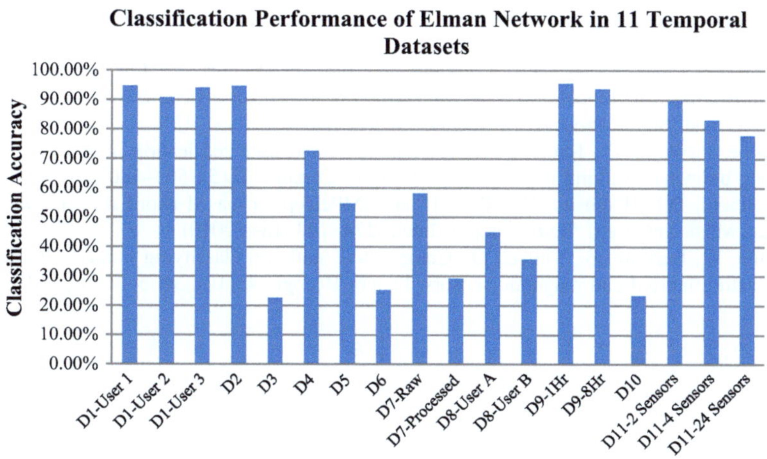

Fig. 2 Classification accuracies of Elman network in 11 temporal datasets

4 Conclusion

Elman network has been tested on 11 public temporal datasets to examine its robustness as a temporal classifier. It demonstrated its ability in handling temporal classification. From the observation of the experiments, there are still some rooms to improve the Elman network. The two main challenges facing in Elman network are its limitation in handling a large number of attributes and large number of classes.

References

1. Elman, J.: Finding structure in time. Cogn. Sci. **14**(2), 179–211 (1990)
2. Liou, C.Y., Huang, J.C., Yang, W.C.: Modeling word perception using the Elman network. Neurocomputing **71**(16–18), 3150–3157 (2008)

3. Tong, X., Wang, Z., Yu, H.: A research using hybrid RBF/Elman neural networks for intrusion detection system secure model. Comput. Phys. Commun. **180**(10), 1795–1801 (2009)
4. Ardalani-Farsa, M., Zolfaghari, S.: Chaotic time series prediction with residual analysis method using hybrid Elman-NARX neural networks. Neurocomputing **73**(13–15), 2540–2553 (2010)
5. Liu, H., Zhang, J., Lu, C.: Performance degradation prediction for a hydraulic servo system based on Elman network observer and GMM-SVR. Appl. Math. Model. **39**(19), 5882–5895 (2015)
6. Liu, H., Tian, H.Q., Liang, X.F., Li, Y.F.: Wind speed forecasting approach using secondary decomposition algorithm and Elman neural networks. Appl. Energy **157**, 183–194 (2015)
7. Wang, F., Wang, M.-Y., Tian, F.-S., Liu, Y.-F., Li, L., Zhao, J.: Study on two-dimensional distribution of X-ray image based on improved Elman algorithm. Radiat. Meas. **77**, 1–4 (2015)
8. Chong, S., Rui, S., Jie, L., Xiaoming, Z., Jun, T., Yunbo, S., Jun, L., Huiliang, C.: Temperature drift modeling of MEMS gyroscope based on genetic-Elman neural network. Mech. Syst. Signal Process. **72–73**, 897–905 (2016)
9. Casale, P., Pujol, O., Radeva, P.: Personalization and user verification in wearable systems using biometric walking patterns. Pers. Ubiquit. Comput. **16**(5), 563–580 (2012)
10. Radicioni, D.P., Esposito, R.: BREVE: an HMPerceptron-based chord recognition system. Adv. Music Inf. Retr. Stud. Comput. Intell. **274**, 143–164 (2010)
11. Altun, K., Barshan, B., Tunçel, O.: Comparative study on classifying human activities with miniature inertial and magnetic sensors. Pattern Recogn. **43**(10), 3605–3620 (2010)

Monotone Data Samples Do Not Always Generate Monotone Fuzzy If-Then Rules

Chin Ying Teh and Kai Meng Tay

Abstract The Wang–Mendel (WM) method is one of the earliest methods to learn fuzzy If-Then rules from data. In this article, the WM method is used to generate fuzzy If-Then rules for a zero-order Takagi–Sugeno–Kang (TSK) fuzzy inference system (FIS) from a set of multi-attribute monotone data. Convex and normal trapezoid fuzzy sets are used as fuzzy membership functions. Besides that, a strong fuzzy partition strategy is used. Our empirical analysis shows that a set of multi-attribute monotone data may lead to non-monotone fuzzy If-Then rules. The same observation can be made, empirically, using adaptive neuro-fuzzy inference system (ANFIS), a well-known and popular FIS model with neural learning capability. This finding is important for the modeling of a monotone FIS model, because it shows that even with a "clean" data set pertaining to a monotone system, the generated fuzzy If-Then rules may need to be preprocessed, before being used for FIS modeling. In short, it is imperative to develop methods for preprocessing non-monotone fuzzy rules from data, e.g., monotone fuzzy rules relabeling, or removing non-monotone fuzzy rules, is important (and is potentially necessary) during the course of developing data-driven FIS models.

Keywords Fuzzy If-Then rules · The Wang–Mendel method · ANFIS · Monotonicity property · Multi-attribute monotone data · Monotone fuzzy rule relabeling · Interval-valued fuzzy rules

1 Introduction

Fuzzy inference systems (FISs) with fuzzy If-Then rules have been successfully applied to many real-life applications [1, 2]. Traditionally, fuzzy If-Then rules are obtained from human experts. Owing to the difficulty in soliciting fuzzy rules from

C.Y. Teh · K.M. Tay (✉)
Faculty of Engineering, Universiti Malaysia Sarawak, Kota Samarahan, Malaysia
e-mail: kmtay@feng.unimas.my

A. Bhatti et al. (eds.), *Emerging Trends in Neuro Engineering and Neural Computation*, Series in BioEngineering,
DOI 10.1007/978-981-10-3957-7_15

human experts, many approaches have been proposed to automatically generate fuzzy If-Then rules from numerical data, i.e., data-driven FIS models. Most of the data-driven FIS models involve neural-like iterative learning procedures, or other rule generation methods, e.g., gradient descent learning [3–6], genetic algorithm-based method [7], least square method [8, 9], fuzzy c-means clustering [10], and fuzzy neural method [11]. The focus of this article is on *ad hoc data-driven linguistic rule learning methods* [12], which include those by Wang et al. [13, 14] and Ishibuchi et al. [4]. *Ad hoc data-driven linguistic rule learning methods* are easy to understand and simple to implement, and yet they perform with a rapid learning process, as compared with other methods [3–11]. In [13, 14], Wang et al. used a table-lookup scheme to generate fuzzy rule bases directly from numerical examples (hereafter abbreviated as the Wang–Mendel (WM) method). The fuzzy If-Then rule(s) with the highest firing strength is selected for learning [13, 14]. In [4] and [12], Ishibuchi et al. presented a heuristic method for generating Takagi–Sugeno–Kang (TSK) fuzzy If-Then rules from numerical examples. All firing fuzzy rules are selected for learning [4, 12].

Despite the popularity of data-driven FIS models, investigation on monotone data-driven FIS models is still new. The importance of the monotonicity property in FIS modeling has been highlighted in a number of recent publications [15–23], and a number of studies to preserve the monotonicity property have been proposed. These include the development of mathematical conditions as a set of governing equations [15, 18, 21], application of the developed mathematical conditions to real-world problems [19, 22], as well as extension and synthesis of the mathematical conditions with some advanced FIS modeling techniques [16, 23, 24]. In this article, we focus on analyzing the monotonicity property of the fuzzy If-Then rules generated from multi-attribute monotone data using the WM method. Specifically, convex and normal trapezoid fuzzy sets are used as the fuzzy membership functions (see Definition 2). Besides that, a strong fuzzy partition strategy is used (see Definition 3). Our analytical results suggest that the fuzzy If-Then rules generated do not always monotone, even though the multi-attribute data used are monotone. A simulated example is used for illustration. The same observation can be made, empirically, using adaptive neuro-fuzzy inference system (ANFIS) [5], a well-known fuzzy system with neural learning. Worth mentioning, various studies pertaining to the monotonicity property and neural-computing have also been reported in [25–27].

The organization of this article is as follows. In Sect. 2, the background and related work are described. In Sect. 3, an analysis of the WM method is presented, and a simulation study is conducted. In Sect. 4, the simulation results using ANFIS are presented. Finally, concluding remarks are presented in Sect. 5.

2 Background

Definition 1 Consider an n-dimensional input space, X, and a one-dimensional output space, Y. Variables $\bar{x} = [x_1, x_2, \ldots, x_n] \in X_1 \times X_2 \times \cdots \times X_n$ and y are the elements of X and Y, i.e., $\bar{x} \in X$, and $y \in Y$, respectively. The lower and upper bounds of X_i, where $i \in [1, 2, 3, \ldots, n]$, are denoted as $\underline{x_i}$ and $\overline{x_i}$, respectively. Similarly, the lower and upper bounds of Y are denoted by $\underline{y}$ and $\bar{y}$, respectively.

Definition 2 Consider a trapezoidal fuzzy set, i.e., A_{X_i}, on the X_i domain, as follows (see Fig. 1 as an example).

$$\mu_{X_i}(x_i\colon a_i, b_i, c_i, d_i) = \begin{cases} \frac{x_i - a_i}{b_i - a_i}, & a_i \le x_i \le b_i, \\ 1, & b_i \le x_i \le c_i, \\ \frac{d_i - x_i}{d_i - c_i}, & c_i \le x_i \le d_i, \\ 0. & \text{otherwise.} \end{cases} \tag{1}$$

Definition 3 The X_i domain is divided into k_i fuzzy partitions, and represented by k_i normal trapezoidal fuzzy sets. The fuzzy sets are denoted as $\mu_{X_{i,j_i}}(x_i)$ for linguistic term $A_{X_{i,j_i}}$, where $j_i = 1, 2, 3, \ldots, k_i$. A strong partition is used, which is defined as follows (see Fig. 2 as an example).

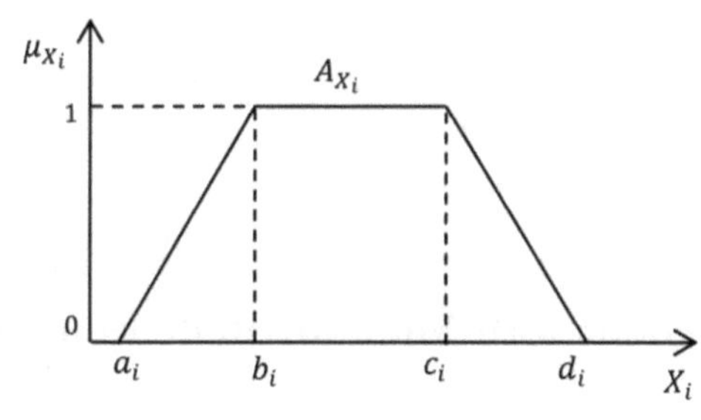

Fig. 1 The membership function of a trapezoidal fuzzy set

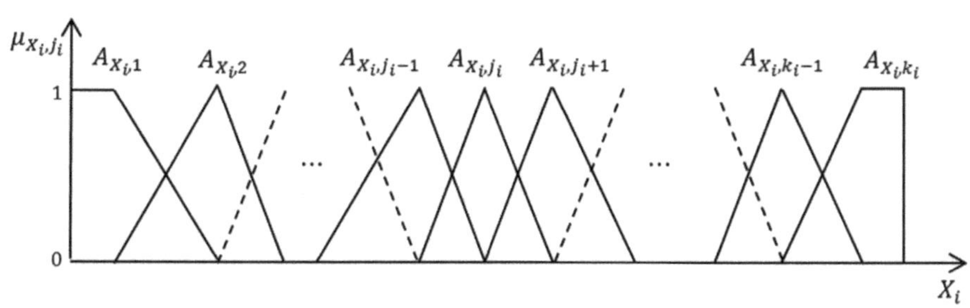

Fig. 2 Fuzzy sets with k_i partitions

3.1. $\mu_{X_i,1}(\underline{x_i}) = 1$ and $\mu_{X_i,k_i}(\overline{x_i}) = 1$, are always true.
3.2. All k_i trapezoidal fuzzy sets are normal.
3.3 $\forall x_i \in X_i$: $\sum_{j_i=1}^{k_i} \mu_{X_i,j_i}(x_i) = 1$ is always true (i.e., a strong partition).
3.4 $\forall x_i \in X_i$: $A_{X_i,j_i} \leq A_{X_i,j_i+1}$, where $j_i = 1, 2, 3, \ldots, k_i - 1$ is always true. A fuzzy ordering exists among all k_i normal trapezoidal fuzzy sets.
3.5. For any two fuzzy sets, i.e., $A_{X_i,j_i} = [a_{i,j_i}, b_{i,j_i}, c_{i,j_i}, d_{i,j_i}]$ and $A_{X_i,j_i+1} = [a_{i,j_i+1}, b_{i,j_i+1}, c_{i,j_i+1}, d_{i,j_i+1}]$ on X_i, $c_{i,j_i} = a_{i,j_i+1}$ and $d_{i,j_i} = b_{i,j_i+1}$, where $j_i = 1, 2, 3, \ldots, k_i - 1$, are always true.

Definition 4 Consider an antecedent in the input space, i.e., $X_1 \times X_2 \times \cdots \times X_n$. It is denoted as $A_{X_1,j_1} \wedge \cdots \wedge A_{X_n,j_n}$, which is simplified as $A^{j_1,\ldots,j_n}$. Each antecedent is mapped to a real value in the output space, i.e., Y and is denoted as $b^{j_1,\ldots,j_n}$. A fuzzy If-Then rule which maps the input space to the output space is denoted as $\mathrm{R}^{j_1,\ldots,j_n}: A^{j_1,\ldots,j_n} \to b^{j_1,\ldots,j_n}$. A collection of $\mathrm{R}^{j_1,\ldots,j_n}$ forms the fuzzy If-Then rules, as follows:

$$IF\left(x_1 \text{ is } A_{X_1,j_1}\right) AND \cdots AND\left(x_n \text{ is } A_{X_n,j_n}\right), THEN\left(y \text{ is } b^{j_1,\ldots,j_n}\right)$$

4.1. For a *complete* fuzzy If-Then rule, all $b^{j_1,\ldots,j_n}$ are known, $i \in [1, 2, 3, \ldots, n]$, $j_i \in [1, 2, 3, \ldots, k_i]$; otherwise it is *incomplete*. A complete fuzzy If-Then rule set should have M rules, i.e., $M = \prod_{i=1}^{M=n} k_i$.
4.2. For a *monotone* fuzzy If-Then rule, $b^{j_i+1, j_{\bar{s}}} \geq b^{j_i, j_{\bar{s}}}$ is always true, for $i \in [1, 2, \ldots, n]$, $s \in \bar{s} = [1, 2, \ldots n]$ and $i \notin \bar{s}$, $j_i = 1, 2, \ldots, k_i - 1$; otherwise the fuzzy If-Then rule is *non-monotone*.

3 The WM Method with Multi-attribute Monotone Data

Definition 5 Consider a set of input–output data pairs, i.e., $(\overline{x^p}; y^p), p \in [1, 2, 3, \ldots, h]$, where output $y^p \in Y$ is influenced by n input variables, $\overline{x^p} = [x_1^p, x_2^p, \ldots, x_n^p] \in X_1 \times X_2 \times \cdots \times X_n$. For simplicity, $(\overline{x^p}; y^p)$ is denoted as $y^p_{x_1^p, x_2^p, \ldots, x_n^p}$. A data sample, $y^q_{x_1^q, x_2^q, \ldots, x_n^q}$, is further considered, where $q \in [1, 2, 3, \ldots, h]$ and $q \neq p$.

5.1 Two notions are defined as follows:

$$\underline{y^q_{x_1^q, x_2^q, \ldots, x_n^q}} = max\left(\left[y^p_{x_1^p \leq x_1^q, x_2^p \leq x_2^q, \ldots, x_n^p \leq x_n^q}\right]\right). \tag{2}$$

$$\overline{y^q_{x_1^q, x_2^q, \ldots, x_n^q}} = min\left(\left[y^p_{x_1^p \geq x_1^q, x_2^p \geq x_2^q, \ldots, x_n^p \geq x_n^q}\right]\right). \tag{3}$$

5.2 The data set is *monotone* if $\underline{y^q_{x_1^q, x_2^q, \ldots, x_n^q}} \leq y^q_{x_1^q, x_2^q, \ldots, x_n^q} \leq \overline{y^q_{x_1^q, x_2^q, \ldots, x_n^q}}$, is true, $\forall p \in [1, 2, 3, \ldots, h]$, $\forall q \in [1, 2, 3, \ldots, h]$ and $q \neq p$. Otherwise, the data set is *non-monotone*.

Definition 6 *A dominant rule(s)* for $\overline{x^p}$ is denoted as $R^{j_1, \ldots, j_n, p}$. $R^{j_1, \ldots, j_n, p}$ is the fuzzy rule(s) with the highest firing strength, i.e., the highest of $\prod_{i=1}^{n} \mu_{X_{i}, j_i}(x_i^p)$. If more than one fuzzy rule have the same firing strength, all fuzzy rules are considered as the *dominant rules*.

Using the WM method, $b^{j_1, \ldots, j_n}$ can be obtained with Eq. (4) [13, 14]. If $p \in R^{j_1, \ldots, j_n}$ is an empty set, then $b^{j_1, \ldots, j_n}$ is unknown.

$$b^{j_1, \ldots, j_n} = \frac{\sum_{p \in R^{j_1, \ldots, j_n}} \prod_{i=1}^{n} \mu_{X_i, j_i}(x_i^p) \times y^p}{\sum_{p \in R^{j_1, \ldots, j_n}} \prod_{i=1}^{n} \mu_{X_i, j_i}(x_i^p)} \tag{4}$$

Proposition 1 *Consider a* $n > 1$ *monotone (see Definition* 5.2*) input data set.*

Remark 1 With Proposition 1, an analysis of the WM method [13, 14] is summarized as follows:

1.1 The lower limit for $b^{j_1, \ldots, j_n}$, denoted as $\underline{b^{j_1, \ldots, j_n}}$, can be obtained with $\underline{b^{j_1, \ldots, j_n}} = min(y^p)$, for $p \in R^{j_1, \ldots, j_n}$.

Proof From Eq. (4), the lower limit for $b^{j_1, \ldots, j_n}$ is equal to $min(y^p)$.

1.2 The upper limit for $b^{j_1, \ldots, j_n}$, denoted as $\overline{b^{j_1, \ldots, j_n}}$, can be obtained with $\overline{b^{j_1, \ldots, j_n}} = max(y^p)$, for $p \in R^{j_1, \ldots, j_n}$.

Proof From Eq. (4), the upper limit for $b^{j_1, \ldots, j_n}$ is equal to $max(y^p)$.

1.3 Using Eq. (4), the resulting fuzzy rules is not always *monotone*.

Proof Any two neighboring fuzzy rules, i.e., $b^{j_i, j_{\bar{i}}}$ and $b^{j_i+1, j_{\bar{i}}}$ for $j_i = 1, 2, \ldots, k_i - 1$ with Eq. (4) are considered. The upper limit for $b^{j_i, j_{\bar{i}}}$ is $\overline{b^{j_i, j_{\bar{i}}}} = max(y_{j_i, j_{\bar{i}}})$ for $y_{j_i, j_{\bar{i}}} \in R^{j_i, j_{\bar{i}}}$. The lower limit for $b^{j_i+1, j_{\bar{i}}}$ is $\underline{b^{j_i+1, j_{\bar{i}}}} = min(y_{j_i+1, j_{\bar{i}}})$ for $y_{j_i+1, j_{\bar{i}}} \in R^{j_i+1, j_{\bar{i}}}$. Therefore, $\underline{b^{j_i+1, j_{\bar{i}}}} \geq \overline{b^{j_i, j_{\bar{i}}}}$ is not always true.

3.1 Simulation Studies

Example 1 Consider a two-attribute *monotone* data set with 100 input–output pairs, i.e., $(\overline{x^p}; y^p), p \in [1, 2, 3, \ldots, 100]$. $\overline{x^p} = [x_1^p, x_2^p] \in X_1 \times X_2$ and single output $y^p \in Y$

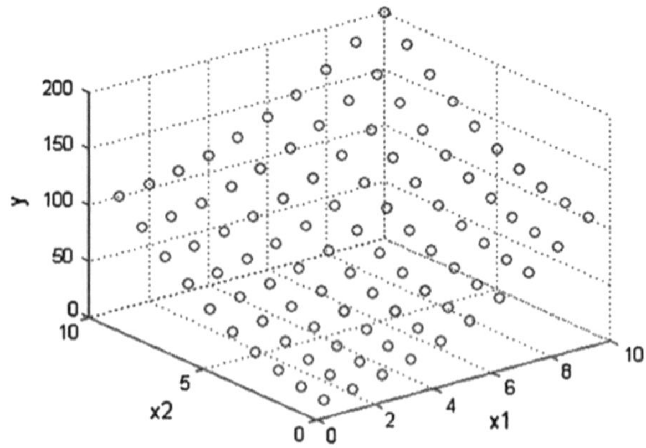

Fig. 3 *Monotone* data for Example 1

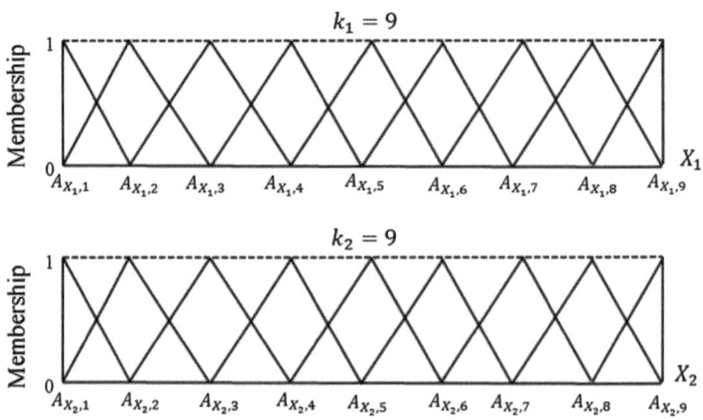

Fig. 4 Fuzzy partition used for simulation

is considered. The data set is generated using a simple quadratic function, i.e., $y = x_1^2 + x_2^2$, in the input space of $[1, 10] \times [1, 10]$, as shown in Fig. 3.

The fuzzy partition illustrated in Fig. 4 is used, i.e., $k_1 = k_2 = 9$. With Eq. (4), a *monotone* fuzzy rule base, as summarized in Fig. 5, is obtained. In this example, *monotone* fuzzy If-Then rules (see Definition 4.2) are produced.

Example 2 A total of 88 data samples from Example 1 are chosen, as depicted in Fig. 6. The same fuzzy partition, as illustrated in Fig. 4, is used.

The obtained fuzzy If-Then rule set using Eq. (4) is summarized in Fig. 7. In this example, *non-monotone* and *incomplete* fuzzy If-Then rules are produced. As an example, for $A^{1,5}$, $b^{j_1=1,j_2=5} = 37.000$; for $A^{3,5}$, $b^{j_1=3,j_2=5} = 34.000$; however, $b^{j_1=3,j_2=5}$ is smaller than $b^{j_1=1,j_2=5}$, i.e., $b^{j_1=3,j_2=5} < b^{j_1=1,j_2=5}$. In short, a *non-*

$A_{x_1,9}$	101.000	104.000	109.000	116.000	130.500	149.000	164.000	181.000	200.000
$A_{x_1,8}$	82.000	85.000	90.000	97.000	111.500	130.000	145.000	162.000	181.000
$A_{x_1,7}$	65.000	68.000	73.000	80.000	94.500	113.000	128.000	145.000	164.000
$A_{x_1,6}$	50.000	53.000	58.000	65.000	79.500	98.000	113.000	130.000	149.000
$A_{x_1,5}$	31.500	34.500	39.500	46.500	61.000	79.500	94.500	111.500	130.500
$A_{x_1,4}$	17.000	20.000	25.000	32.000	46.500	65.000	80.000	97.000	116.000
$A_{x_1,3}$	10.000	13.000	18.000	25.000	39.500	58.000	73.000	90.000	109.000
$A_{x_1,2}$	5.000	8.000	13.000	20.000	34.500	53.000	68.000	85.000	104.000
$A_{x_1,1}$	2.000	5.000	10.000	17.000	31.500	50.000	65.000	82.000	101.000
	$A_{x_2,1}$	$A_{x_2,2}$	$A_{x_2,3}$	$A_{x_2,4}$	$A_{x_2,5}$	$A_{x_2,6}$	$A_{x_2,7}$	$A_{x_2,8}$	$A_{x_2,9}$

Fig. 5 Generated fuzzy If-Then rules for Example 1

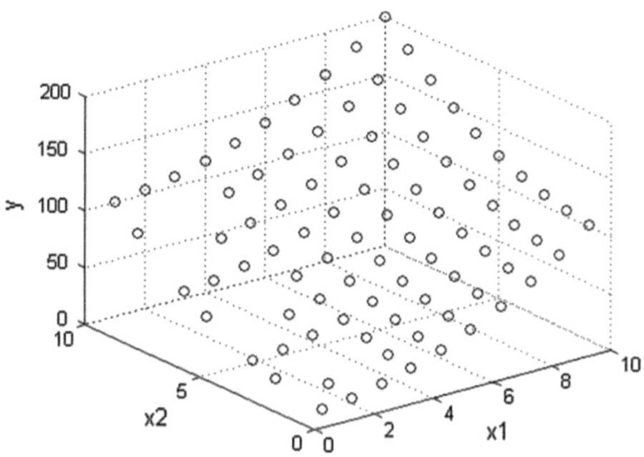

Fig. 6 *Monotone* data for Example 2

$A_{x_1,9}$	101.000	104.000	109.000	116.000	130.500	149.000	164.000	181.000	200.000
$A_{x_1,8}$	82.000	85.000	90.000	97.000	111.500	130.000	145.000	162.000	181.000
$A_{x_1,7}$	65.000	68.000	73.000	80.000	94.500	113.000	128.000	145.000	164.000
$A_{x_1,6}$	50.000	53.000	58.000	65.000	79.500	98.000	113.000	130.000	149.000
$A_{x_1,5}$	31.500	34.500	39.500	46.500	61.000	79.500	94.500	111.500	130.500
$A_{x_1,4}$	17.000	20.000	25.000	32.000	46.500	65.000	80.000	97.000	116.000
$A_{x_1,3}$	10.000	NaN	NaN	25.000	**34.000**	58.000	73.000	NaN	109.000
$A_{x_1,2}$	5.000	8.000	NaN	20.000	NaN	53.000	NaN	NaN	104.000
$A_{x_1,1}$	2.000	NaN	10.000	17.000	**37.000**	50.000	NaN	82.000	101.000
	$A_{x_2,1}$	$A_{x_2,2}$	$A_{x_2,3}$	$A_{x_2,4}$	$A_{x_2,5}$	$A_{x_2,6}$	$A_{x_2,7}$	$A_{x_2,8}$	$A_{x_2,9}$

Fig. 7 The generated fuzzy If-Then rules for Example 2

monotone fuzzy If-Then rule set is obtained. On the other hand, for $A^{2,3}$, $b^{j_1=2,j_2=3} = NaN$ (i.e., $b^{j_1=2,j_2=3}$ is unknown), resulting in an *incomplete* fuzzy If-Then rule set.

$A_{x_{1,9}}$	101.000	104.400	110.300	118.800	129.900	143.600	159.800	178.600	200.000
$A_{x_{1,8}}$	79.620	83.000	88.950	97.460	108.600	122.200	138.400	157.200	178.600
$A_{x_{1,7}}$	60.820	64.200	70.140	78.660	89.750	103.400	119.600	138.400	159.800
$A_{x_{1,6}}$	44.590	47.960	53.910	62.430	73.520	87.180	103.400	122.200	143.600
$A_{x_{1,5}}$	30.930	34.300	40.250	48.770	59.860	73.520	89.750	108.600	129.900
$A_{x_{1,4}}$	19.840	23.210	29.160	37.680	48.760	62.430	78.650	97.470	118.800
$A_{x_{1,3}}$	11.320	14.700	20.640	**29.160**	40.260	53.900	**70.160**	88.940	110.300
$A_{x_{1,2}}$	5.375	9.184	**2.786**	**29.170**	**29.500**	**38.730**	**82.670**	**18.090**	104.400
$A_{x_{1,1}}$	**2.000**	**1.903**	**12.310**	**19.340**	**16.490**	**62.640**	**24.720**	**79.620**	101.000
	$A_{x_{2,1}}$	$A_{x_{2,2}}$	$A_{x_{2,3}}$	$A_{x_{2,4}}$	$A_{x_{2,5}}$	$A_{x_{2,6}}$	$A_{x_{2,7}}$	$A_{x_{2,8}}$	$A_{x_{2,9}}$

Fig. 8 The generated fuzzy If-Then rules using ANFIS [5]

4 Simulation with ANFIS

Example 2 is revisited. The same fuzzy partition, as illustrated in Fig. 4, is adopted. The obtained fuzzy If-Then rule set using ANFIS [5] is summarized in Fig. 8. In this simulation, it is observed that the fuzzy MFs at antecedent are not adjusted or learned. Worth mentioning, ANFIS with fixed MFs can be viewed as a functional link network [2, 5]. Again, *non-monotone* fuzzy If-Then rules are produced. As an example, for $A^{1,1}$, $b^{j_1=1,j_2=1}=2.000$; for $A^{1,2}$, $b^{j_1=1,j_2=2}=1.903$, while $b^{j_1=1,j_2=2}$ is smaller than $b^{j_1=1,j_2=1}$, i.e., $b^{j_1=1,j_2=2}<b^{j_1=1,j_2=1}$.

5 Summary

In this article, the WM method [13, 14] has been used to generate fuzzy If-Then rules from *monotone* data. A simulated "clean" *monotone* data set, which is free from noise, pertaining to a *monotone* system is considered. The "clean" *monotone* data set is used for learning using an FIS model. Our study shows that while a *monotone* multi-attribute data set is used, the fuzzy If-Then rules generated using the WM method may not always be *monotone*. The same observation occurs when the ANFIS model is used [5]. This finding is important for modeling a *monotone* FIS model, as it shows that even with a "clean" data set, the generated fuzzy If-Then rules may need to be preprocessed, in order to satisfy the monotonicity property. Developing useful methods to preprocess *non-monotone* fuzzy rules from data samples is, therefore, a key step toward producing data-driven *monotone* TSK FIS models.

For future work, we will develop useful methods to preprocess *non-monotone* fuzzy rules from data. The possible methods include relabeling *non-monotone* fuzzy rules [28] or transforming *non-monotone* fuzzy rules to *monotone* interval-valued fuzzy rules [29].

References

1. Mendel, J.: Fuzzy logic systems for engineering: a tutorial. Proc. IEEE **83**(3), 345–377 (1995)
2. Jang, J.S.R., Sun, C.T., Mizutani, E.: Neuro-fuzzy and Soft Computing: A Computational Approach to Learning and Machine Intelligence. Prentice-Hall, Upper Saddle River, NJ (1997)
3. Ichihashi, H., Watanabe, T.: Learning control system by a simplified fuzzy reasoning model. In: Proceeding of IPMU, pp. 417–419 (1990)
4. Ishibuchi, H., Nozaki, K., Tanaka, H., Hosaka, Y., Matsuda, M.: Empirical study on learning in fuzzy systems by rice taste analysis. Fuzzy Sets Syst. **64**, 129–144 (1994)
5. Jang, J.S.R.: ANFIS: adaptive-network-based fuzzy inference system. IEEE Trans. Syst. Man Cybern. **23**, 665–685 (1993)
6. Nomura, H., Hayashi, I., Wakami, N.: A learning method of fuzzy inference rules by descent method. In: Proceeding FUZZ-IEEE, pp. 203–210 (1992)
7. Nomura, H., Hayashi, I., Wakami, N.: A self-tuning method of fuzzy reasoning by genetic algorithm. In: Proceeding International Fuzzy Systems and Intelligent Control Conference, pp. 239–245 (1992)
8. Sugeno, M., Kang, G.T.: Structure identification of fuzzy model. Fuzzy Sets Syst. **28**, 15–33 (1988)
9. Takagi, T., Sugeno, M.: Fuzzy identification of systems and its applications to modeling and control. IEEE Trans. Syst. Man Cybern. **15**, 116–132 (1985)
10. Sugeno, M., Yasukawa, T.: A fuzzy-logic-based approach to qualitative modelling. IEEE Trans. Fuzzy Syst. 7–31 (1993)
11. Takagi, H., Hayashi, I.: NN-driven fuzzy reasoning. Approx. Reason. **5**, 191–212 (1991)
12. Nozaki, K., Ishibuchi, H., Tanaka, H.: A simple but powerful heuristic method for generating fuzzy rules from numerical data. Fuzzy Sets Syst. **86**(3), 251–270 (1997)
13. Wang, L.X., Mendel, J.M.: Generating fuzzy rules by learning from examples. IEEE Trans. Syst. Man Cybern. **22**, 1414–1427 (1992)
14. Wang, L.X.: The WM method completed: a flexible fuzzy system approach to data mining. IEEE Trans. Fuzzy Syst. **11**(6), 768–782 (2003)
15. Won, J.M., Park, S.Y., Lee, J.S.: Parameter conditions for monotonic Takagi-Sugeno-Kang fuzzy system. Fuzzy Sets Syst. **132**(2), 135–146 (2002)
16. Kouikoglou, V.S., Phillis, Y.A.: On the monotonicity of hierarchical sum-product fuzzy systems. Fuzzy Sets Syst. **160**(24), 3530–3538 (2009)
17. Lindskog, P., Ljung, L.: Ensuring monotonic gain characteristics in estimated models by fuzzy model structures. Automatica **36**(2), 311–317 (2000)
18. Seki, H., Ishii, H., Mizumoto, M.: On the monotonicity of fuzzy-inference methods related to T-S inference method. IEEE Trans. Fuzzy Syst. **18**(3), 629–634 (2010)
19. Tay, K.M., Lim, C.P.: On the use of fuzzy inference techniques in assessment models: Part I: theoretical properties. Fuzzy Optim. Decis. Mak. **7**(3), 269–281 (2008)
20. Tay, K.M., Lim, C.P.: On the use of fuzzy inference techniques in assessment models: Part II: industrial applications. Fuzzy Optim. Decis. Mak. **7**(3), 283–302 (2008)
21. Broekhoven, E.V., Baets, B.D.: Only smooth rule bases can generate monotone Mamdani-Assilian models under center-of-gravity defuzzification. IEEE Trans. Fuzzy Syst. **17**(5), 1157–1174 (2009)
22. Tay, K.M., Lim, C.P.: A fuzzy inference system-based criterion-referenced assessment model. Expert Syst. Appl. **38**(9), 1112–11136 (2011)
23. Tay, K.M., Lim, C.P.: Optimization of Gaussian fuzzy membership functions and evaluation of the monotonicity property of fuzzy inference systems. In: Proceeding of the 20th International Conference on Fuzzy Systems, pp. 1219–1224 (2011)
24. Tay, K.M., Lim, C.P.: An evolutionary-based similarity reasoning scheme for monotonic multi-input fuzzy inference systems. In: Proceeding of the 20th International Conference on Fuzzy Systems, pp. 442–447 (2011)

25. Ameloot, T.J., Van den Bussche, J.: Positive neural networks in discrete time implement monotone-regular behaviors. Neural Comput. **27**(12), 2623–2660 (2015)
26. Li, S.T., Chen, C.C.: A regularized monotonic fuzzy support vector machine model for data mining with prior knowledge. IEEE Trans. Fuzzy Syst. **23**(5), 1713–1727 (2015)
27. Fernández-Navarro, F., Riccardi, A., Carloni, S.: Ordinal neural networks without iterative tuning. IEEE Trans. Neural Netw. Learn. Syst. **25**(11), 2075–2085 (2014)
28. Pang, L.M, Tay, K.M., Lim, C.P.: Monotone fuzzy rule relabeling for the zero-order TSK fuzzy inference system. IEEE Trans. Fuzzy Syst. **24**(6), 1455–1463 (2016)
29. Kerk, Y.W, Tay, K.M., Lim, C.P.: An analytical interval fuzzy inference system for risk evaluation and prioritization in failure mode and effect analysis. IEEE Syst. J. doi:10.1109/JSYST.2015.2478150

MIX
Papier aus verantwortungsvollen Quellen
Paper from responsible sources
FSC® C105338

If you have any concerns about our products,
you can contact us on
ProductSafety@springernature.com

In case Publisher is established outside the EU,
the EU authorized representative is:
Springer Nature Customer Service Center GmbH
Europaplatz 3, 69115 Heidelberg, Germany

Printed by Libri Plureos GmbH
in Hamburg, Germany